Frontiers in Fluid Mechanics

A Collection of Research Papers
Written in Commemoration of the 65th Birthday
of Stanley Corrsin

Editors: S. H. Davis and J. L. Lumley

With 132 Figures

Springer-Verlag Berlin Heidelberg New York Tokyo

Professor Stephen H. Davis, Ph. D.
The Technological Institute, Northwestern University,
Evanston, IL 60201, USA

Professor John L. Lumley, Ph. D.
Sibley School of Mechanical and Aerospace Engineering, Cornell University, 238 Upson Hall,
Ithaca, NY 14853, USA

ISBN 978-3-642-46545-1 ISBN 978-3-642-46543-7 (eBook)
DOI 10.1007/978-3-642-46543-7

Softcover reprint of the hardcover 1st edition 1985

This work relates to Department of Navy Grant N00014-84-G-0127 issued by the Office of Naval Research. The United States Government has a royalty-free permission to reproduce all or portions of the above work, and to authorize others to do so, for US Government purposes.

Stanley Corrsin

Dedication

This volume is dedicated to *Stanley Corrsin*, Theophilus Halley Smoot Professor of Fluid Mechanics at the Johns Hopkins University, in anticipation of his 65th birthday on April 3, 1985.

Stan obtained his Bachelor's degree in Mechanical Engineering in 1940 from the University of Pennsylvania. Being interested in aerodynamics, he went to the California Institute of Technology to study in the Guggenheim Laboratory with von Karman. He completed his M.S. in 1942 and his Ph.D. in 1947, both in Aeronautics, being the first advisee of Liepmann. He was hired in 1947 on the Faculty of Aeronautics at Hopkins by F. Clauser and has remained there under a variety of titles ever since. He is Fellow of the American Academy of Arts and Sciences, the American Physical Society, and the American Society of Mechanical Engineers. He is a member of the National Academy of Engineering and Docteur Honoris Causa, Université Claude Bernard (Lyon). He was awarded the 1983 Fluid Dynamics Prize by the American Physical Society. His research interests have been mainly in the areas of turbulence, turbulent transport, and biomechanics.

Stan discovered in 1943 the phenomenon of intermittency at the edge of a turbulent region, which results in distinct regions of vortical and non-vortical fluid being swept past a fixed probe. With Kistler in 1954 he found that the interface between such regions is very sharp, spanning a length of the order of the smallest scales of the flow. The recognition of these distinct regions led to a whole new concept of turbulence measurement, conditional sampling, in which account is taken separetely of field measurements at a point according to whether that point is instantaneously on one side of the interface or the other. This measurement technique is widely used for the identification of "coherent structures" in turbulence.

Stan made with Kistler in 1954 a first theoretical attempt to study the entrainment process by posing a mechnism based on molecular diffusion. His study of the interface in 1955 was one of the first in a small, select subject called "random geometry".

Stan's insistence on a strong interplay between theory and experiment has led to his devoting a major effort to the study of simplified turbulent-flow models free of unnecessary complication. Stan has studied extensively isotropic turbulence and the transport of scalars in isotropic turbulence. One of the earliest contributions was in 1951 in which Stan gave for the first time the form of the spectrum of a passive scalar in isotropic turbulence.

By an elaboration of Onsager's model, he extended this in 1958 to a scalar undergoing a first-order chemical reaction. This initial work on chemical reactions in turbulence was further developed in 1961 and 1965.

The dispersion of scalars in uniform shear flows was explored in 1953. Here the equation for the dissipation of scalar variance was given for the first time.

The transport of scalars in flow fields involves the relationship between Eulerian and Lagrangian quantities, a topic that has interested Stan his entire career. He first suggested in 1963 a form for the Lagrangian time spectrum and a relation between Lagrangian and Eulerian integral scales. He gave a simple proof in 1972 of the old conjecture that fluid line elements grow on the average. In collaboration with his students he carried out extensive experimental work on this and related random-geometry problems, including the 1972 work with Karweit on the mixing of scalar stripes and angular dispersion of line elements.

In the late 1960's Stan became interested in biomechanics. He has investigated topics that range from the efficiency of human walking and the characteristics of maternal blood flow in the human placenta, to the flutter frequency of flexible tubes conveying fluid. Stan gave with Ross in 1974 a quantitative theory of the pumping of mucous in the airways by the beating beds of cilia. He formulated with Berger in 1974 a mechanism for the restoration of the pre-corneal tear film after the lid blinks. He examined with Higdon in 1978 the aerodynamic efficiency of the formation flying of bird flocks.

Research and teaching have been the focus of Stan's activities at Hopkins for almost 38 years. These activities are blended and indivisible. Hundreds of students took Stan's course "Fluid Mechanics" and first learned fluid mechanics as a field of research. Dozens of graduates took Stan's course on turbulence where they learned by example to give open-ended research problems for examinations. Stan developed an undergraduate course on the "Mechanics of Animal Motion" in which many pre-meds saw their first, and perhaps last, quantitative view of the life sciences. The laboratory component of "Animal Motion" saw students measuring lizards walking on water and Hopkins faculty hopping around the track, aiming to relate the physiological responses to the number of legs used. Stan's leadership shows that teaching is a multifaceted endeavor including one-on-one research supervision and collaboration, research-conference development and organization as well as classroom lecturing on the undergraduate and graduate level. His availability to workers in all fields is an invaluable, though unmeasurable, gift to Hopkins.

Finally, Stan has set an example of style and tone. In his research he has established a point of view in fluid mechanics in which problems are freed of their frills, attacked with rigor and the result explained with clarity. This approach has been conveyed to his 25 Ph.D. students and dozens of associates. Scores of students and associates have participated in his 10 AM coffee hours in which science and religion, among other things, are all discussed with large doses of humor; he has shown many of us that we need to view our own work seriously, but light-heartedly. It was here that the first test was made of contact lenses for the eyes of potatoes! It was here and at the wine and cheese parties, toasting newly awarded Ph.D.'s, that undergraduates, graduates, post-doctoral associates and faculty mingled with no hint of a "class" sytem. This serves as a model for many a group around the country.

The editors would like to take this opportunity to thank the Office of Naval Research, in the person of Bob Whitehead, for providing generous support for the undertaking from which this volume resulted.

Evanston - Ithaca
December 1984

S. H. Davis
J. L. Lumley

Contents

On the Approach to Isotropy of Homogeneous Turbulence: Effect of the Partition of Kinetic Energy Among the Velocity Components

L. Le Penven, J.N. Gence, and G. Comte-Bellot

Laboratoire de Mécanique des Fluides, associé au CNRS, Ecole Centrale de Lyon et Université de Lyon I, 36 Avenue Guy de Collongue, F-69131 Ecully Cedex, France

1. Introduction

It has long been admitted by folk wisdom rather than by analytical proof that in the absence of mean velocity gradients, homogeneous anisotropic turbulence evolves towards isotropy. The first experimental findings were due to UBEROI (1957) for the velocity and vorticity components, and to MILLS and CORRSIN (1959) for the velocity components, the temperature fluctuations and their associated skewness factors (signals and derivatives). Experimental confirmation was later given by COMTE-BELLOT and CORRSIN (1966), TUCKER and REYNOLDS (1968), WARHAFT (1980), and GENCE and MATHIEU (1980) for velocity components. Numerical studies made in wave-number space supported also the same trend (SCHUMANN and PATTERSON, 1978).

The question of how fast the return to general isotropy (i.e. on all scales) takes place, is however of fairly long standing. Recently GENCE and MATHIEU (1980) (see also GENCE, 1983) made an interesting suggestion by taking advantage of the invariants II and III of the deviatoric tensor b_{ij} introduced by ROTTA (1951) and extensively used by LUMLEY and NEWMANN (1977) in turbulence modeling:

$$b_{ij} = \frac{\overline{u_i u_j}}{q^2} - \frac{\delta_{ij}}{3} \qquad \left(q^2 = \overline{u_i u_i} = (u_1')^2 + (u_2')^2 + (u_3')^2\right) \tag{1}$$

$$II = b_{ij} b_{ji} \tag{2}$$

$$III = b_{ik} b_{kj} b_{ji} \quad . \tag{3}$$

A time scale τ_R for the return to isotropy was defined from invariant II, by:

$$\frac{1}{\tau_R} = - \frac{1}{II} \frac{dII}{dt} \tag{4}$$

and compared to the usual time scale τ_D for the turbulence decay:

$$\frac{1}{\tau_D} = - \frac{1}{q^2} \cdot \frac{dq^2}{dt} \quad . \tag{5}$$

The return to isotropy was then observed to be relatively rapid, at least at the beginning of the process, (i.e., τ_R between 0.2 τ_D and 0.4 τ_D) in most of the cases where the invariant III was negative, whereas the return to isotropy was found to be fairly slow (i.e. $\tau_R \simeq 0.6\ \tau_D$) in GENCE and MATHIEU (1980), a case for which the invariant III was positive. In Table 1 (upper part) are listed all the results available presently. The Reynolds number Re_L has been also indicated, although it was rather large in all the experiments (Re_L is defined by $(q^2)^2/9\bar{\varepsilon}\nu$ which gives $Re_L \simeq u'L/\nu$ for isotropic turbulence). In Fig. 1 the curves of $\log(q^2/\bar{U}^2)$ are plotted versus log II (a suggestion made by J.L. LUMLEY, private communication). The slopes of the linear curves which are obtained are then simply related to τ_R and τ_D by:

$$\frac{d(\log q^2)}{d(\log II)} = \frac{(1/q^2) \cdot (dq^2/dt)}{(1/II) \cdot (dII/dt)} = \frac{\tau_R}{\tau_D} \quad . \tag{6}$$

We recall that invariant III specifies the shape of the ellipsoid associated with the Reynolds stress tensor. It is positive when only one principal component is relatively large and it is negative when two principal components are large. This is illustrated in Fig. 2 for two cases of axisymmetric turbulence

The possible role of the invariant III on the return to isotropy was an incentive for two recent investigations. At Cornell University,

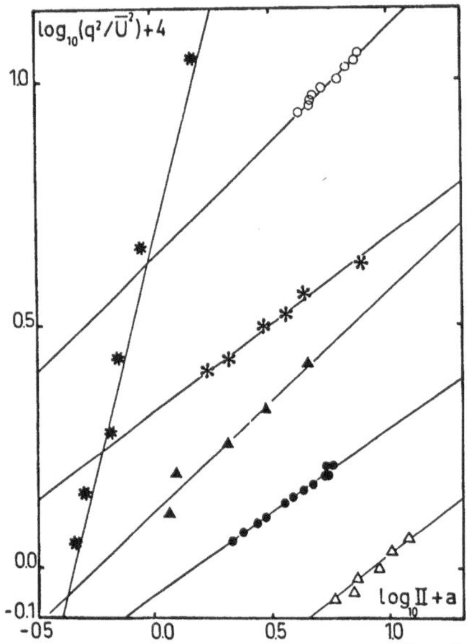

Fig. 1 - Evolution of $\log(q^2/\bar{U}^2)$ against log II in different experiments of return towards isotropy ∗ Uberoi (1957); ● Mills and Corrsin (1959); ▲ Tücker and Reynolds (1968); ○ Gence and Mathieu (1980); △ Warhaft (1980); in all these experiments a = 2; ✱ Comte-Bellot and Corrsin (1966) with a = 3.

- Table 1 -

Experiment	Initial values of the parameters: Re_L	II	III	τ_D/τ_R	Mean values of τ_D/τ_R derived from Fig. 1
UBEROI (1957) Axisymmetric contraction	160	0.047	$-4\ 10^{-3}$	2.4	2.8
MILLS & CORRSIN (1959) Axisymmetric contraction	60	0.056	$-5.3\ 10^{-3}$	4.3	2.8
COMTE-BELLOT & CORRSIN (1966) (Grid without contraction)	420	0.0015	$+2.4\ 10^{-5}$	0.48	0.5
TUCKER & REYNOLDS (1968) (plane distorsion)	460	0.08	$-5.2\ 10^{-3}$	4.5	2
WARHAFT (1980) Axisymmetric contraction	150	0.116	$-16.1\ 10^{-3}$	1.3	2.6
GENCE and MATHIEU (1980) (plane distortion)	450	0.08	$+4\ 10^{-3}$	1.6	2
CHOI (1983) (plane distortion)	180	0.061	$+0.75\ 10^{-3}$	5.1	1.22
CHOI (1983) (Axisymmetric expansion)	260	0.062	$+5.8\ 10^{-3}$	1.3	≈ 1
Present experiment	350	0.054	$+3.1\ 10^{-3}$	1	≈ 1
Present experiment	450	0.062	-5.10^{-3}	2.4	2.6

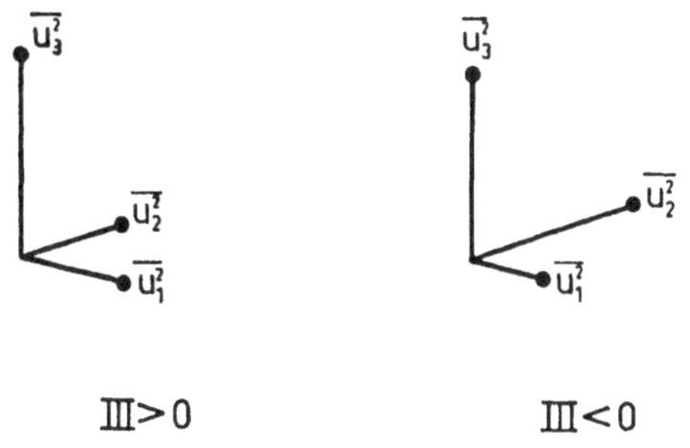

Fig. 2 - Physical meaning of the sign of invariant III.

CHOI (1983) examined the evolution of two anisotropic turbulence fields with positive invariant III. At the Ecole Centrale de Lyon, LE PENVEN and GENCE (1983) generated two anisotropic turbulence fields with nearly opposite values of III and equal values of II and Re_L.

The present paper will describe the new results we have obtained. A physical interpretation for the role of the invariant III will also be attempted. Finally, since the return to isotropy is of great interest for turbulence modeling, mainly for the pressure-strain terms which are responsible for the intercomponent energy interchange, several actual models will be compared with our experimental findings.

2. Experimental Arrangements

1°) Distorting ducts

Two kinds of irrotational three-dimensional strains were used to create, from the same isotropic grid-generated turbulence, the two required anisotropic turbulence fields with opposite invariants III and equal invariants II. Indeed, a three-dimensional deformation offers two adjustable parameters, $\bar{U}_{1,1} = \partial\bar{U}_1/\partial x_1$, $\bar{U}_{2,2} = \partial\bar{U}_2/\partial x_2$ and not just one as for a plane or an axisymmetric deformation where $\bar{U}_{1,1} = \pm\bar{U}_{2,2}$. As usual, the fluid is assumed incompressible so that

$$\bar{U}_{3,3} = -(\bar{U}_{1,1} + \bar{U}_{2,2}) \quad . \tag{7}$$

Additional useful requirements were also specified :

- the length of the distorting ducts should be of the order of 1.5 m to offer a gradual straining able to smooth out the discontinuities at the inlet and outlet of the distorting ducts. This length also suited the wind-tunnel facilities available in the Laboratory.
- the shortest side of the rectangular outlet section of the ducts should be at least of the order of 7 integral length scales to insure an acceptable lateral homogeneity of the turbulence field.
- the mean velocity at the inlet of the two distorting ducts should be the same to insure identical turbulence Reynolds numbers for the initially isotropic grid turbulence.
- along the distorting duct, positive values of $\bar{U}_{1,1}$ are advantageous for reducing flow separation.

The stream surfaces limiting the distorting ducts are given by :

$$x_2 = l_0 \left(1 + \frac{\bar{U}_{1,1}}{U_0} x_1\right)^{\bar{U}_{2,2}/\bar{U}_{1,1}} \tag{8}$$

$$x_3 = h_0 \left(1 + \frac{\bar{U}_{1,1}}{U_0} x_1\right)^{-(1+\bar{U}_{2,2}/\bar{U}_{1,1})} \quad . \tag{9}$$

$2l_0$ and $2h_0$ are the lateral and vertical dimensions of the inlet rectangular section (Fig. 3) where U_0 is the streamwise mean velocity. Moreover, if T is the residence time of the turbulence in the duct of length L, it can be written

$$\bar{U}_{1,1}T = \mathrm{Ln}\left(1 + \frac{\bar{U}_{1,1}}{U_0} L\right) , \tag{10}$$

so that for a given shape of the distorting duct, the ratio $\bar{U}_{1,1}/U_0$ and the dimensionless gradients $\bar{U}_{1,1}T$ and $\bar{U}_{2,2}T$ are determined. The values of $\bar{U}_{1,1}$ and $\bar{U}_{2,2}$ are imposed when U_0 is given.

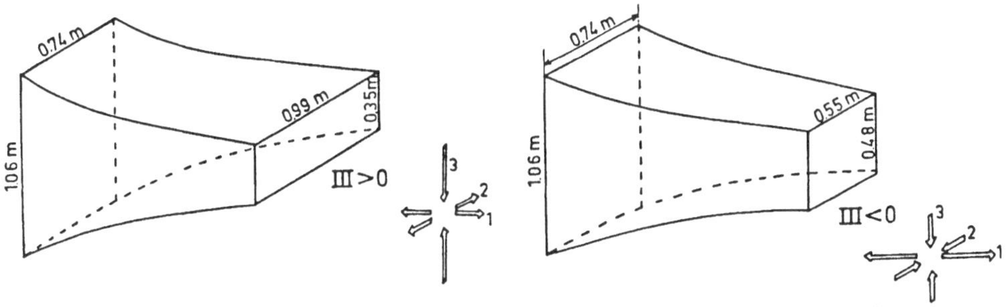

Fig. 3 - The two different distorting ducts used in the experiments.

A small time linear theory with isotropic initial conditions indicates that one can write

$$II(T) = \alpha^2 \bar{U}_{i,k}\bar{U}_{k,i}T^2 + O(T^3) \tag{11}$$

$$\left(\alpha = \frac{4}{15}\right) .$$

$$III(T) = -\alpha^3 \bar{U}_{i,k}\bar{U}_{k,j}\bar{U}_{j,i}T^3 + O(T^4) \tag{12}$$

Then, as a first step $\bar{U}_{1,1}T$ and $\bar{U}_{2,2}T$ were chosen to give the same II and opposite III in this linear approximation, and to be realized in ducts satisfying the above mentioned constraints.

As a second step, several runs were made to include non-linear effects. The spectral method developed by CAMBON, JEANDEL and MATHIEU (1981) was used with the three-dimensional energy spectrum (at 42 meshes) of COMTE-BELLOT and CORRSIN (1971) as initial condition. The values of $\bar{U}_{1,1}T$ and $\bar{U}_{2,2}T$ which were retained are listed in Table 2 with the predicted levels of the invariants II and III and the time T used in the computations.

For the ducts given in Fig. 3, whose lengths L are equal to 1.5 m, the values of time T correspond to a streamwise inlet velocity U_0 of 6.7 m/s. In the experiments U_0 was equal to 6.06 m/s for $III > 0$ and to 7.2 m/s for $III < 0$. The predicted values of II are slightly smaller than those observed in the present experiments and indicated in Table

- Table 2 -

	III > 0	III < 0
$\bar{U}_{1.1}$ T	0.8	1.09
$\bar{U}_{2.2}$ T	0.29	-0.29
II	0.045	0.045
III	0.0021	-0.0024
T	0.146 s	0.123 s

1. This is probably due to the initial spectrum used in the calculations which is taken at the same distance downstream of the grid as in the experiments, but at a higher convection velocity U_G so that the mesh Reynolds number is too large by a factor of about 2. The non-linear effects reducing the anisotropy level are then stronger in the computations than in the experiments.

2°) The wind-tunnel

The distorting ducts are included in a flow facility which is of the open return type and where the basic elements are (Fig. 4) :

- a collector which is 4 m x 4 m in cross section followed by a settling chamber which is 2 m long and equipped with dust filters and screens.
- two contractions with a overall ratio of 16 : 1.
- a constant area duct (0.83 m x 1.19 m) in which the grid which generates the turbulence is located at 1 m from the inlet.
- the grid is biplane with square meshes (M = 5.08 cm) and square rods. (solidity σ =0.33).
- an additional 1.27 : 1 contraction placed at 18 meshes downstream of the grid, to improve the turbulence isotropy as suggested by COMTE-BELLOT and CORRSIN (1966).

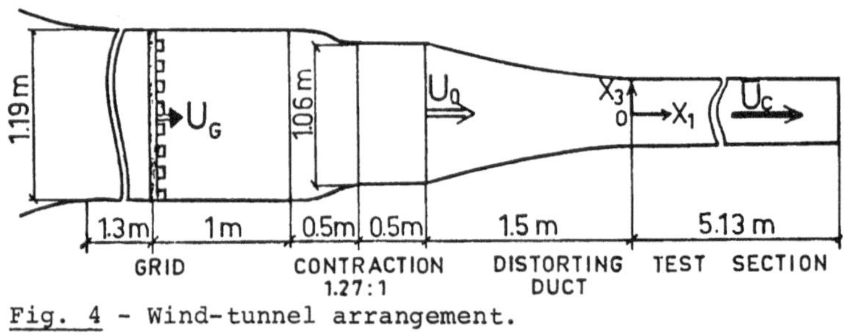

Fig. 4 - Wind-tunnel arrangement.

- the distorting duct, either that for invariant III > 0 or that for invariant III < 0.
- a 5.13 m constant area duct in which the return to isotropy is investigated ; its cross section is 0.99 m x 0.35 m when III is > 0 and 0.55 m x 0.48 m when III is < 0.

3. Measuring Equipment and Procedure

All turbulence data were taken with standard X-meter probes DISA 55 P 51, with hot wires 5 μm in diameter. They were connected to DISA 55 D 01 constant-temperature anemometers and operated at an overheat ratio of about 0.8. The output signals were then passed through DISA 55 D 25 band filters (1 Hz - 20 kHz). Finally, the root mean square values of the signals were obtained by a DISA D 35 r.m.s. voltmeter.

The wire sensitivities were determined empirically from the first derivatives of the heat transfer law with respect to the longitudinal velocity component and the wire angle with the x_1 axis. Calibrations were made in the empty wind tunnel (no grid). The weighted sum and difference, $e_1 + Ke_2$ and $e_1 - K'e_2$, were formed from the two hot wire signals of the X-meter to produce outputs directly proportional to the longitudinal and lateral velocity fluctuations. The probes were first used to get u_1 and u_2, then turned by 90° to obtain u_1 and u_3. The inevitable scatter from one run to the other, from day to day because of drift in the ambiant temperature, or due to change of hot wire probes, was reduced by taking arithmetic means over 6 measurements for all the data.

The mean velocity in the test section increases slightly along the axis due to boundary-layer growth (Fig. 5). The resulting effect of this strain rate ($\simeq 0.2\ s^{-1}$) on the dynamics of $\overline{u_i u_j}$ is fortunately

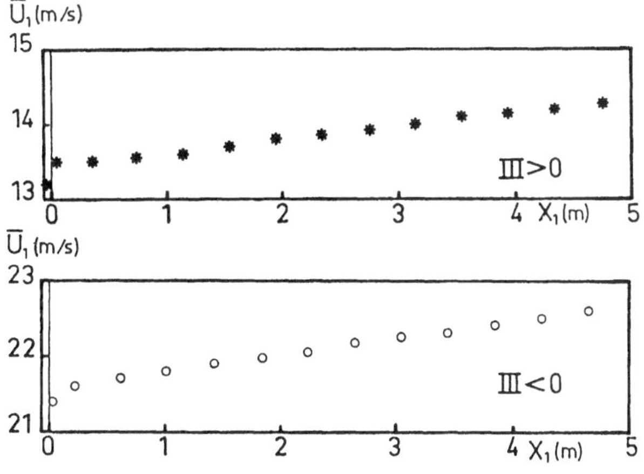

Fig. 5 - Longitudinal mean streamwise velocity along the axis of the test section.

very small (see section 5). The local mean velocity is however taken into account when computing the travel time t downstream of the test section.

Background (no grid) turbulence was measured for each component, at various locations in the test section (Table 3). An appreciable part of it seemed due to a slight unsteadiness of the fan (u_1' is larger than u_2' and u_3' and its spectrum is large for low frequencies). All the turbulence results were corrected for such noise, by its subtraction from the mean squared values. The correction was down to 0.2 % in the best situation, for u_3' at the beginning of the test section for III positive, and reached a maximum of 10% for u_1' at the end of the test section for III positive or negative.

Table 3 - Variation ranges for background turbulence.

	III > 0	III < 0
$(u_1'/U_c)^2$	$4.1\ 10^{-6}$	$1.1\ 10^{-6}$ to $2.4\ 10^{-6}$
$(u_2'/U_c)^2$	$2.7\ 10^{-7}$ to $6.6\ 10^{-7}$	$4.3\ 10^{-7}$ to $6.5\ 10^{-7}$
$(u_3'/U_c)^2$	$2.8\ 10^{-7}$ to $1.15\ 10^{-6}$	$4.3\ 10^{-7}$ to $6.5\ 10^{-7}$
U_c	13.5 m/s	21.5 m/s

Lateral homogeneity of the turbulence field was considered as acceptable even close to the end of the test section and along its shortest (vertical) side. Fig. 6 shows that the central plateau covers about half of the height, which approximately corresponds to 6 times the average integral length scale of turbulence estimated as $M(h/h_0)$ ($2h_0$ height of the distorting duct at its inlet, and 2h at its outlet).

4. Experimental Results

In all the results presented in this section for the return to isotropy, the elapsed time t is computed from

$$t = \int_{\Delta x_1}^{x_1} \frac{dx_1}{\bar{U}_1(x_1)} , \tag{13}$$

where the space origin is slightly shifted downstream of the beginning of the constant area duct. The reason, which is apparent from Fig. 5 and also from Fig. 7, is that the distortion continues to act just after the duct area has changed (the mean velocity on the axis continues to increase rapidly and the turbulent quantities do not change noticeably. The shift estimated from Fig. 7 is $\Delta x_1 = 0.03$ m for the

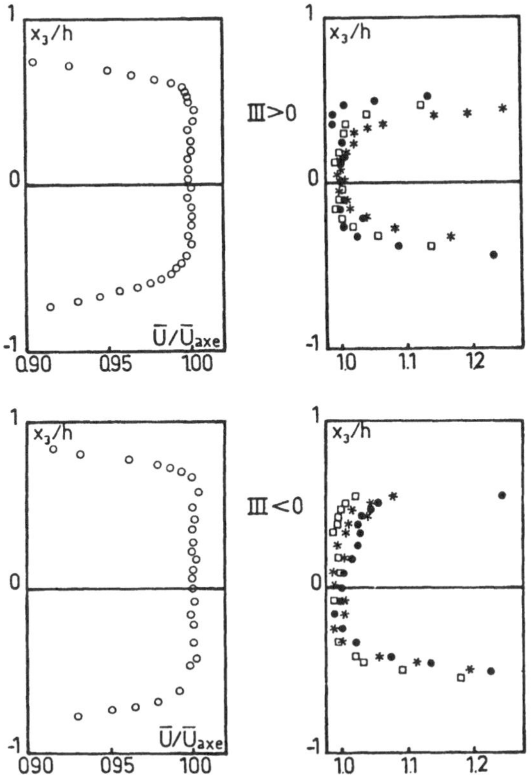

Fig. 6 - Vertical profiles of the mean velocity and of the turbulence intensities divided by their axial value: $*u'_1$; $\square u'_2$; $\bullet u'_3$, at 0.45 m upstream of the outlet section and in the plane $x_2 = 0$

runs at positive invariant III and $\Delta x_1 = 0.22$ m for the runs at negative invariant III. Use of this corrected elapsed time will also enable us to make more direct comparisons with the turbulence modeling results reported in sections 5 and 6.

The time evolution of $(u'_1)^2$, $(u'_2)^2$ and $(u'_3)^2$ are given in Fig.7 for III > 0 and III < 0. The most interesting fact is that the smaller component, $(u'_1)^2$, is slightly increasing at the beginning of the return to isotropy when invariant III is negative, whereas it decays like all the other components when invariant III is positive. This implies that the pressure strain correlation terms have a large and even visible action for exchanging energy among the three velocity components when one of them is very small.

Figure 8 illustrates the time evolution of invariant II and of the total kinetic energy q^2 for positive and negative values of invariant III. The effect of invariant III appears very clearly. When III is positive, the time evolutions of II and q^2 (normalized by their

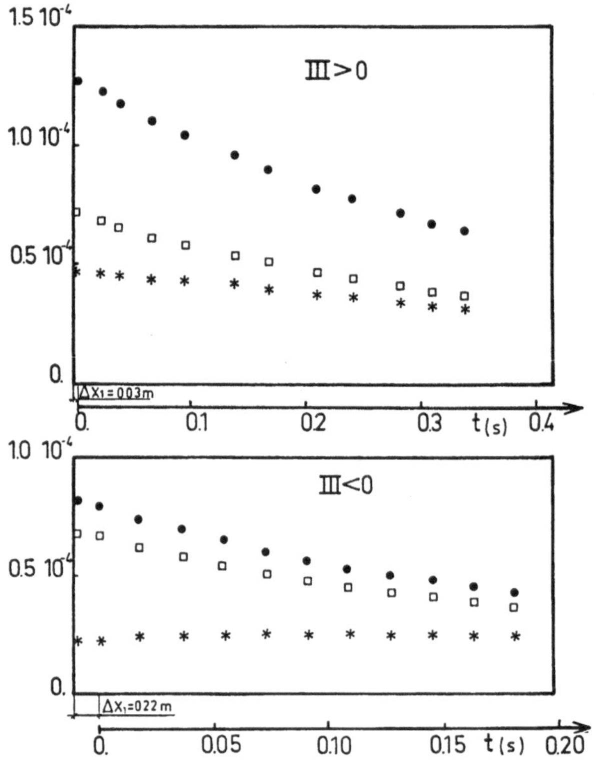

Fig. 7 Caption see opposite page

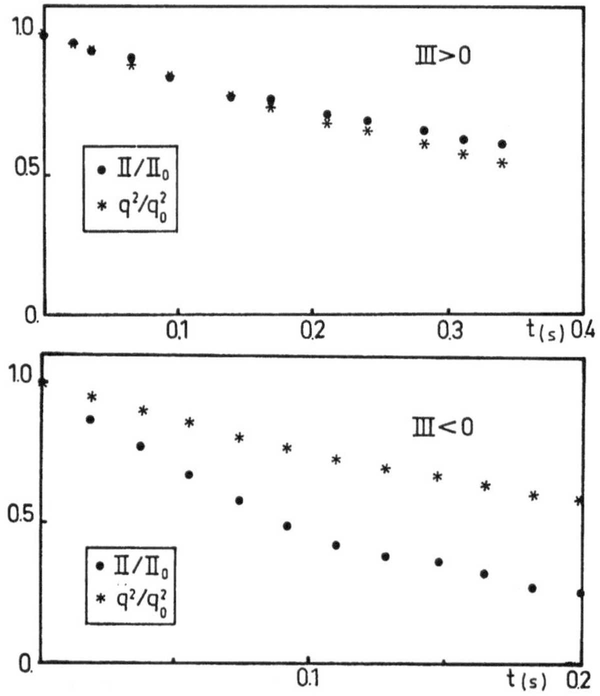

Fig. 8 Caption see opposite page

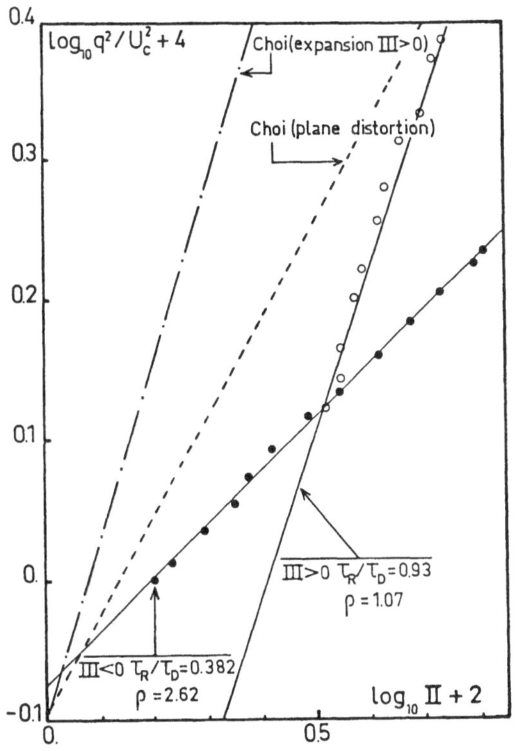

Fig. 9 - Evolution of $\log(q^2/U_c^2)$ against log II in the two present experiments

respective values at t = 0) are almost identical so that $\tau_R \simeq \tau_D$, whereas when III is negative, II evolves more rapidly than $\bar{q}^2$, so that $\tau_R < \tau_D$ (in fact $\tau_R \simeq 0.38\ \tau_D$).

Another presentation of this result is given in Fig. 9 where $\log(q^2/U_c^2)$ is plotted against log II. We recall that the slope of the approximately linear curves gives τ_R/τ_D (see section 1). In other words, the larger the slope, the slower is the return to isotropy. The recent results of CHOI (1983) are also included in Fig. 9. His first case ($III_o = +5.8\ 10^{-3}$, axisymmetric expansion) agrees very well with our experiment at $III_o = 3.1\ 10^{-3}$). The results of his second case ($III_o = +\ 0.75\ 10^{-3}$, plane distorsion) are located between our two sets of data for positive and negative invariant III, which shows that, as expected, a monotonic effect of invariant III exists on the return to isotropy mechanism.

Fig. 7 - Time evolution of the partition of kinetic energy among the velocity components. $*\ (u_1')^2/U_c^2$; $\square\ (u_2')^2/U_c^2$; $\bullet\ (u_3')^2/U_c^2$ (U_c = 13.5 m/s for III > 0 and U_c = 21.5 m/s for III < 0)

Fig. 8 - Time evolution of q^2/q_0^2 and II/II_0 ($q_0^2 = 44.5 \times 10^{-3}\ m^2s^{-2}$; II_0 = 0.055 for III > 0; $q_0^2 = 7.8 \times 10^{-2}\ m^2s^{-2}$; II_0 = 0.0621 for III < 0)

5. Consequences for Turbulence Modeling in the Physical Space

The evolution of homogeneous anisotropic turbulence in the absence of a mean velocity gradient is a basic situation for modeling workers because it is the simplest case in which the pressure strain correlation terms are different from zero and the dissipation terms possibly not isotropic.

The major attempt to close in a rational way the equations governing the Reynolds stress tensor, and the dissipation, was proposed by LUMLEY and NEWMAN in 1977. First they wrote these equations as follows :

$$\frac{d\overline{u_i u_j}}{dt} = \frac{1}{\rho}\overline{P\,(u_{i,j} + u_{j,i})} - 2\nu\,\overline{u_{i,k}u_{j,k}} \tag{14}$$

$$= \left[\frac{1}{\rho}\overline{P\,(u_{i,j} + u_{j,i})} - 2\nu\,\overline{u_{i,k}u_{j,k}} + \frac{2}{3}\,\bar{\varepsilon}\delta_{ij}\right] - \frac{2}{3}\,\bar{\varepsilon}\delta_{ij} \tag{15}$$

$$= -\,\bar{\varepsilon}\Phi_{ij} - \frac{2}{3}\,\bar{\varepsilon}\delta_{ij} \qquad \text{and} \tag{16}$$

$$\frac{d\bar{\varepsilon}}{dt} = -\,2\nu\,\overline{u_{i,k}u_{k,j}u_{j,i}} - 2\nu^2\,\overline{u_{i,kj}u_{i,kj}} \tag{17}$$

$$= -\psi\,\frac{\bar{\varepsilon}^2}{q^2} \tag{18}$$

with $\bar{\varepsilon} = \nu\overline{u_{i,k}u_{i,k}}$ as usual. Then, they regarded these equations as equations for Φ_{ij} and ψ and assumed that Φ_{ij} and ψ can be approached by functions of the present state defined by $\overline{u_i u_j}$, $\bar{\varepsilon}$ and ν or b_{ij}, q^2, $\bar{\varepsilon}$ and ν (and not by functionals of these arguments which would involve the history of b_{ij}, q^2 and $\bar{\varepsilon}$). Finally, the use of invariant theory permits them to obtain:

$$\Phi_{ij} = \beta(II,III,Re_L)\,b_{ij} + \gamma\,(II,III,Re_L)\,(b_{ik}b_{kj} - \frac{II}{3}\,\delta_{ij}) \tag{19}$$

and

$$\psi = \psi\,(II,III,Re_L) \quad . \tag{20}$$

At that stage, modeling concerns the unknown functions β, γ and ψ. Use is made of realizability conditions (SCHUMANN, 1977) and of existing experimental data. Cumbersome expressions are sometimes elaborated and LUMLEY (1978) suggested simplifications with no apparent loss of efficiency.

The simplification $\gamma = 0$ and ψ independent of invariant III, proves to be satisfactory for predicting experimental situations in which the invariant III is negative (LUMLEY, 1978). We made, therefore, numerical computations based on this simplification for both

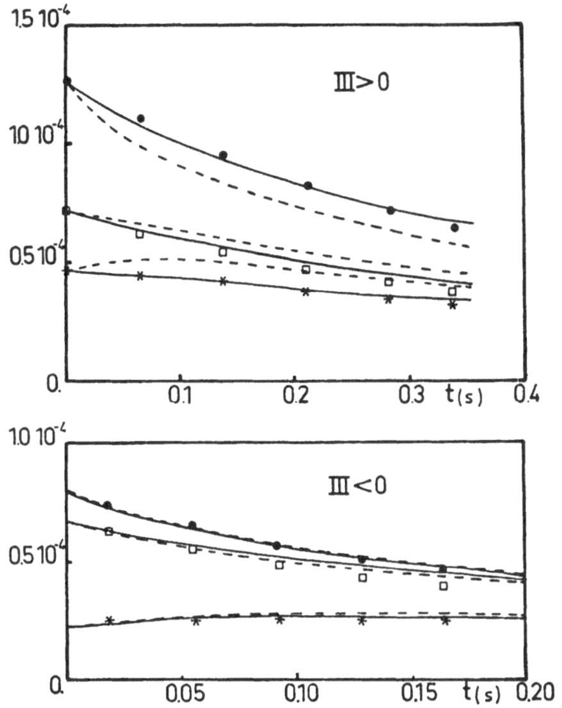

Fig. 10 - Comparison of the numerical predictions with the present experimental data [(---) model of Lumley; (——) model of Choi]; symbols for experimental data are the same as in Fig. 7

our situations. From figure 10 it is clear that the case of negative invariant III is again well predicted, but that a too fast return to isotropy is computed when invariant III is positive. In order to gain physical insight into the modeling, it is worth examining more closely what is implied by the assumption $\gamma = 0$. CHOI (1983) has shown explicitly that $\gamma = 0$ implies, from equation (16), whatever the form of β, that:

$$\frac{1}{2II}\frac{dII}{dt} = \frac{1}{3III}\frac{dIII}{dt} \quad , \tag{21}$$

or, after integration:

$$II^{1/2} = \alpha|III|^{1/3} \quad . \tag{22}$$

where α is a numerical factor (see appendix). By the way, we recall that β was simply chosen as a constant by ROTTA (1951). Now, it happens that expression (22) is necessarily satisfied by any axisymmetric turbulence for which α is: $(3/2)^{1/2}(4/3)^{1/3}$, or more simply $6^{1/6}$, a value which is also apparent from the plot of limiting values of the second and third invariants for turbulence (LUMLEY and NEWMAN, 1977).

It is, therefore, most interesting to considerer experimental situations for which the turbulence is not axisymmetric in order to draw conclusions about the validity of the simplified model $\gamma = 0$. This is done in Fig. 11 where $II^{1/2}$ is plotted versus $III^{1/3}$ for the few experiments available in the literature and also for the present investigation. The main observation is that the plotted curves are not, a priori, straight lines converging to the origin of the axes. More precisely it seems, when invariant III is positive, that turbulence evolves towards axisymmetry before returning to isotropy. When invariant III is negative, the same phenomenon does not seem to occur, but the data are scarce or exhibit a large scatter.

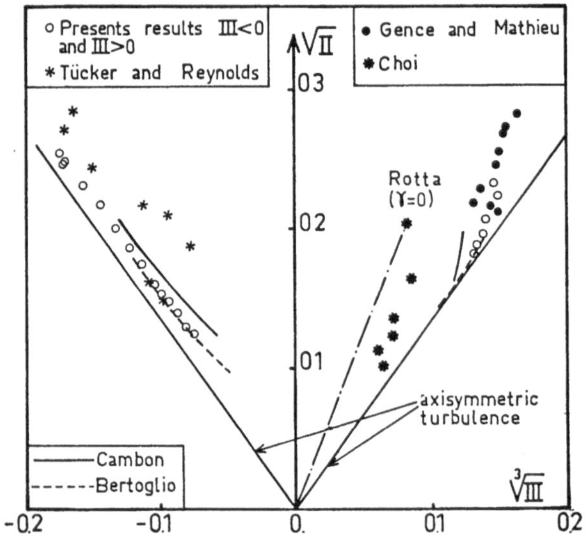

Fig. 11 - Evolution of $II^{1/2}$ versus $III^{1/3}$ for different experiments of return towards isotropy in non-axisymmetric cases.
(---) Prediction of Bertoglio } (Private communication)
(——) Prediction of Cambon }

In his newly suggested model, CHOI (1983) takes into account this evolution towards axisymmetry by setting γ different from zero. The non-linear terms in b_{ij} are therefore kept in Φ_{ij}. Adjustment of the dependance of γ on the invariants II and III was then made, as usual, from realizability conditions and with respect to experimental results. Use of CHOI's model for the present investigation predicts very well our data for positive invariant III (Fig. 10). This is a valuable improvement of LUMLEY's work (1978). However, when the invariant III is negative the superiority of CHOI's method over LUMLEY's is not so clear. CHOI's method seems to give by its nature a slightly too fast return to axisymmetry which is not apparent in the experimental data.

Now, we can return to the effect on the Reynolds stress of the slight mean velocity gradient which exists in the constant area duct used to study the return to isotropy. We have measured $\bar{U}_{1,1} = 0.17\ s^{-1}$ for III positive and $\bar{U}_{1,1} = 0.20\ s^{-1}$ for III negative. Estimates of the effect on $\overline{u_i u_j}$ was made as follows. For the linear part of the pressure-strain correlation terms, use was made of the model suggested by LUMLEY (1979). For the non-linear part of the pressure-strain correlation terms, use was made of LUMLEY's (1978) model in the case where the invariant III is negative and of CHOI's (1983) model in the case where the invariant III is positive. Comparisons of the computed values of $(u_i')^2$ $(i = 1,2,3)$, q^2 and II were made for the end of every duct between the two cases $\bar{U}_{1,1} = 0$ and $\bar{U}_{1,1}$ equal to the above values. The respective variations obtained are 0.3 %, 3.3 %, 3 %, 2.4 %, 13 % when invariant III is negative and 0.7 %, 4.5 %, 2.6 %, 2.4 %, 4.8 % when invariant III is positive.

6. Anisotropy of the Spectral Tensor

An insight into the physics involved by the sign of invariant III can be gained from an inspection of what happens in the wave-number space for the anisotropy of the spectral tensor $\psi_{ij}(\mathbf{K},t)$. Of course, measurements of $\psi_{ij}(\mathbf{K},t)$ would be interesting but they are almost beyond reach in non-isotropic turbulent fields for which all the components of $R_{ij}(\mathbf{r},t)$, the two points space velocity correlation tensor, should be measured.

Help can fortunately come from modeling in spectral space, and two methods are available for non-isotropic cases. The first one is the approach of CAMBON, JEANDEL and MATHIEU (1981) where the equation governing $\psi_{ij}(\mathbf{K},t)$ is averaged over the sphere of radius K and the closure made by an E.D.Q.N.M. technique. The second approach is due to BERTOGLIO (1982) who retains $\psi_{ij}(\mathbf{K},t)$ and models the three-dimensional transfer terms directly by an E.D.Q.N.M. assumption, and who made recent improvements to his numerical procedure (3-D computations are now possible for every value of the wave number, BERTOGLIO, private communication). The anisotropic spectrum $\psi_{ij}(\mathbf{K},0)$ needed for the initial condition was deduced by submitting the isotropic grid-generated turbulence spectrum (COMTE-BELLOT and CORRSIN (1971)), to the irrotational straining of the distorting ducts. These initial data do not correspond exactly to those in the present experiments (as already pointed out in section 2), but closely related trends can be obtained since the turbulence Reynolds number differs only by a factor of about two.

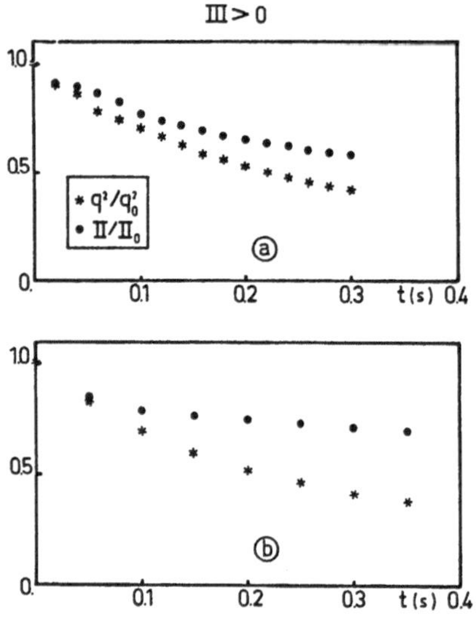

Fig. 12 - Numerical spectral simulations (a) Bertoglio (b) Cambon, Case III > 0.

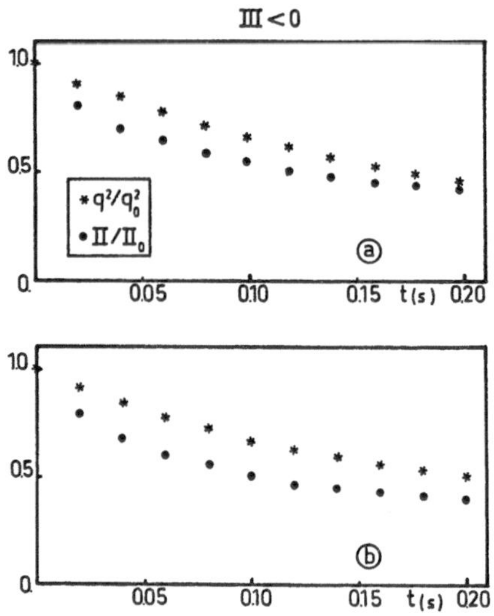

Fig. 13 - Numerical spectral simulations (a) Bertoglio (b) Cambon, Case III < 0.

As a preliminary check let us look at the results obtained after an integration of $\psi_{ij}(\mathbf{K},t)$ over the wave vector $\mathbf{K}$, which gives $q^2(t)$ and II(t). Their time evolutions are reported in Figs. 12 and 13 and exhibit the same trend as the experimental findings, i.e., the slowest return to isotropy is when the invariant III is positive. From Fig. 12 it can be seen that even $\tau_R > \tau_D$ (the curve of II is above the curve of q^2).

The respective evolution of the invariants II and III can also be obtained. They are indicated in Fig. 11 under the form $II^{1/2}$ versus $III^{1/3}$. It is worth noting that the two ways of modeling non isotropic turbulence in spectral space predict the experimental findings very well, i.e. that turbulence evolves towards axisymmetry before its return to isotropy when invariant III is positive, and does not do so when this invariant is negative.

Now that these global checks have been done, we can focus our attention on the anisotropy of the spectral tensor $\psi_{ij}(\mathbf{K},t)$ itself. This anisotropy can be considered along two complementary ways by taking advantage of the local coordinate system built on the wave number $\mathbf{K}$ and two axes in the plane perpendicular to $\mathbf{K}$ (the usual reference system of spherical coordinates). For an incompressible fluid the spectral tensor expressed in these new axes reduces to four non zero components ψ'_{ij} related to that plane (i and j = 1 or 2) with Hermitian symmetry. Imaginary parts of ψ'_{ij} are zero (in usual tur-

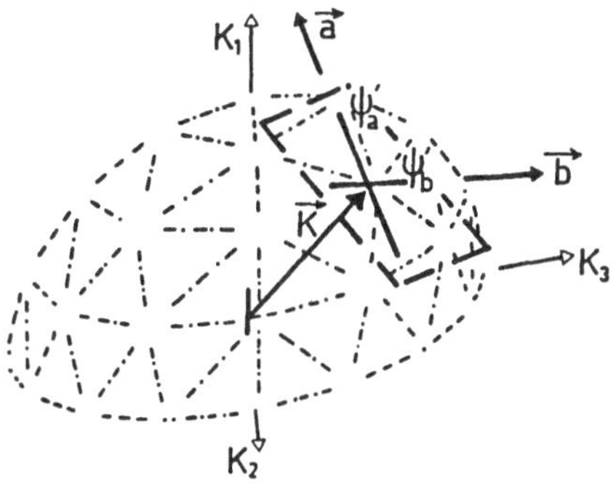

Fig. 14 - In the plane normal to the wave vector **K** (also tangent to the sphere of radius $K = \|\mathbf{K}\|$: **a**, **b** are the principals axes of ψ'_{ij}; ψ'_a, ψ'_b are segments proportional to the principal values of ψ'_{ij} (for isotropic turbulence $\psi'_a = \psi'_b$). (CRAYA, 1958; Bertoglio, 1982)

bulence there is no helicity), so that the principal axes and the principal values of ψ'_{ij} can be easily visualized in the plane normal to **K**, for each vector **K** (Fig. 14).

For a given wave number modulus K, it is then of interest (i) to determine if all the wave number directions are uniformly fed, as they would be for isotropic turbulence, and (ii) to compare the two principal values of ψ'_{ij}, which would be identical for isotropic turbulence.

Figure 15 visualizes $\psi'_{ij}(\mathbf{K},0)$ at the beginning of the return to isotropy when invariant III is positive, for a relatively small value of the wave number modulus (close to that at which the maximum of the kinetic energy spectrum is located, $K = 0.3\ cm^{-1}$) and for a values 10 times larger ($K = 3.47\ cm^{-1}$). For $K = 0.3\ cm^{-1}$ depleted zones are clearly visible in the "artic" and "antartic" regions along K_3 on the left and right (large velocity components in direction 3 preclude wave-numbers in the same direction). The principal values ψ'_a, ψ'_b are also very different, one being almost zero. On the other hand, the spectral contributions at $K = 3.67\ cm^{-1}$ are much more uniformly distributed, as expected for small structures from the local isotropy concept.

Finally, let us try to relate the anisotropy of the spectral tensor to the invariants II and III used in the physical space. General expression are probably beyond reach, but fortunately simple relations can be obtained for axisymmetric turbulence :

$$II(t) = \frac{2}{3}\frac{1}{(q^2)^2}\left\{\int_0^\infty [\varphi_{11}(K,t) - \varphi_{33}(K,t)]dK\right\}^2 \qquad (23)$$

$$III(t) = \frac{2}{9}\frac{1}{(q^2)^3}\left\{\int_0^\infty [\varphi_{11}(K,t) - \varphi_{33}(K,t)]dK\right\}^3 \quad ;$$

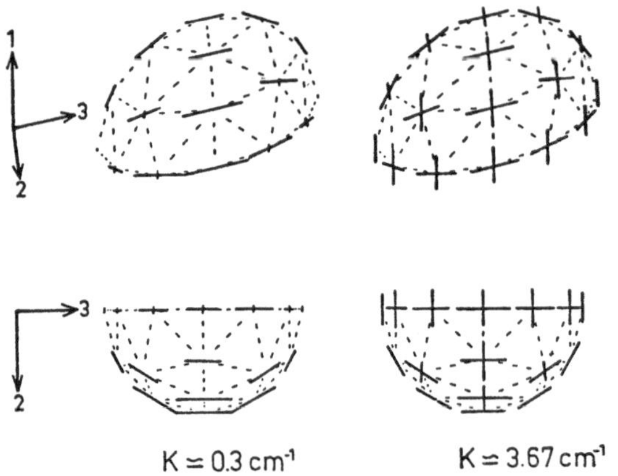

Fig. 15 - Visualization of the anisotropy of the spectral tensor for two values of K (time t = 0)

$$\left[\overline{u_i u_j}(t) = \int_0^\infty \varphi_{ij}(K,t)\,dK\right] \tag{24}$$

where x_1 is the axis of symmetry. Then, it is clear that the quantity

$$\Delta(K,t) = |\varphi_{11}(K,t) - \varphi_{33}(K,t)|/q^2(t) \tag{25}$$

is a spectral measure of the anisotropy and the sign of $\int_0^\infty \Delta(K,t)\,dK$ is that of invariant III.

The numerical results obtained by CAMBON's method for two turbulence fields specifically generated by axisymmetric distorsions such that invariants II are equal and invariants III are opposite, are indicated in Fig. 16 for t = 0 (end of the distortion and beginning of the return to isotropy).

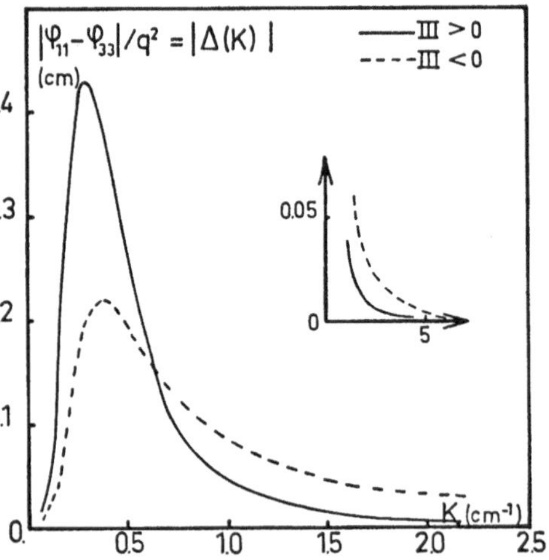

Fig. 16 - Anisotropy distribution among wave number modulus (t = 0)

The areas under the two curves of Fig. 16 are almost the same. The main result is therefore that the anisotropy is concentrated, in structures which are larger for positive rather than for negative invariant III. It is then tempting to conclude that these large and energetic structures will need a relatively long time for their return to isotropy.

7. Conclusions

In the present investigation we have tried to illustrate the effect of unequal kinetic energy partition among the three fluctuating velocity components, on the rate of return to isotropy of homogeneous turbulence. This unequal partition of energy was particularly reflected in the third invariant, III, of the deviatoric tensor of the Reynolds stress.

In the situation where one component only is very energetic (positive values of III), the return to isotropy was found to be slow and possibly preceeded by an evolution towards axisymmetry for the two small components. In the situation where two components are initially large (negative values of III), a case for which the energy was already expected to be more uniformly distributed in the physical space, the return to isotropy was indeed observed to be faster, compared to the decay time, and with no visible tendency to axisymmetry. For brievety, here, the limit between the fast and the slow rates was located at III = 0, but, of course, a slightly different value is certainly possible.

The third invariant is directly involved in the modeling of turbulence in the physical space, along with several numerical factors which have to fit experimental data. We therefore believe that attention has to be paid to the class of data to be taken into account, according to the sign and value of their invariant III.

Turbulence modeling in wave number space seemed to work whatever the sign of invariant III. The reason could be that one numerical factor only has to be adjusted and that this can be done by comparison with theoretical models (test field) rather than with experimental data. In the present investigation spectral modeling permitted us to visualize the anisotropy of the spectral tensor which is beyond reach in experiments.

The present work was limited to non-isotropic turbulence generated by irrotational strains. Other types of anisotropy would certainly be of interest as intermediate steps towards usual turbulent shear flows in which invariant III can be expected to be positive, since in most cases the longitudinal velocity component is the largest, by its direct gain of energy from the mean flow.

Acknowledgments. This work was supported by the Direction des Recherches Etudes et Techniques under contract no. 83/286.

Hearty thanks are addressed to J.P. BERTOGLIO and C. CAMBON who promptly made all the numerical turbulence modeling needed for sections 2 and 6. The assistance of O. VITALI who drew most of the figures and of F. MAUPAS who did the typescript is also gratefully acknowledged.

Appendix: Relation between $II^{1/2}$ and $III^{1/3}$ when $\gamma = 0$

From expression (1) and (16), and the definition of $\bar{\varepsilon}$, i.e.:

$$b_{ij} = \frac{\overline{u_i u_j}}{q^2} - \frac{\delta_{ij}}{3}$$

$$\frac{d}{dt}\overline{u_i u_j} = -\bar{\varepsilon}\Phi_{ij} - \frac{2}{3}\bar{\varepsilon}\delta_{ij}$$

$$\frac{d}{dt} q^2 = -2\bar{\varepsilon}$$

one obtains:

$$\frac{d}{dt} b_{ij} = \frac{d}{dt}\left(\frac{\overline{u_i u_j}}{q^2}\right)$$

$$= -\frac{1}{q^2}\left(\bar{\varepsilon}\Phi_{ij} + \frac{2}{3}\bar{\varepsilon}\delta_{ij}\right) + \frac{\overline{u_i u_j}}{q^2}\frac{2\bar{\varepsilon}}{q^2}$$

$$= -\frac{\bar{\varepsilon}}{q^2}(\Phi_{ij} - 2b_{ij}) \quad .$$

Now, if $\gamma = 0$ in expression (19), Φ_{ij} reduces to:

$$\Phi_{ij} = \beta b_{ij} \quad ,$$

so that:

$$\frac{d}{dt} b_{ij} = -\frac{\bar{\varepsilon}}{q^2}(\beta - 2)b_{ij} \quad .$$

The time derivatives of II and III are then simply:

$$\frac{dII}{dt} = 2b_{ij}\frac{d}{dt}b_{ij} = -2\frac{\bar{\varepsilon}}{q^2}(\beta - 2)II \quad \text{and}$$

$$\frac{dIII}{dt} = 3b_{ik}b_{kj}\frac{d}{dt}b_{ij} = -3\frac{\bar{\varepsilon}}{q^2}(\beta - 2)III \quad ,$$

which gives,

$$\frac{1}{2II}\frac{dII}{dt} = \frac{1}{3III}\frac{dIII}{dt} \quad ,$$

and hence:

$$II^{1/2} = \alpha|III|^{1/3}$$

where α is a numerical factor. This result holds whatever the dependence of β on II, III and Re_L.

References

BERTOGLIO J.P., 1982, A model of three-dimensional transfer in non-isotropic homogeneous turbulence, Turbulent Shear Flows 3, Springer Verlag, p. 253-261.

CAMBON C., BERTOGLIO J.P. & JEANDEL D., 1981, Spectral closure of homogeneous turbulence, Comparison of computations with experiment, AFOSR-HTTM, Stanford Conf. on complex turbulent flows, p. 1307-1311.

CAMBON C., JEANDEL D. & MATHIEU J., 1981, Spectral modelling of homogeneous non-isotropic turbulence, J. Fl. Mech. 104, p. 247-262.

CHOI K.S., 1983, A study of the return to isotropy of homogeneous turbulence, Ph. D. Thesis, Cornell Univ.

COMTE-BELLOT G. & CORRSIN S., 1966, The use of a contraction to improve the isotropy of grid-generated turbulence, J. Fl. Mech. 25, p. 657-682.

COMTE-BELLOT G. & CORRSIN S., 1971, Simple Eulerian time correlation of full and narrow-band velocity signals in grid-generated "isotropic" turbulence, J. Fl. Mech. 48, p. 273-337.

CRAYA A., 1958, Contribution à l'analyse de la turbulence associée à des vitesses moyennes, Publi. Sci. Tech. Ministère Air, Paris, n° 345.

GENCE J.N. & MATHIEU J., 1980, The return to isotropy of an homogeneous turbulence having been submitted to two successive plane strains, J. Fl. Mech. 101, p. 555-566.

GENCE J.N., 1983, Homogeneous turbulence, Ann. Rev. Fl. Mech. 15, p. 201-222.

LE PENVEN L. & GENCE J.N., 1983, Quelques nouveaux résultats concernant l'étude expérimentale du retour à l'isotropie du tenseur de Reynolds dans le cas où le troisième invariant de son déviateur est positif, C.R. Acad. Sciences Paris, série II, 297, p. 389-392.

LUMLEY J.L. & NEWMAN G.R., 1977, The return to isotropy of homogeneous turbulence, J. Fl. Mech. 82, p. 161-178.

LUMLEY J.L., 1978, Computational modeling of turbulent flows. In "Advances in Applied Mechanics", 18, 123. Academic Press.

LUMLEY J.L., 1979, Second order modeling of turbulent flows. Lecture series "Prediction methods for turbulent flows" at the Von Karman Institut for Fluid Dynamics, Belgium.

MILLS R.R. Jr. & CORRSIN S., 1959, Effect of contraction on turbulence and temperature fluctuations generated by a warm grid, NASA Memo 5-5-59 W.

ROTTA J.C., 1951, Statistische theorie nichthomogener turbulenz, Mitteilung-Zeitschrift für Physik, 129, p. 547-572.

SCHUMANN U. & PATTERSON G.S., 1978, Numerical study of the return of axisymmetric turbulence to isotropy, J. Fl. Mech., 88, p. 711-735.

SCHUMANN U., 1977 , Realizability of Reynolds-stress turbulence models, Phys. Fluids 20, p. 721-725.

TUCKER H.J. & REYNOLDS A.J., 1968, The distortion of turbulence by irrotational plane strain, J. Fl. Mech., 32, p. 657-673.

UBEROI M.S., 1957, Equipartition of energy and local isotropy in turbulent flows, J. Appl. Phys. 28, p. 1165-1170.

WARHAFT Z., 1980, An experimental study of the effect of uniform strain on thermal fluctuations in grid-generated turbulence, J. Fl. Mech. 99, p. 545-573.

Random Incompressible Motion on Two and Three-Dimensional Lattices and Its Application to the Walk on a Random Field

Michael Karweit

Department of Chemical Engineering, The Johns Hopkins University
Baltimore, MD 21218, USA

1. Introduction

One of the more important features of turbulent flow is its ability to disperse contaminants. But, because the most convenient way of describing the turbulent field is in terms of Eulerian or "laboratory" coordinates, and particle dispersion is naturally expressed in Lagrangian or "material" coordinates, dispersion is not well predicted quantitatively. The principal difficulty is that the two-point Eulerian statistics which characterize the velocity field do not admit to unique two-point Lagrangian displacement statistics which describe particle dispersion.

A way to gain insight into this problem is to work with a process which is conceptually simpler, and which exhibits some of the same difficulties. In this paper we take this approach and develop an analog to turbulent dispersion based on an extension to the principle of the random walk.

The classical random walk was first used by Einstein (1905) as a means by which to understand the relatively simple problem of brownian motion. In the random walk technique, hypothetical particles are given randomly selected, discrete movements and the statistics of the motions of an ensemble of particles are accumulated. The approach has since been used by others for more complex problems, e. g. Bugliarello and Jackson (1964) for laminar convective diffusion. Karweit and Corrsin(1972) expanded the idea of a random walk from "simple" to "compound" to simulate material line growth in turbulence. However, such classical random walks are only peripherally related to turbulent dispersion because particle movement is assigned to the particle and no consideration is given to a field in which the particle might be moving. In the classical scheme only statistics following the particle, i.e., Lagrangian statistics, are calculable.

To more completely model turbulent dispersion, Lumley and Corrsin (1959) introduced the concept of a "walk on a random field". Here, the motion of fluid parcels (or particles) is prescribed by random velocities attached to a field of lattice points in space rather than to the particles themselves; and the movement of particles proceeds from point to point as dictated by the velocity at the point in the field where each is located. Both the Eulerian statistics of the underlying velocity field and the Lagrangian statistics of particle motion are obtainable. Patterson and Corrsin (1966) used this idea to model the dispersion of particles moving on time-evolving, one-dimensional, latticed, binary-velocity fields. Their experiments considered dispersion on fields which had a variety of spatial and temporal correlations. A fundamental limitation in their work, however, is that one-dimensional, random velocity fields are necessarily compressible; and consequently their results have marginal applicability to "real" turbulent dispersion.

The extension of those experiments to the more applicable case of two and three space dimensions, however, proved to be difficult. The problem lay in producing acceptable fields. What is needed for two dimensions are fields which are random, bi-directional-binary, and incompressible; and they must have some "persistence" in both space and time, i.e., correlation. (Our definition of "bi-directional-binary" is +1 or -1 in the direction of either of the two coordinates). Further, their properties should be statistically homogeneous and "isotropic". Patterson (1958) derived the equations of constraint for incompressibility, but did not develop an algorithm for producing satisfactory fields. It has been only recently that Karweit (1984) was able to demonstrate a possible method for generating acceptable fields. In that work he also used those fields to carry out several dispersion experiments.

A geometric portrayal of these latticed fields is given in Figure 1. Here we show an example of a two-dimensional, bi-directional-binary random velocity field--in this case a Markovian one, i.e., one having no spatial correlation.

It is this class of random fields and walks on these random fields that we explore in this paper. Partially replicating the experiments of Karweit (1984), we generate correlated two-dimensional, bi-directional-binary fields, (from here on referred to simply as discrete fields) and investigate some of their unique properties. Although we demonstrate the extension of our procedures to three-dimensional, tri-directional-binary ones, our principal treatment will be for the two-dimensional case. We show the results of several numerical dispersion experiments. These experiments consist of following the trajectories of fluid parcels whose motions are prescribed by the discrete fields. We calculate Eulerian statistics of the fields; and by accumulating the details of the particle

Figure 1. A typical two-dimensional Markovian discrete velocity field.

trajectories over an ensemble of realizations, we produce dispersion statistics for the particles. The relation between the field statistics and the dispersion statistics are presented. Also, we compare these statistics from experiments made on discrete fields to those obtained from analogous experiments performed on continuous, two-dimensional, incompressible fields.

2. Description and Generation of Discrete Fields

A discrete field consists of a two-dimensional, square lattice of bi-directional-binary velocity vectors. The vectors can take on only one of four values: +1 or -1 in the x-direction, or +1 or -1 in the y-direction. Thus a velocity vector always points to one of the four adjacent lattice points(nodes). A field is incompressible if at every node exactly one vector is leaving and one vector (from another node) is entering. We disallow an exchange of exit(entry) vectors between two nodes. So, in a unit of time a parcel of fluid leaves every point in the lattice, and a different parcel of fluid enters every point from an adjacent node.

As indicated above, we are interested in producing not just arbitrary discrete fields but discrete fields which are statistically homogeneous and "isotropic", and have particular spatio-temporal characteristics, e.g., specified spatial and temporal correlations. Earlier attempts to produce such fields by Patterson(1958) and later by Maxey(1979) focused on generating incompressible discrete fields--constrained fields--in terms of preassigned unconstrained fields. Their algorithms were designed to translate the easily-generated unconstrained fields into correct incompressible fields. Their method

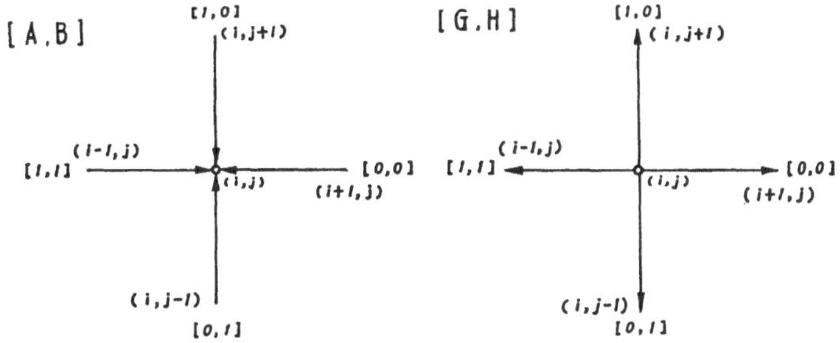

Figure 2. Notation for describing the velocity at lattice node (i,j). Values for the two ordered pairs of binary variables [A,B] and [G,H] define the directions for the entry and exit vectors, respectively. Lattice nodes from (to) which entry (exit) vectors point are also given.

was successful in that correct incompressible fields could be produced, but the fields thus produced were spatially biased, and hence not acceptable. In this paper we proceed somewhat differently and manage to overcome the difficulty.

Consider the notation presented in Figure 2 for describing the entry and exit vectors at a point. [G,H](i,j) identifies the exit vector at the point (i,j) and [A,B](i,j) identifies the entry vector at (i,j)--A, B, G, H are all binary variables with permissible values of zero or one. The two vectors taken together comprise what we will call the flow vector [G,H,A,B](i,j). This vector completely specifies the flow through node (i,j). It is with this flow vector that we will develop a procedure for producing incompressible discrete fields.

Provided all four variables A, B, G, H are defined at every point in the lattice, incompressibility is automatically satisfied, i.e., there is exactly one entry and one exit vector passing through the point. The constraint of not allowing an exchange of exit(entry) vectors between two nodes is satisfied provided $[G,H](i,j) \neq [A,B](i,j)$. The complication is that one cannot arbitrarily specify four values at a given point, because entry(exit) vectors attached to one node are exit(entry) vectors for an adjacent node. Denoting the conjugate of a binary variable, say G, as $G' = 1 + G \pmod 2$, we can list the consistency relations between nodes for the exit vector [G,H](i,j) in terms of the algebraic relations:

$$G'H'(i,j) = AB\ (i+1,j)$$
$$GH'\ (i,j) = A'B\ (i,j+1)$$
$$GH\ \ (i,j) = A'B'(i-1,j)$$
$$G'H\ (i,j) = AB'\ (i,j-1).$$

An analogous set of consistency relations also exists for the entry vector [A,B](i,j). Thus for every entry(exit) vector explicitly assign to one node, we implicitly assign an exit(entry) vector at an adjacent node.

We now describe a procedure for systematically defining the complete flow vector [G,H,A,B](i,j) which is consistent with flow vectors at adjacent nodes. Suppose that one is allowed to prescribe explicitly exit(entry) elements of the flow vector at a point with the restriction that H (or B) can be assigned only the value "0". G(or A) can take on either of the permissible values. This corresponds to defining an exit(entry) vector pointing toward(away from) nodes (i,j+1) or (i+1,j). (See Figure 2.) Such an assignment implicitly defines an entry(exit) vector in the flow vector of an adjacent node as follows:

	[G,H,A,B]		[G,H,A,B]
a)	[1,0, ,](i,j)	=	[, ,0,1](i,j+1)
b)	[0,0, ,](i,j)	=	[, ,1,1](i+1,j)
c)	[, ,1,0](i,j)	=	[0,1, ,](i,j+1)
d)	[, ,0,0](i,j)	=	[1,1, ,](i+1,j).

Note that, although H and B can be prescribed explicitly as only "0", they carry the value "1" as elements in the flow vector of the adjacent nodes. Each of these four possible assignments a) -- d) is permissible provided that the corresponding elements of the flow vector in the specified adjacent node have not been previously assigned. For example, were elements A and B at node (i,j+1) defined by a previous assignment (not necessarily related to node (i,j)), then alternative a) would not be permitted.

The procedure begins at node (0,0). Since we would like to minimize edge effects in producing our finite, discrete fields we must assume some "previous" exit(entry vector(s) as partially defining the four elements of the flow vector at (0,0). Then we proceed to complete the flow vector with alternatives corresponding to a) -- d). If we continue to nodes (0,1) and (1,0), and then to nodes (0,2), (1,1), and (2,0), and so on, along diagonals from (0,k) to (k,0), an incompressible field will be produced a diagonal at a time. Only the first node along any diagonal (0,k) requires an assumption of "previous" assignments from hypothetical nodes outside the finite field. As the flow vector for each node is defined, the implied G,H,A,B values for the flow vector at the relevant adjacent node is also assigned. Then, when the flow vector of that node is to be specified, it will already contain some assigned elements

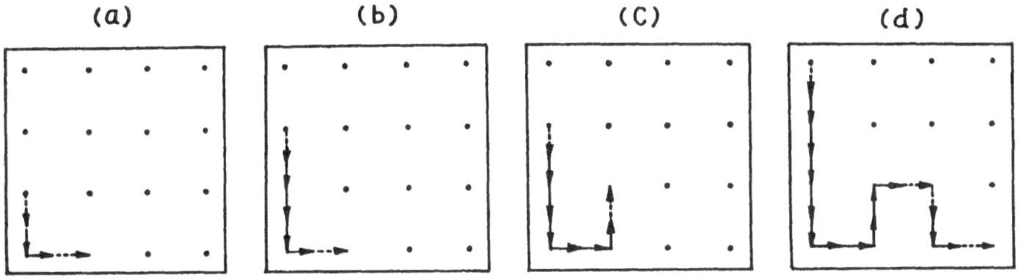

Figure 3. Constructing a discrete field. Panels (a), (b), and (c) illustrate the successive completing of flow vectors at nodes (0,0), (0,1), and (1,0), respectively. Both entry and exit vectors are shown. Dotted vectors are constraints for nodes not yet addressed. (d) shows the results after operating on the next diagonal, (0,2), (1,1), and (2,0).

corresponding to portions of the flow vector defined earlier at an adjacent node. Figure 3 illustrates several steps in this process.

As one moves from node to node along a diagonal completing flow vectors, depending on previous assignments, there will be 0 to 4 remaining alternatives. It is through these remaining alternatives that the spatial characteristics of the discrete field can be dictated.

Note: either a complete field of exit vectors [G,H](i,j) or a complete field of entry vectors [A,B](i,j) is sufficient to fully describe the discrete velocity field, so our using both the entry and exit information is redundant. However, this redundancy permits us to describe the complete flow through a node in terms of information at that node.

3. Production of Markovian and Correlated Discrete Fields

Our main objective is to produce a discrete analog to twodimensional turbulent flow. Assuming that time as well as space is defined in discrete units, we can emulate a time-dependent velocity field as a sequence of latticed fields, whose change from timestep to timestep is characterized by temporal correlation.

In the last section we discussed a process by which discrete fields might be generated. We noted that particular spatial or temporal properties could be incorporated into the fields through the assignment of elements of the flow vector at each point. In this section we describe the production of discrete, time-evolving fields on which our random walk experiments will be based. In each case we produce an ensemble of 64 realizations of sequences of fields, each sequence consisting of 15 discrete fields (timesteps), each field in the sequence being generated on a 32 x 32 lattice.

Two classes of sequences are produced: one, Markovian--having neither spatial nor temporal persistence; the other, "correlated"--having both spatial and temporal persistence. The spatial properties of these two classes will be characterized in terms of the Karman-Howarth longitudinal and lateral velocity correlation functions f(r) and g(r) [see e.g., Hinze (1975)]. Purely temporal properties will be given in terms of R_E (T)--the Eulerian temporal correlation function. "r" and T are separations in space and time and do not depend on absolute position or time. Note that r and T can take on only discrete values corresponding to gridspacing and timestepping. Further, f(r) and g(r) are definable only along the two axis directions. Use of these functions requires that the considered fields be statistically homogeneous and "isotropic" in space and statistically stationary in time. Consequently, we restrict our attention to fields having these attributes. Of the two classes, the "correlated" sequences are, of course, more relevant to turbulent dispersion. However, we treat Markovian fields as a basis for comparison. We begin with a discussion of the Markovian fields.

Markovian fields are produced with our algorithm by configuring the flow vector at each point with randomly selected elements, i.e.,randomly selected vector alternatives a) -- d), as described in the last section. We discovered it necessary to assign slightly non-equal probabilities to these alternatives to compensate for a directional preference of the procedure. But the ensembles of fields thus produced were found to be empirically homogeneous and "isotropic". Even a Karman-Howarth-like two-point velocity statistic designed to test the diagonal directions yielded no irregularities. Potential "edge" effects--associated with having to preassign some elements of the flow vector at the leftmost column of lattice points--were minimized by creating 34 x 34 discrete fields and discarding the two lower and leftmost rows to yield the final 32 x 32 fields. Refer again to Figure 1 for a typical field.

The production of correlated discrete fields is more complex, because correlation must be introduced both in space and time. Patterson and Corrsin (1966), working with one-dimensional, time-dependent discrete fields, introduced correlation by smoothing Markovian fields with various low-pass filters. But their fields were compressible. With two space dimensions and the constraint of incompressibility, their technique does not ensure that an initially-incompressible field remains so after filtering. So we require a different scheme.

In our algorithm, proper definition of the elements of the flow vector at each point in the field automatically generates correct incompressible fields. To introduce correlation, we use "preferred" flow directions at each point in space and time rather than randomly-selected flow directions as in the Markovian fields. We begin by defining a field of "preferred" directions which contains the desired space and time correlations. Then we proceed with the algorithm to generate the discrete field, using the preferred directions as much as possible to determine the flow vectors. The extent to which the "preferred" directions are admissable as flow vectors is the degree to which the discrete field matches the field of preferred directions and exhibits the same correlation properties.

The field of "preferred directions" is the mechanism by which correlation is introduced into the discrete fields and may be generated in any one of a number of ways. Directions may be continuously variable and incompressibility need not be satisfied. For example, spatial and temporal smoothing of a perfectly random field will yield a satisfactory preferred field.

In this work we attempt to produce correlated, discrete fields which are suggestive of "real" flows. So, our "preferred" fields are derived from the Euler (inviscid Navier-Stokes) equations. We define initial two-dimensional velocity fields in terms of Fourier modes and integrate the Fourier form of the equations over time, throwing out the higher order modes introduced by the non-linear terms. (This simplification was first used by Lorenz (1960) and by Kraichnan (1963) to study interactions between velocity wavevectors). We limit the complexity of our fields to six band-limited velocity wavevectors $k_1, \ldots, k_6$: $k_1, \ldots, k_4$ are chosen randomly, and $k_5 = -k_1 - k_2$ and $k_6 = -k_3 - k_4$ to produce "triad interaction". All velocity amplitudes and wavevector phases are chosen randomly. Results are then interpreted on a square lattice to yield the "preferred" fields mentioned above. In producing the discrete fields from the Euler fields only the velocity direction is considered; velocity amplitude is ignored. (Hereafter, we will refer to the Euler fields as "continuous" fields.) An example of a continuous field and the resulting discrete field is given in Figure 4. The fields at two points in time are shown. Compare the spatial persistence of these correlated discrete fields with that of the Markovian discrete field in Figure 1.

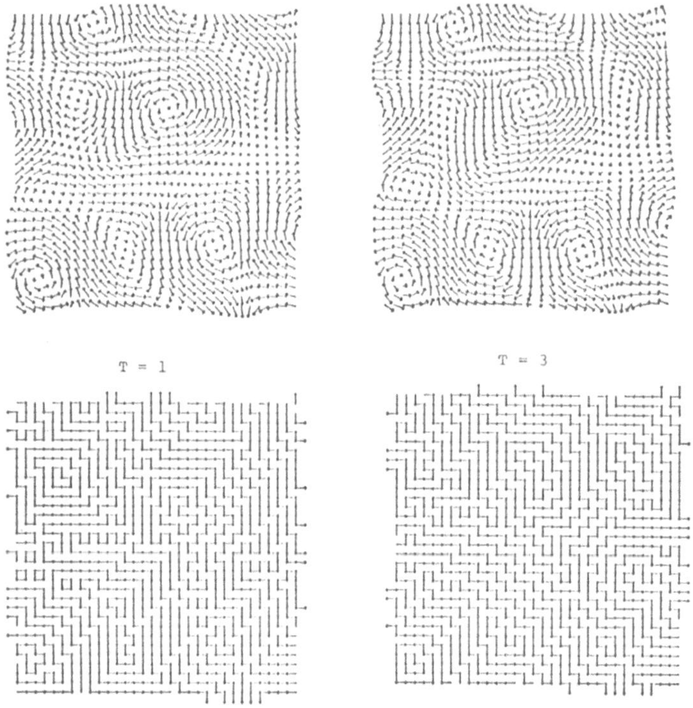

Figure 4. A continuous field at two different times; and the corresponding discrete field.

4. Statistics of the Fields

Based on the exit vectors for the discrete fields and the complete velocity vectors for the continuous fields, we calculate the correlations f(r), g(r), and R_E (T) for all of our ensembles.

The Markovian fields exhibit some characteristics unique to incompressible discrete fields. For example, the longitudinal correlation function f(r) is not zero for all $r \neq 0$. Refer to Figure 5. Rather, it oscillates between zero and a positive envelop which decreases with r. This result is a manifestation of incompressibility. Consider the calculation of $f(1) = \langle u(i,j)u(i+1,j) \rangle$ where $\langle \; \rangle$ denotes ensemble averaging over all realizations and all i and j, and u is the velocity in the i (or X) direction. Let the velocity at a node (i,j) be defined solely in terms of its exit vector. Then node (i,j) has only two permissible values of u: +1 , corresponding to an exit vector pointing to node (i+1,j), and -1, corresponding to an exit vector pointing to node (i-1,j). For two adjacent nodes (i,j) and (i+1,j) there are two ways that the u's can be positively correlated: u(i,j) = u(i+1,j) = 1, and u(i,j) = u(i+1,j) = -1. However, there is only one way that the u's can be negatively correlated: u(i,j)

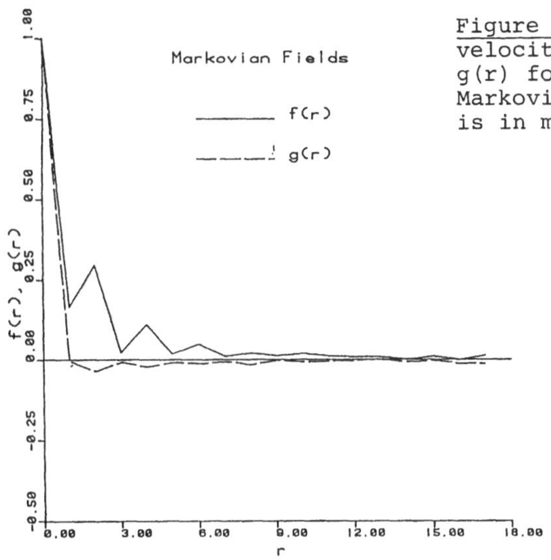

Figure 5. Longitudinal and lateral velocity correlation functions f(r) and g(r) for an ensemble of 64 discrete Markovian fields over 15 timesteps; r is in mesh lengths.

= -1, u(i+1,j) = 1. The only other possibility, u(i,j) = 1 and u(i+1,j) = -1, is precluded by the constraint of incompressibility. Thus, in calculating f(1), on the average there will be twice as many +1 contributions to the correlation as there will be -1 contributions; and the expected value of f(1) is positive, not zero as might be anticipated from a Markovian field. Carrying this analysis to larger r's confirms the oscillatory behavior of f(r) for these discrete, "uncorrelated" fields. An analogous oscillatory result is obtained for the lateral velocity correlation g(r), as well. Note that these results are only semi-quantitative; because we consider just the involved nodes and not the potential constraints due to surrounding nodes.

Spatial correlations for the correlated fields are shown in Figure 6. Curves for both the discrete and continuous cases are given. First, note that the oscillatory behavior of the correlations for the discrete fields is retained as in the Markovian case. Second, a substantial degradation of correlation between the continuous and discrete fields exists. Since the discrete fields were derived from the continuous ones, we might expect that the spatial correlations would be similar. But on reflection, we find that the smaller correlation extent for the discrete fields is not so surprising.

A "preferred" velocity at a lattice point in a discrete field is produced from the velocity at the same point in the continuous field. But virtually none of these lies along one of the four permitted directions. So a decision must be made as to how the preferred velocity should be assigned. A deterministic assignment

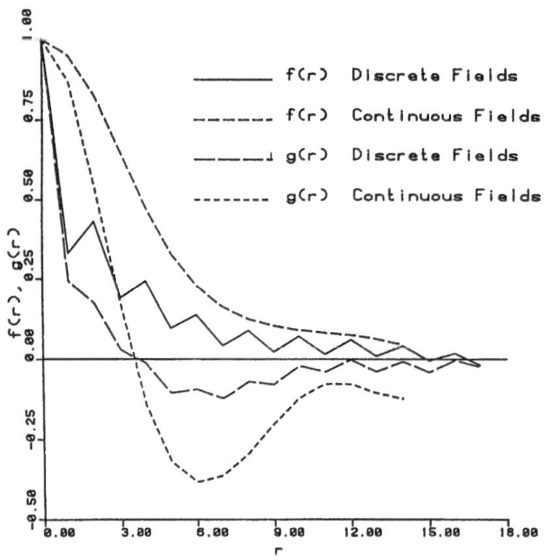

Figure 6. Longitudinal and lateral velocity correlation functions f(r) and g(r) for discrete correlated fields and their continuous counterparts; r is in mesh lengths; results based on 64 realizations over 15 timesteps.

based on closest permitted direction yields discrete fields whose streamlines are relatively unrelated to those in the continuous fields. A probabilistic assignment weighted by closest permitted direction yields fields whose streamlines are more commensurate with the continuous case, but whose spatial correlations are degraded.

To illustrate the effect of these two methods for selecting a preferred direction, consider a continuous velocity field consisting of uniform flow in a direction of 43 degrees with respect to the x-axis. A deterministic assignment would produce a discrete velocity field uniform along the x-axis--the closest permitted direction. A probabilistic assignment, on the other hand, would produce a random zig-zag flow whose average orientation would be 43 degrees. The spatial correlation of the former is unity--the same as the continuous field; but the correlation of the latter is substantially less. Since one of our goals is to explore particle dispersion, and particle dispersion is most influenced by streamline geometry we choose the latter method to generate the discrete fields.

Figure 7 shows R_E (T) for the continuous and correlated discrete fields. Again, the correlation for the discrete fields is smaller, because from timestep to timestep the discrete fields are manufactured probabilistically from the continuous ones. R_E (T) for the Markovian fields is not shown, but of course is zero for $T \neq 0$, because sequential fields are produced entirely independently.

An incompressible, bi-directional-binary, latticed field is simply an analogy to a "real" field. Here we would like to carry that analogy further to include the concepts of vorticity and strain

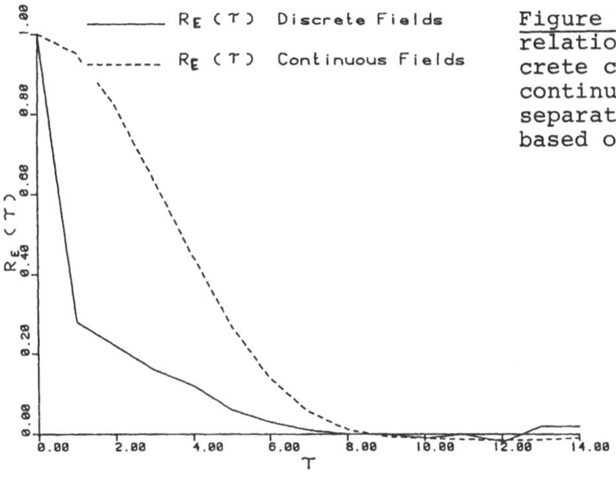

Figure 7. Eulerian temporal correlation function R_E (T) for discrete correlated fields and their continuous counterparts; T is separation in timesteps; results based on 64 realizations.

rate. They are traditionally defined as

$$\text{vorticity:} \quad \tfrac{1}{2}\left(\frac{\partial u}{\partial y} - \frac{\partial v}{\partial x}\right)$$

$$\text{strain rate:} \quad \tfrac{1}{2}\left(\frac{\partial u}{\partial y} + \frac{\partial v}{\partial x}\right)$$

To implement these definitions for discrete fields we define non-binary velocities u and v in terms of the flow vector [G,H,A,B], where u = G'H' - GH - A'B' + AB and v = GH' - G'H - AB' + A'B. The velocities u and v can take on values from -2 to 2 in unit increments. Vorticity and strain rate are most conveniently evaluated not at the already-existing lattice points, but rather at pseudo-lattice points centered in each grid-square of the field. Then partial spatial derivatives may be calculated based on nearest-neighbor velocity differences, e.g., $\partial u/\partial y = [u(i+1,j+1)+u(i,j+1)-u(i+1,j)-u(i,j)]/2$.

With these definitions we can compare the nature of the Markovian and correlated discrete fields with respect to vorticity and strain rate. Figures 8 and 9 show the distributions of values. Not surprisingly, the correlated fields are characterized by lower strain rate and vorticity indicated by smaller tails and a larger central peak in the distributions. But, unlike their quantification in continuous fields, vorticity and strain rate here depend on a fixed length scale--the mesh size, and constant magnitude velocities--one.

An interesting feature of discrete fields is that they admit to characterization in ways that are not possible with conventional fields: for example, in terms of distribution of streamline lengths. Streamlines on latticed fields are unambiguously defined along flow vectors from node to node and fill the lattice. There are a countable number of them with integer lengths. With finite lattices a streamline

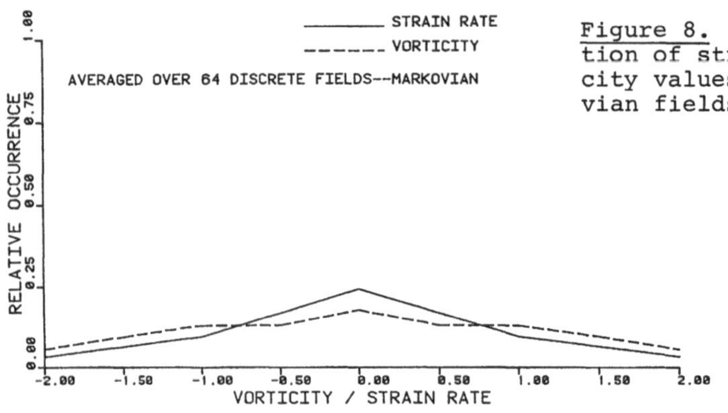

Figure 8. Frequency distribution of strain rate and vorticity values in discrete Markovian fields.

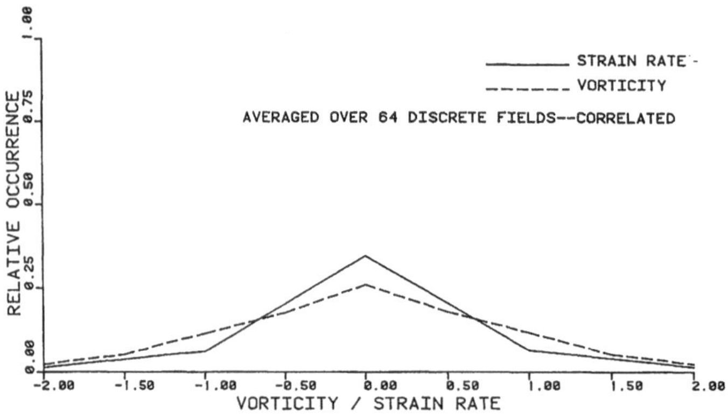

Figure 9. Frequency distribution of strain rate and vorticity values in discrete correlated fields.

must either close upon itself (a closed loop) or enter and leave the domain as an open path. Figures 10 and 11 show the relative occurrence of streamlines by streamline length for Markovian and correlated discrete fields. Both closed loops and open paths are presented.

These results are difficult to interpret quantitatively because the distributions are strongly influenced by the boundaries of the finite fields, i.e., an open path may have become a closed path were the field slightly larger. However, we can make some qualitative comments. Markovian fields yield a higher incidence of short paths--especially of the shortest closed path, length four. Conversely, correlated fields typically produced longer paths. (The longest was 196 lattice lengths!) Note also that the ratio of occurrences between closed and open paths differs significantly between the two cases. Because of spatial scale differences, there is much more of an opportunity for a streamline in a Markovian field to complete its loop within the extent of the finite field than there is for a streamline in a correlated field.

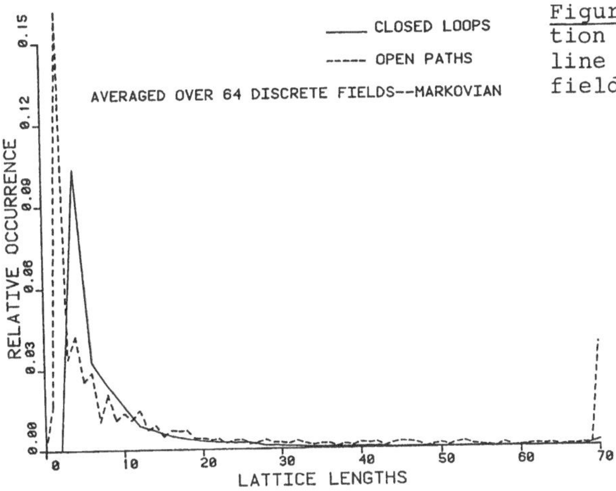

Figure 10. Frequency distribution of open and closed streamline lengths in discrete Markovian fields.

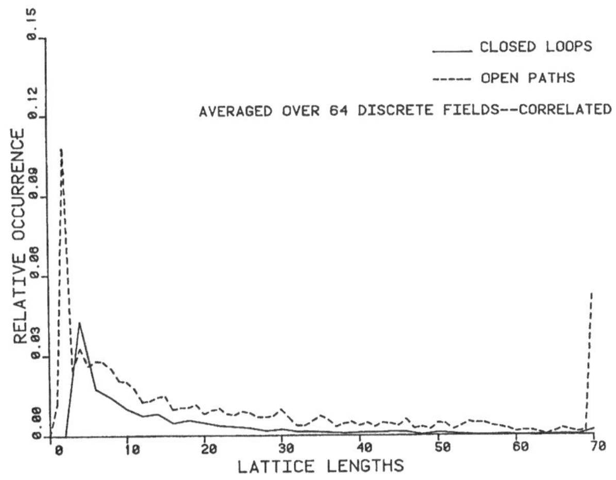

Figure 11. Frequency distribution of open and closed streamline lengths in discrete correlated fields.

5. Statistics of Particle Dispersion

We now address particle dispersion. On each of our 64 sets of initial fields, both continuous and discrete, we identify fluid parcels at relative separations of 1, 2, and 4 mesh lengths. Each parcel (particle) then moves according to the velocity at its instantaneous location prescribed by the sequence of time-evolving fields. Through 15 timesteps we follow the motions of these particles and record their successive positions. Over the ensemble of realizations we thus can calculate r.m.s. single particle dispersion $\langle D^2(t)\rangle^{1/2}$ and r.m.s. two-particle relative dispersion $\langle D^2_{rel}(t)\rangle^{1/2}$.

Although the principal focus here is on dispersion in incompressible, latticed fields, it is interesting to compare results with comparable continuous fields; especially since it is from continuous fields that our discrete fields are derived. The most appropriate comparison of dispersion would be on fields whose statistics were nearly the same. But, as described in the last section, our discrete fields are not really comparable to their parent continuous ones, especially with respect to integral space and time scales. Jin (1984) is carrying out dispersion experiments on more closely related fields, but the discrete fields he uses are not directly produced from his continuous ones.

Figure 12 presents r.m.s. single-particle dispersion for four cases of particle motion: classical random walk, walks on discrete Markovian fields, walks on discrete correlated fields, and movement on continuous fields. The particle "walks" by definition have a velocity scale of unity. To obtain a rudimentary level of comparison for particle motion on the continuous fields, particles were followed not on the original continuous fields, but rather on those fields rescaled to have r.m.s. velocities of one. Thus, as seen in Figure 12, all four cases yield an average dispersion of one after one timestep.

Taylor (1921) deduced asymptotic dispersion statistics for homogeneous turbulent fields: for small t, $\langle D^2 (t)\rangle^{1/2}$ is proportional to t; for large t, $\langle D^2 (t)\rangle^{1/2}$ is proportional to $t^{1/2}$ with a coefficient depending on the Lagrangian autocorrelation function. Our dispersion

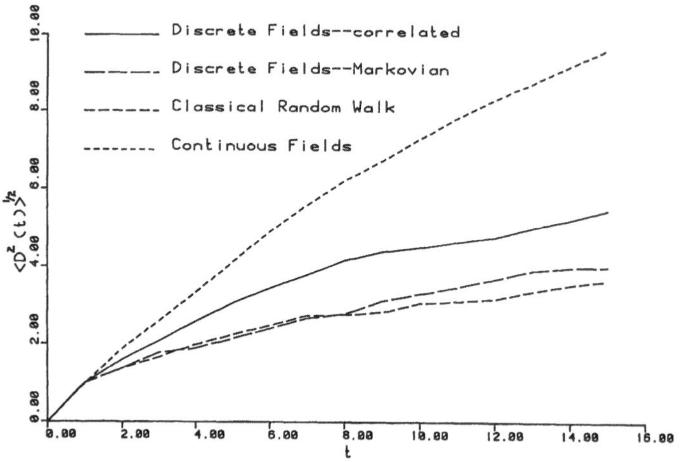

Figure 12. R.m.s. single-particle dispersion $\langle D^2 (t)\rangle^{1/2}$ for four cases: classical random walk, walks on discrete Markovian fields, walks on discrete correlated fields, and movements on related continuous fields; dispersion in mesh lengths and t in timesteps.

curves for the correlated discrete fields exhibit that form. Initially $\langle D^2(t)\rangle^{1/2}$ increases linearly with t, and then slows toward the expected $t^{1/2}$ limit. Dispersion on the continuous fields proceeds in a similar manner, but takes more time to reach the asymptotic regime because of the continuous fields' larger time scale. Consequently, dispersion is greater for the continuous case. Dispersions for the Markovian fields and the random walk are identical. Because there is no temporal persistence in the Markovian fields, single-particle dispersion coincides with that of a random walk. [Einstein (1905) showed that a random walk also disperses particles at an r.m.s. rate of $t^{1/2}$ for large t.]

Twice differentiating the dispersion curves for the correlated discrete and continuous cases would give the respective Lagrangian autocorrelation functions [Taylor (1921)]. But since our results are based on so few realizations we do not perform these calculations.

In contrast, relative dispersion of two particles separated initially by a finite distance is much more dependent on the spatial properties of the fields. Figure 13 gives relative dispersion results for pairs of particles separated initially by 1, 2, and 4 mesh lengths on continuous and correlated discrete fields. For the first timestep the statistics of relative particle motion are dependent on only the Eulerian velocity correlations f(r) and g(r). But, for later timesteps the statistics depend on a non-stationary Lagrangian autocorrelation function [Corrsin (1962)]. [See also Brier (1950) for a treatment of relative dispersion in terms of a sequence of single, discrete timesteps.]

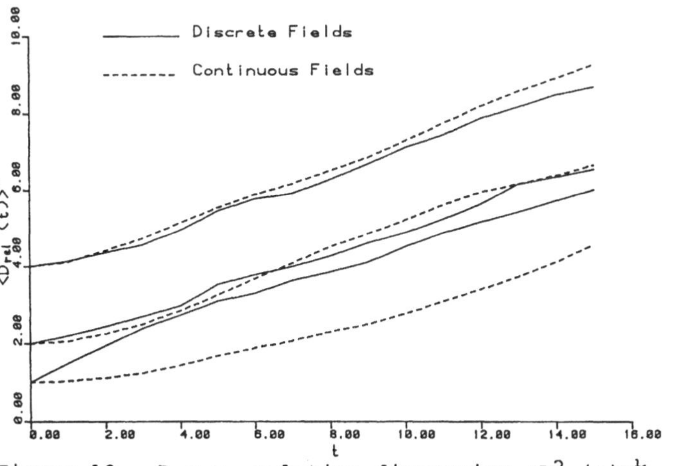

Figure 13. R.m.s. relative dispersion $\langle D^2(t)\rangle^{1/2}$ between two particles on discrete correlated fields and their continuous counterparts; dispersion in mesh lengths and t in timesteps. Initial separations are given by t=0.

In the continuous fields for small t, $\langle D^2_{rel}(t)\rangle^{1/2}$ grows linearly with t with a coefficient proportional to $[1-f(r_0)]$, where r_0 is the initial separation. At long times (and at large separations) the particles move independently, and relative dispersion reduces to $\sqrt{2}$ times the single particle dispersion rate [Corrsin (1962)].

Relative dispersion on discrete fields behaves differently and demonstrates the unique properties of latticed, incompressible, binary velocity fields. Whereas particles on continuous fields move with variable speed and direction, particles on discrete fields move with unit speed and only orthogonal directions. The solid curves in Figure 13 shows relative dispersion on discrete fields. At small times the rate of relative dispersion is greatest for small initial separations and least for large initial separations--contrary to the continuous case. (This phenomenon does not extrapolate to zero separation where, of course, the particles will always move together.) The result can be explained as follows. When particles are close together, say one mesh length, and must move only along lines of the grid, any relative movement will increase their separation to at least $\sqrt{5}$ mesh lengths. Incompressibility precludes their approaching one another; geometry defines the minimum change in separation. At larger initial separation distances, however, particles can approach one another; in these cases the geometry admits to lesser changes in separation. These dispersion characteristics cannot be inferred directly from the Eulerian spatial correlations f(r) and g(r) as was the case with the continuous fields. Consequently we find that an additional factor plays a role in dispersion on discrete fields--at least for small separations. Once particles are far enough apart, their dispersion characteristics are at least qualitatively similar to those on continuous fields.

6. Extending the Work to Three-Dimensional Discrete Fields

Two dimensional discrete fields are relatively easy to display, relatively inexpensive to generate, and relatively convenient to experiment with. However, the applicability of two-dimensional velocity fields to real flows (let alone the applicability of discrete two-dimensional velocity fields) is tenuous. Insofar as turbulence is a fully three-dimensional process, a more appropriate analogy is obtained in three dimensions. Further, some of the idiosyncracies found in the present discrete fields may be reduced when working with the three-dimensional case.

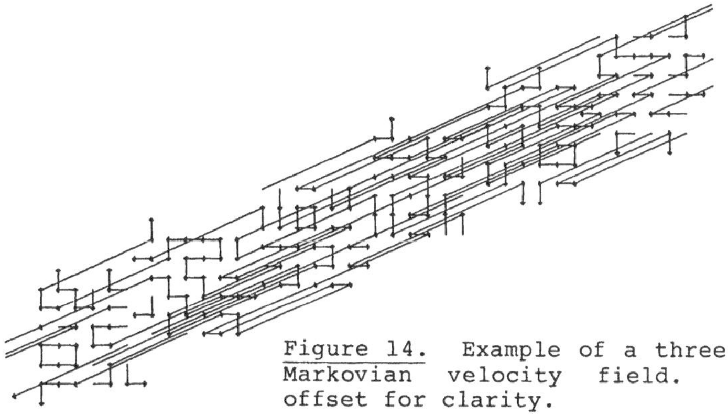

Figure 14. Example of a three-dimensional discrete Markovian velocity field. Lattice planes are offset for clarity.

As a prelude to future work, we demonstrate the potential extension of our current work to three dimensions with Figure 14, a portrayal of a 6 x 6 x 6 discrete Markovian field.

7. Conclusions

This work represents a first attempt at using discrete fields as a research tool for turbulent dispersion. Working with a wider range of Eulerian space and time scales, estimating additional Lagrangian properties of particle motion, and, of course, expanding to the three-dimensional case are obvious extensions of this work.

Possibly basic to this research is the study of the mathematical properties of discrete fields. Currently, we cannot fully separate the properties of the flow and particle dispersion from the idiosyncracies of the latticed velocity fields. The relationship between a "walk on a random field" and a random walk needs study--especially with respect to the constraint of incompressibility. Treating long-chain polymers as linked particles moving on discrete fields might be a non-turbulent application.

The author wishes to thank S. Corrsin for suggestions, to thank G. Jin for considerable assistance in developing this work; and to acknowledge the Society for Statistical Geometry.

This work was supported by the National Science Foundation, Atmospheric Sciences Division, ATM 77-04901.

References

Brier, G. W. (1950), "The statistical theory of turbulence and the problem of diffusion in the atmosphere." J. of Meteorology, 7-4, pp. 283-290.

Bugliarello, G. and E. D. Jackson, III (1964), "Random walk study of convective diffusion." J. Engrg Mech. Div., Proc. ASCE, **90, EM4**, pp. 49-77.

Corrsin, S. (1962), "Theories of turbulent dispersion." Mecanique de la Turbulence, Colloques Int'l du C.N.R.S., No. 108, Marseille, 1961, pp.27-52.

Einstein, A. (1905), "Uber die von der molekularkinetischen Theorie der Warme geforderte Bewegung von in ruhenden Flussigkeiten suspendierten Teilchen." Ann. d. Physik, **17**, p. 549.

Hinze, J. O. (1975), Turbulence, McGraw-Hill, New York.

Jin, G. (1984), "Particle dispersion on bi-directional-binary two-dimensional, incompressible velocity fields." Master's Essay, The Johns Hopkins University, Baltimore.

Karweit, M. J. and S. Corrsin (1972), "Simple and compound line growth in random walks." Lecture Notes in Physics, **12**, Springer-Verlag, Berlin, pp. 317-326.

Karweit, M. (1984), "Dispersion of particles in bi-directional-binary, two-dimensional, incompressible velocity fields: some numerical experiments." Comp. Methods and Exp. Measurements, Proc. 2nd Int'l Conf., New York/Southampton, in press.

Kraichnan, R. H. (1963), "Direct-interaction approximation for a system of several interacting simple shear waves." Phys. Fluids, **6**-11, pp. 1603-1609.

Lorenz, E. N. (1960), "Maximum simplification of the dynamic equations." Tellus, **12**-1, pp. 243-254.

Lumley, J. L., and S. Corrsin (1959), "A random walk with both Lagrangian and Eulerian statistics." Advances in Geophysics, **6**, pp. 179-184.

Patterson, G. S., Jr. (1958) Private communication.

Patterson, G. S. Jr. and S. Corrsin (1966) "Computer experiments on random walks with both Eulerian and Lagrangian Statistics." Dynamics of Fluids and Plasmas, Academic Press, New York, pp. 275-307.

Taylor, G. I. (1921), "Diffusion by continuous movements." Proc. London Math. Soc., **A**, vol. 20, p. 196.

Transition and Turbulence in Fluid Flows and Low-Dimensional Chaos

K.R. Sreenivasan

Department of Mechanical Engineering, Yale University
New Haven, CT 06520, USA

Recent studies of the dynamics of low-dimensional nonlinear systems with chaotic solutions have produced very interesting and profound results with several implications in many disciplines dealing with nonlinear equations. However, the interest of fluid dynamicists in these studies stems primarily from the expectation that they will help us understand better the onset as well as dynamics of turbulence in fluid flows. At this time, much of this expectation remains untested, especially in 'open' or unconfined fluid flows. This work is aimed at filling some of this gap.

Measurements made in the wake of a circular cylinder, chiefly in the Reynolds number range of about 30-10^4, have been analyzed to show aspects of similarity with low-dimensional chaotic dynamical systems. In particular, it is shown that the initial stages of transition to turbulence are characterized by narrow windows of chaos interspersed between regions of order. The route to the first appearance of chaos is much like that envisaged by Ruelle & Takens; with further increase in Reynolds number, chaos disappears and a return to three-frequency quasiperiodicity occurs. This is followed in turn by the reappearance of chaos, a return to four-frequency quasiperiodicity, reappearance of chaos yet again, and so on. We have observed several alternations between order and chaos below a Reynolds number of about 200, and suspect that many more exist even in the higher Reynolds number region. Each window of chaos is associated with a near-discontinuity in the vortex shedding frequency and the rotation number, as well as a dip in the amplitude of the vortex shedding mode. It is further shown that the dimension of the attractor constructed using time delays from the measured velocity signals is truly representative of the number of degrees of freedom in the ordered states interspersed between windows of chaos; it is fractional within the windows of chaos, and is higher than those in the neighbouring regions of order. Our measurements suggest that the dimension is no more than about 20 even at a moderately high Reynolds number of 10^4, and that it probably settles down at about that value.

1. Introduction

a. General remarks

The principal parameter of incompressible viscous flows, in situations free of body forces, is the Reynolds number, Re. Observations show that for given (fixed or time-independent) boundary conditions (and external forces if applicable), the flow is unique and steady for Re < Re_{cr}, where Re_{cr} is a certain critical value of Re; this is the steady laminar motion. As Re increases, the fluid motion may first become periodic, quasiperiodic, and 'eventually' chaotic. (Chaos is defined better in section 3 and in the appendix, but we shall also loosely use the word to designate a state in which the details of motion are not reproducible.) This chaotic state is not necessarily turbulence as generally understood — and we shall discuss this shortly — but it is believed that one attains the turbulent state if the Reynolds number is taken to a sufficiently high value. The goal of the stability theory is to understand how the evolution from the laminar to the turbulent state occurs, while turbulence theories aim at unearthing and predicting the mysteries of the (fully) turbulent state itself.

It is generally believed that the key to both these problems lies in the Navier-Stokes (NS) equations, and that no additional hypotheses of fundamental nature are required for describing either the onset of turbulence or its dynamics. Much effort has thus been spent on mastering the NS equations. However, the difficulties, both analytical and computational (at high enough Reynolds numbers), remain intimidating.

In the recent past, claims have been made that autonomous dynamical systems with small number of degrees of freedom, typified by

$$\frac{db_i}{dt} = f(b_i;\ \varepsilon_i), \tag{1.1}$$

where the b_i characterize the state of the system (the so-called 'state variables'), i is a small integer, and ε_i are the so-called control parameters (analogous to Re in the NS equations), help us towards attaining both the goals mentioned above. It is to a discussion of aspects of these claims, via an example of fluid flow behind circular cylinders, that this paper is devoted.

b. Remarks on degrees of freedom, genericity, and spatial chaos

Several questions arise immediately. One natural question concerns the relevance to fluid flows of low-dimensional dynamical systems. To give some meaning to the concept of degrees of freedom in fluid flows, let us *approximate* the velocity vector u_j appearing in the NS equations as

$$u_j = \sum_{\underset{\sim}{k}} a_j(\underset{\sim}{k};t)\,e^{i\underset{\sim}{k}\cdot\underset{\sim}{x}} \quad (j = 1,2,3), \tag{1.2}$$

where the wave number vector $\underset{\sim}{k}$ is an element of a discrete (finite or infinite) set. The NS equations can then be written formally as

$$\frac{\partial a_i(\underset{\sim}{k};t)}{\partial t} = F(a_i;\ Re),\ i = 1,2,\ldots.N\ \text{(large)}. \tag{1.3}$$

The number of the coefficients a_i which, for given boundary conditions for the fluid flow, are capable of variation in time can now be called the degrees of freedom of the fluid flow governed by the NS equations (to within the approximation implied in (1.2) and (1.3)). Since the laminar flow is uniquely specified by the boundary (and external force) conditions, this number is zero. If Re increases just past

Re_{cr}, only a few degrees of freedom are excited, and hence it appears that, at least in the positive neighbourhood of Re_{cr} (to be called transcritical region henceforth), consideration of these few degrees of freedom is adequate.

An interesting hypothesis (which we shall examine in this paper) is that the number of degrees of freedom (not necessarily in the sense described above) remains small even in (certain type of) high Reynolds number turbulence.

Assuming that the number of degrees of freedom excited in the transcritical region is indeed small, we must ask whether the behavior in this transcritical region does not depend on the broad nature of the right hand side of equations (1.1) and (1.3). The most often cited justification for the belief that this dependence is in some sense of secondary importance comes from the work of Ruelle & Takens [1] and Newhouse, Ruelle & Takens [2] which indicates that chaos sets in abruptly following a few Hopf bifurcations, and that this behavior is 'generic' or 'typical'.

The words 'generic' and 'genericity' find their frequent use in the literature on dynamical systems, and so, it is perhaps useful to discuss the concept briefly. Ruelle & Takens make this concept quite specific for the vector fields they were considering, but we shall be content with a rather loose qualitative description. Consider as an *example*, a class of functions possessing continuous derivatives up to a certain order, and satisfying differential equations of the type (1.1). Properties of this class of functions which are the rule and not the exception, and which do not depend on the precise nature of the right hand side of (1.1), are called generic. The conclusions of Ruelle & Takens strictly hold for an idealized mathematical system, and whether the concept of genericity is powerful enough to embrace fluid systems is not clear. One should attempt to answer this question by looking at the specific form of F in (1.3) and/or by observing the actual bifurcations in experiments on laminar-turbulent transition.

Even if the concept of genericity does hold for fluid flows, it is not obvious that interesting nongeneric phenomena do not occur. To make this notion specific, let us consider the following rather far-fetched example. Suppose we link (as in our example above) genericity to the existence of velocity fields possessing continuous derivatives of a certain order. Those generic properties may be irrelevant to a turbulent boundary layer since one cannot exclude the possibility that at some moment during bursting near the wall (a key event sustaining turbulence production) this smoothness condition is destroyed in spite of viscosity. It is therefore sensible to keep in mind that nongeneric behavior is neither uninteresting nor unlikely, especially when conditions such as configurational symmetry, vicinity to wall, play an important role in the evolution of the flow.

Finally, one must mention the predominant role played by spatial chaos (and order!) in turbulent flows of fluids. An important characteristic of fluid turbulence is random vorticity, whose presence necessarily implies that the velocity vector is a random function of *position*. Autonomous dynamical systems of the type (1.1), on the other hand, do not contain any space information. While temporal chaos in fluid turbulence may in some sense be symptomatic of spatial chaos, it is clear that autonomous dynamical systems have little to say directly about the latter, at least at the current state of development.

c. 'Closed' and 'open' flow systems

Notwithstanding these remarks, it is necessary to note that several beautiful experiments now exist in the Taylor-Couette flow (e.g., Refs. 3, 4 and 5) and the convection box (e.g., Refs. 5 and 7) which have lent support to the notion that the behavior of fluid flows in the transcritical region could be similar to that of low-dimensional dynamical systems. This in itself is undoubtedly remarkable, but it should be remembered that these two flows are special in the following sense. In all 'closed flow' systems — of which the convection box and the Taylor-Couette flow are two popular examples — the boundary is fixed so that only certain class of eigenfunctions can be selected by the system; this does not hold for another class of flows we may call 'open flow systems' — for example, boundary layers, wakes, jets — in which the flow boundaries are continuously changing with position. Thus, while in closed flow systems each value of the control parameter (for example, the rotation speed of the inner cylinder in the Taylor-Couette problem) characterizes a given state of the flow globally, this is not true of open systems. Consider as an example the near field of a circular jet. For a given set of experimental conditions, the flow can be laminar at one location, transitional at another and turbulent at yet another (downstream) location. This usually sets up a strong coupling between different phenomena in different spatial positions in a way that is peculiar to the particular flow in question. Secondly, the nature and influence of external disturbances (or the 'noise', or the 'background or freestream turbulence') is more delicate and difficult to ascertain in open flows: the noise, which is partly a remnant of complex flow manipulation devices and partly of the 'long range' pressure perturbations, is not 'structureless' or 'white', no matter how well controlled. Finally, it is well known that closed flow systems can be driven to different states by means of different start-up processes; for example, different number of Taylor vortices can be observed in a Taylor-Couette apparatus depending on different start-up accelerations [8]. This type of path-sensitivity in a temporal sense does not apply to open systems, where the overriding factor is the path-sensitivity in a spatial sense (i.e., the 'upstream influence').

d. Scope of the paper

On balance, all these considerations suggested to us that it is desirable to look at some open flows to determine the extent to which dynamical systems can assist us in our goals of understanding transition and turbulence in fluid flows. This is the motivation for the work described in this paper, which is to be viewed more as a progress report than as a complete account; obviously much more remains to be done. Our approach is to select well-known flows and follow the bifurcations as closely as possible. (We reported some of our earlier work in pipe flows in [9] and wake work in [10].) Surprisingly, while much work has been done in these flows in the past, an amazing amount of new information can still be acquired that will facilitate clarifying the relation between low-dimensional chaotic systems and fluid flow transition and turbulence. Part of the reason for this is undoubtedly that the details one looks for are often dictated by contemporary concerns.

2. Experiments

a. Experimental conditions

Although we have conducted experiments in wakes, jets and pipe flows, we choose to discuss here only our wind tunnel experiments in two-dimensional wakes behind circular cylinders. The Reynolds number range covered is from about 30 (slightly below the vortex shedding value) to about 10^4. Two wind tunnels — one of the blower type and one of the suction type — were used. Nylon threads, stainless steel wires and aluminium tubes, stretched tightly across the width of the wind tunnels, were used as wake generators. The aspect ratio varied between about 70 and 2000. The basic experimental conditions are summarized in Table 1.

Table 1. The flow configuration and experimental conditions

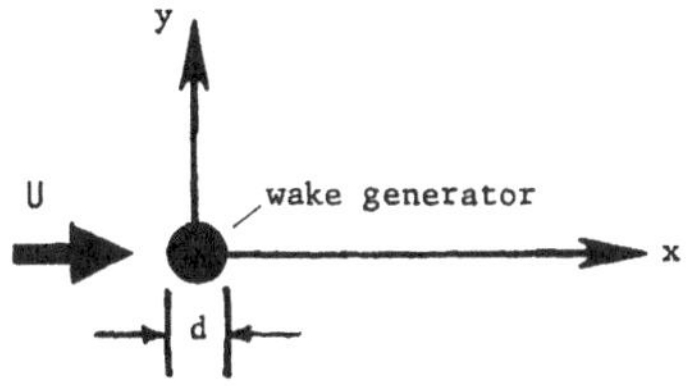

d (mm)	≈x/d	≈y/d	aspect ratio	wind tunnel characteristics
0.24	5	1	2000	suction type; turbulence level ≈ 0.2% at speeds of interest
0.24	50	1	2000	
0.36	5	1	1330	
4.0	5	1	170	
0.36	11	1	70	blower type; turbulence level varied from 0.68% at speeds ≈ 1 m/s to 0.06% at speeds ≈ 10 m/s

All velocity signals were obtained with a hot-wire operated on a DISA 55M01 constant temperature anemometer. The speed of the tunnel was monitored with a Pitot tube connected to a calibrated MKS Baratron with adequate resolution (and an averager). The hot-wire and the Pitot tube were mounted on a specially designed slim holder.

Some of the data to be presented in this and later sections is in the form of power spectral density of the streamwise velocity component, u. Nearly all the signals were digitized at sufficiently high frequency (60 kHz or more) to ensure that, whenever the signal was periodic, at least 30 digitized points were contained in one period of the basic frequency (so that it was a good representation of the analog signal). Further, the entire length of the signal (which contained at least 100 cycles of the basic frequency) was Fourier transformed at once using the Cooley-Tukey FFT algorithm. The overriding criterion was that the spectral resolution should be as good as possible (here, between 0.5 Hz and 2 Hz compared with shedding frequencies of the order of 2000 Hz or more) and that one must not miss any low frequency modulations.

b. The background turbulence

We have worked with varying levels of background turbulence, and found that the occurrence of different stages of transition reported here is in itself not terribly sensitive to the turbulence level as long as it is not too high; larger turbulence levels blur the distinction between different stages and alter the details somewhat erratically. One should, however, strive to eliminate all strong discrete frequency components in the background turbulence structure.

Figure 1a shows a typical power spectral density of u in the freestream at Re = 60. (The ordinate is the logarithm to base 10 of the power.) The 'noise' (though

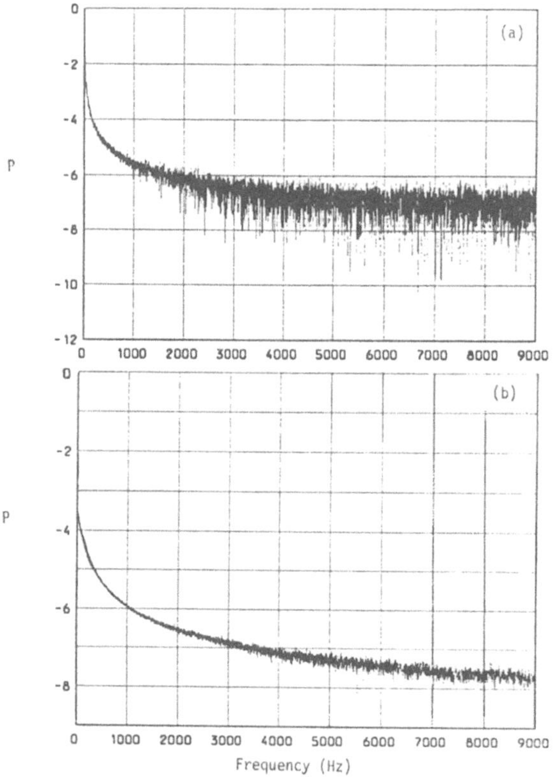

FIGURE 1: Normalized power (or frequency) spectrum of (a) noise of the instrumentation and digitizer, plus freestream disturbances, Re = 60; (b) instrumentation and digitizer noise only with no flow.

devoid of any discrete peaks) does not appear to be 'white' but has a much larger low frequency component. Figure 1b shows the power spectral density measured with the flow completely shut off, but the hot-wire and other electronic instruments operating the same way as before. It is clear that the anomalously high low frequency content is not representative of the flow itself, but of electronic and computer noise. Allowance should thus be made for this fact in the interpretation of the spectral data to follow.

3. Results from Spectral Measurements

a. Route to chaos: the first appearance

Figure 2 shows the logarithm (to base 10) of the normalized power spectral density of u at a Reynolds number (based on the freestream velocity and the diameter of the cylinder) of about 36, which is approximately the onset value for vortex shedding. Notice that the instrumentation and other noise level is around 10^{-8}, while the peak of the spectrum (marked f_1), corresponding to the basic vortex shedding frequency behind the cylinder, is at round $10^{-0.5}$, about 7½ orders of magnitude higher than the noise level! The sharpness of the peak (as well as of the other peaks to the right of f_1 which are the harmonics of f_1) is excellent.

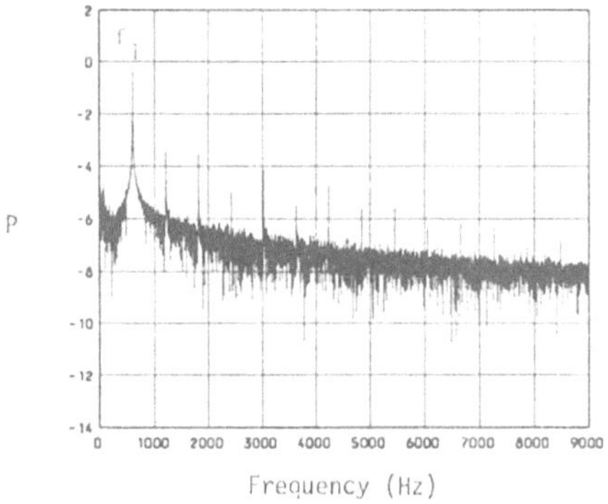

FIGURE 2: Normalized frequency spectrum of u at Re ≈ 36. Note that the power P is plotted on a logarithmic scale (to base 10). The peak at $f_1 \approx 590$ Hz corresponds to the vortex shedding, and the subsequent strong peaks above the noise level are simply harmonics of f_1.

At a somewhat higher Reynolds number of 54, there appear a number of peaks in the spectrum (figure 3a); as shown in the expanded version (figure 3b) all the peaks can be identified precisely in terms of the interaction of the two frequencies — the basic vortex shedding frequency f_1 and another incommensurate frequency f_2.

At an Re = 66 the spectrum (figure 4) shows broadened peaks with no overwhelmingly strong discrete components — quite a different situation from that of figures 2 and 3. One might say, in the language of dynamical systems, that chaos has set in!

The sequence of events leading to chaos are so far literally like that envisaged in the Ruelle-Takens-Newhouse (RTN) picture of transition to chaos, and so, a brief digression *roughly* describing this picture is quite useful. (The appendix is an introduction to the basic terminology.) With increasing Re, the steady laminar motion

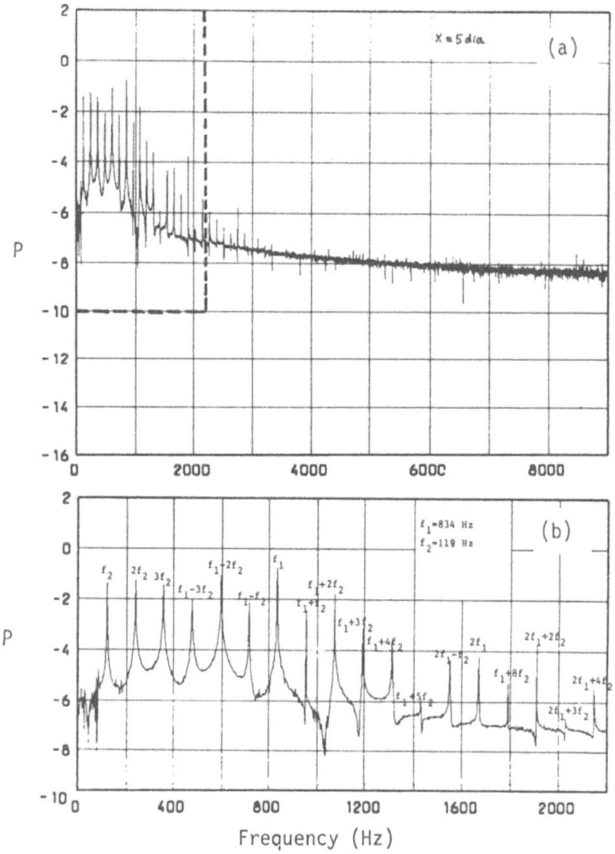

FIGURE 3: (a) Normalized frequency spectrum of u at Re = 54. In (b), the frequency range 0-2200 Hz is expanded. All significant peaks in (b) are simple combinations of the vortex shedding frequency f_1 (corresponding to the most dominating peak), and another incommensurate frequency f_2. After satisfying ourselves that there are no subharmonics of f_1 (and that 119.02 Hz is unrelated to the line frequency or spurious oscillations of the cylinder) we have picked f_2 by hypothesizing that the peaks nearest f_1 must be $f_1 \pm f_2$. The value of f_2 thus obtained accounts for every other significant peak as shown in (b) — actually to 4 or 5 decimal places for reasons we do not understand! At least part of the reason for the relatively low noise level (compared with figure 2) is the increased signal level.

loses stability and becomes periodic with frequency f_1 (say); the power spectral density will have (as in figure 2) a peak at f_1 (and its harmonics), and the phase diagram will show a limit cycle behavior. Loss of stablility of this new state yields a quasiperiodic motion with two independent frequencies, f_1 and (say) f_2. The spectral density will now show f_1, f_2 and various combinations $mf_1 \pm nf_2$ (as in figures 3a, b), and the phase portrait will be a two-torus. Further increase in Reynolds number yields a quasiperiodic motion with three frequencies (three-torus). Newhouse, Reulle & Takens [2] argue that even a weak nonlinear coupling (of a certain variety!) among the three frequencies is likely to result in chaos or a strange attractor (see appendix), one of whose symptoms is an increased broadband content (see figure 4). This contrasts the classical picture of Landau, according to which turbulence is the asymptotic state of increasingly higher order quasiperiodicities.

Phase diagrams provide complementary information on the sequence of events leading to chaos. To construct phase diagrams, it would seem that one would require the measurement of N independent variables (in general, a hopeless task!), but embedding theorems like those of Takens [11] justify the use of a single measured variable. From the measured local velocity $u(t)$ — for example — one constructs a d-dimensional diagram from the vectors $\{u(t_i), u(t_i + \tau), \ldots u(t_i + (d-1)\tau)\}$, $i = 1, \ldots\infty$, τ being a time delay whose precise value in a certain wide range seems to be immaterial. According to the embedding theorems, the phase diagrams constructed in the above manner will have essentially the same properties as the one with N independent variables, as long as $d \geq 2N + 1$ (although exceptions to this now commonly assumed philosophy are not hard to concoct). In practice, d is increased by one at a time until the properties of interest become independent of d.*

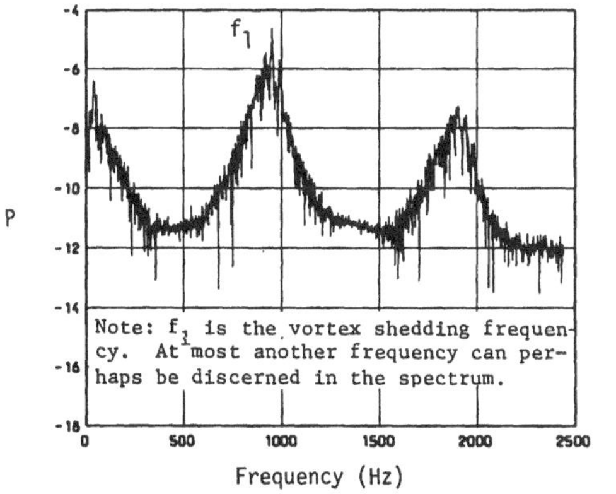

FIGURE 4: The first appearance of chaos at Re = 66. The broadband nature implies chaos; onset of chaos does not rule out the existence of spectral peaks. (Note: This does not signify some high order quasiperiodicity as dimension and entropy calculations of section 4 show.)

Figures 5, 6 and 7 show respectively the plot of $u(t_i+ \tau)$ vs $u(t_i)$ at Re = 36, 54 and 66, and can be considered as projections of the phase diagrams on a two-dimensional plane. The limit cycle behavior at Re = 36 is evident, the scatter visible in the figure being partly due to experimental noise (see figure 2) and partly due tothe jitter in the signal. Further, a Poincaré section reveals no discernible structure. The situation is thus basically periodic.

* About two years ago (October 1982) when we first started constructing phase diagrams in this manner, we were unaware of any literature on embedding theorems, but were guided solely by elementary *ad-hoc* considerations.

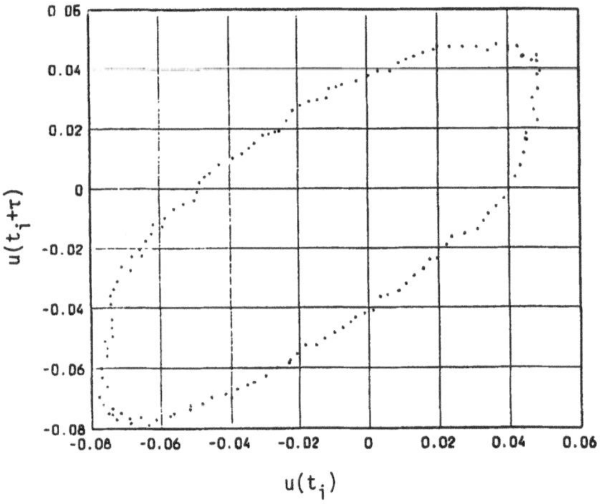

FIGURE 5: The phase plot from the velocity signal u at Re ≈ 36, showing limit cycle behavior. The time delay τ = 10 sampling intervals; the starting point t_1 is arbitrary.

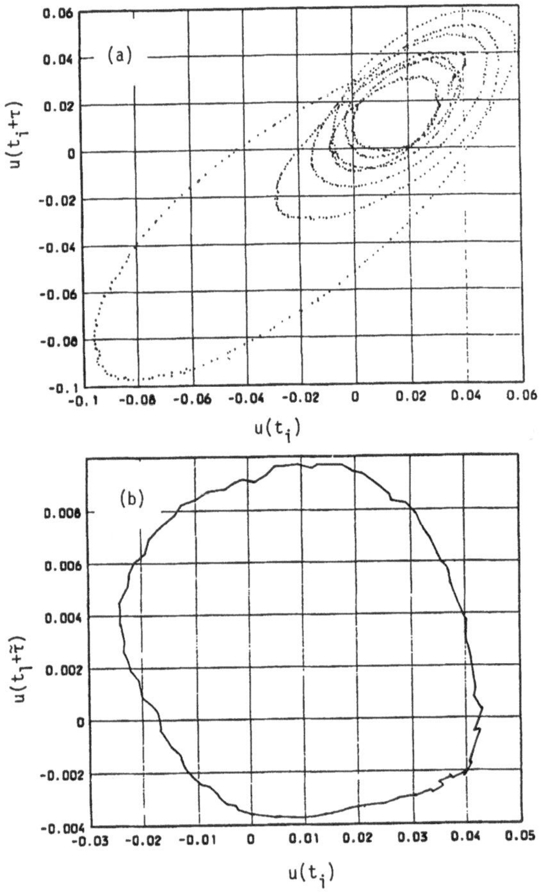

FIGURE 6: (a) The phase plot from u at Re ≈ 54. τ = 10 sampling intervals. (b) Poincaré section for the phase plot of (a). This is simply a plot of $u(t_i+\tilde{\tau})$ vs $u(t_i)$ with $\tilde{\tau}$ spaced exactly $1/f_2$ apart.

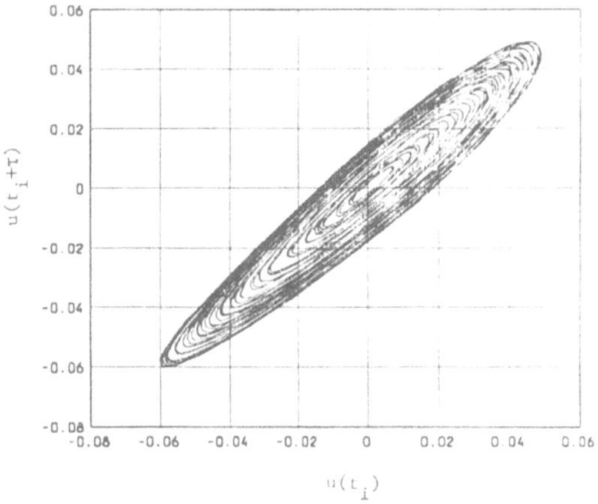

FIGURE 7: The phase diagram for Re = 66. τ = 10 sampling intervals. The continuous curve is now the result of joining successive data points (done for clarity).

At Re = 54, although the projection of the phase diagram is complicated in appearance*(figure 6a), a Poincaré section (figure 6b) yields a limit cycle, reinforcing the fact that only two degrees of freedom are present. On the other hand, not only is the projection of the phase diagram at Re = 66 complex (figure 7), but also its Poincaré sections (not shown), no matter how defined. This, as well as the fractional dimension of the attractor (see section 4a) show that the signal is indeed chaotic.

(As equally valuable measures of chaos, one could evaluate the Lyapunov exponent (characterising the exponential divergence of nearby trajectories) or the Kolmogorov entropy (which, for typical systems, equals the sum of positive Lyapunov exponents). Limitations of various kinds have prevented us from measuring the Lyapunov exponent — such measurements for a Taylor-Couette flow have been made by Brandstäter et al. [5] — but we do discuss some entropy measurements in section 4d.)

This progression towards chaos — underlying the possible presence of a strange attractor — proceeds much like that proposed by Newhouse, Ruelle & Takens [2]. It is thus extraordinary that the 'generic' behavior indicated by Ruelle & Takens for an idealized mathematical system should have a nontrivial bearing on a rather complex fluid dynamical system!

It should be noted that few would feel comfortable in designating as turbulent the *signal* we have recognized as chaotic. Clearly, to the extent that a turbulent *flow* must possess *spatial* randomness, we cannot say much of value as to whether the *flow* at Re = 66 is turbulent or not without a global survey of the flow field at this Reynolds number. Further, if one *defines* turbulence as a high Reynolds number phenomenon (as is often done!), it is tautologically true that the signal does not represent turbulence. Further, a look at the signal (figure 8) would prevent someone with an everyday familiarity with high Reynolds number turbulence from accepting it

* Note that the trajectory resides most often in the upper right quadrant, but only rarely strays away into the lower left quadrant. This behavior in the phase plane can be related to the finite skewness of the signal.

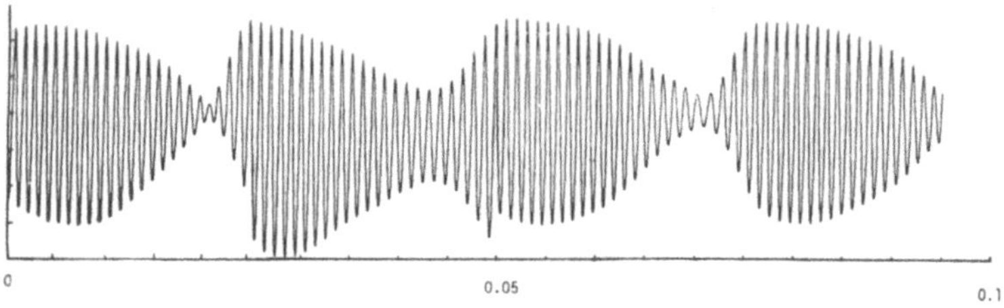

FIGURE 8: The signal u(t) at Re = 66.

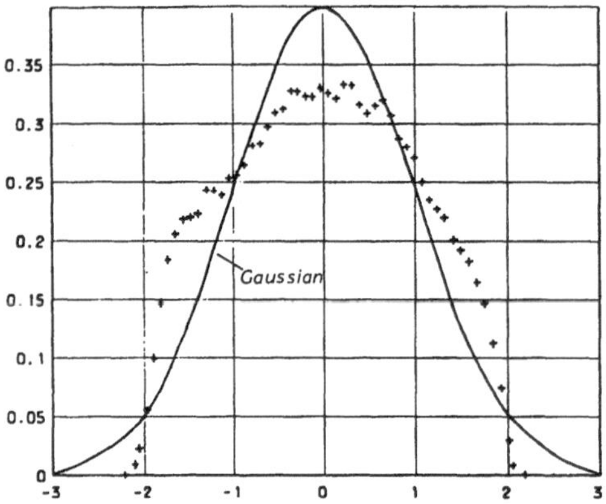

FIGURE 9: The measured probability density of u at Re = 66. The abscissa is the amplitude about the mean normalized by the root-mean-square of the signal, and the ordinate is the probability density. The signal has a skewness of near zero and a flatness of about 2.4.

as turbulent. Nevertheless, we would like to suggest that the signal shown in figure 8 is indeed random (for example, in terms of algorithmic complexity required to specify it [12]) with a well-defined probability density (see figure 9; for a comparison with similar data at 'large' Reynolds numbers in the far wake, see Thomas [13]). What this means is that even at low enough Reynolds numbers, the interaction of only a few degrees of freedom leads to randomness! It is also pertinent to point out that at least in some respects the signal of figure 8 resembles a narrow band pass filtered turbulent signal at high Reynolds numbers. (Perhaps the word 'preturbulence' also used commonly in dynamical systems literature, is sufficiently useful to designate the signal such as the one shown in figure 8, and its dynamics.)

b. Chaos and its aftermaths

No qualitative change occurs between Re = 66 and about 71. Soon thereafter the system becomes reordered. For example, the spectral density at Re = 76 shows (essentially) nothing but discrete peaks again (figure 10a). These peaks, shown in detail in figure 10b, can all be identified with great precision as arising from the interaction of *three* irrational frequencies. (That there are definitely three independent frequencies can also be seen from Poincaré sections (not shown here) and the dimension of the attractor discussed in section 4b). After a small increase in Reynolds number to about 81, one can see the onset of the broadband spectral content (figure 11), and we may consider chaos to have set in again!

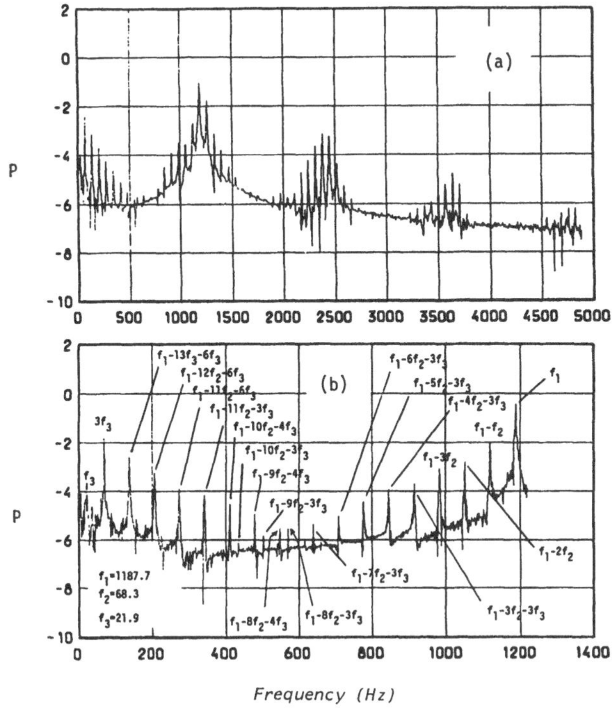

FIGURE 10: Reordering at Re = 76. (a) The measured power spectral density of u, and (b) its details in the frequency range 0-1250 Hz. Note that all peaks above noise level can be represented by combinations of three frequencies f_1, f_2 and f_3. This conclusion can certainly be influenced by the finite FFT resolution, but our belief in the accuracy of this statement comes also from dimension calculations (section 4).

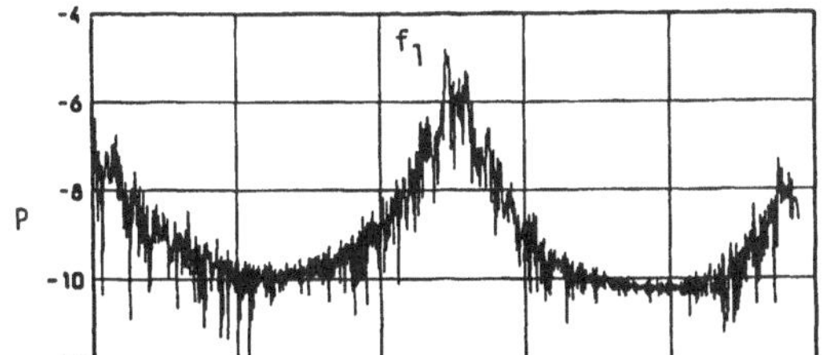

FIGURE 11: Chaos at Re = 81. f_1 is the vortex shedding frequency.

The system reorders itself around an Re of about 90, and we have discussed elsewhere [10] that this reordered state is quasiperiodic with four frequencies. (That this is the case will be demonstrated also by dimension measurements in section 4d.) Chaos sets in again at an Re = 140, followed by yet another reordering around an Re = 143. In fact, this sequence of return to chaos and reordering continues for much higher Reynolds numbers although it becomes progressively more difficult with increasing Re to distinguish experimentally between the two states.

Two related points of importance emerge. First, there do exist quasiperiodic motions with three or four independent frequencies; just like Landau's quasiperiodicities, the Ruelle-Takens picture of transition is also not the whole story. Second, transition to turbulence (at least in the temporal sense) is characterized by regions of chaos interspersed between regions of relative order. Each of these deserves at least a brief discussion.

c. Note on quasiperiodicities with more than two frequencies

We have shown that the route to the lowest Reynolds number chaos occurs in our experiments precisely as postulated in the RTN picture of transition. On the other hand, our experiments also show that quasiperiodicities with three (and possibly four) frequencies do exist. This type of disagreement with the RTN scheme has been noted earlier in the Taylor-Couette flow [14] and the convection problem [15]. It is thus pertinent to inquire whether there are (in some sense) exceptional conditions to be satisfied for the RTN scheme to hold. Greborgi et al. [16], who address this question in a specific numerical experiment, suggest that the three frequency quasiperiodicity is indeed quite likely to occur in practice, and that the special perturbation required to destroy this state (as in the RTN scheme) is unlikely. Haken [17] discusses this issue at some length and concludes that if the frequencies possess a certain kind of irrationality with respect to each other (or, more precisely, the so-called Kolmogorov - Arnold - Moser condition holds), bifurcation from a two-torus toa three-torus is possible. Both these discussions are strictly relevant to systems with no externally imposed noise (or fluctuations), a condition that does not strictly obtain in experiments (especially open systems). Our own experience is that the precise nature of even small amounts of noise (some of which is controllable in our wind tunnels and some of which is not!) has an influence on the evolution of the system (for a brief discussion of this influence, see subsection 3e). It is not hard to visualize that in our experiments the detailed conditions of intrinsic noise itself could have altered from before to after the first occurrence of chaos. Clearly, this is an area for further work, both experimentally and theoretically.

d. Windows of order and chaos

Figure 12 summarizes the changes occurring in the low end of the Reynolds number range we have considered. The shaded regions indicate windows of chaos, and the question marks indicate the uncertainty and difficulty in quantifying what we believe are reordered states.

At least two questions arise: What is the mechanism that permits the reordering of a chaotic state? What determines the length and location of the windows of chaos? Our understanding of these matters is rather limited, but even within these limits, some comments seem called for. Let us consider the first question now, and relegate the second one to the next subsection. The observed alternation between chaos and order has been known to occur in several low-dimensional dynamical systems; for ex-

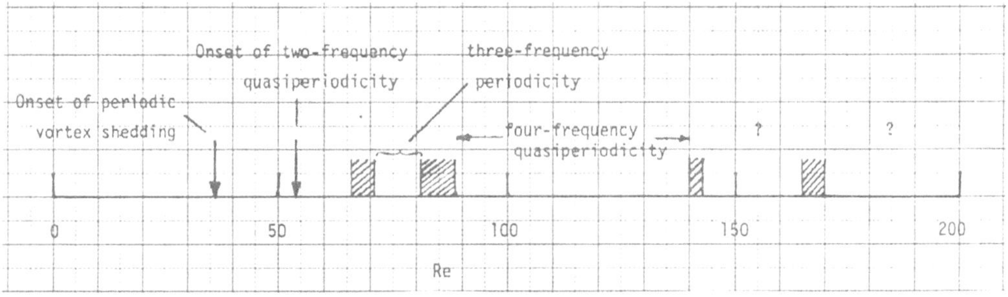

FIGURE 12: Window of chaos and order

ample, Lorenz equations [18], and spherical pendulum [19]. In these systems, the occurrence of reordering is independent of *external* noise. The numerical experiments of Matsumoto & Ysuda [20] show that chaotic orbits could be unstable to external noise, and noise addition to deterministic chaos (i.e., chaos characteristic of deterministic dynamical systems) yields an ordered state in some cases. They specifically consider the so-called Belousov-Zhabotinskii (BZ) reaction and some variants of the logistic model. Roux et al. [21] find windows of chaos and order in their experiments on the BZ reaction.

In experiments on open systems, it is hard to ascertain whether the return to order is tied intimately to external noise or the increased degrees of freedom associated with the appearance of chaos itself. In any case, the analogy between this situation and increased eddy viscosity in turbulent flows appears to be more than superficial: addition of high frequency modes results in a lowering of an effective Reynolds number and increased stability of the flow.

Though we have not made detailed spectral measurements at higher Reynolds numbers, it is our contention that the succession of order and chaos in a wake continues indefinitely even at very high Reynolds numbers (with the caution that order must now be interpreted to mean spectral sharpening). Roshko [22] pointed out several years ago that order reappears in the Reynolds number range of 10^6. More recently, the fluctuating lift force measurements of Schewe [23] on a circular cylinder showed that the spectral density of the lift coefficient was broad at $Re = 3.7 \times 10^6$ (upper end of transition) and became increasingly narrow until, at $Re = 7.1 \times 10^6$, it was quite sharp, rather like a narrow-band-pass filtered signal. Although the fluctuating lift force can at best be related to the squared fluctuating velocity filtered via the transfer function corresponding to the response of the circular cylinder, its behavior is nevertheless indicative of the flow itself in the vicinity of the cylinder.

e. The vortex shedding frequency and windows of chaos

Consider now the variation of the vortex shedding frequency f_1 with Reynolds number (figure 13). The frequency does not vary monotonically with Re but shows several more or less distinct breaks. Such breaks have been noted before [24,25,26], and perhaps most convincingly demonstrated in a beautiful experiment by Friehe [27].

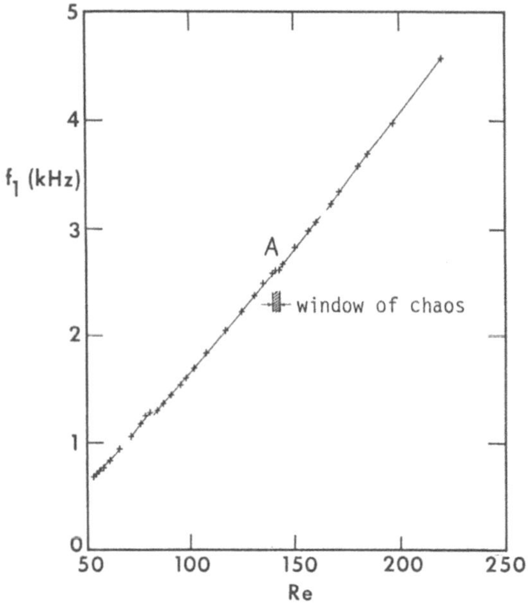

FIGURE 13: Variation of vortex shedding frequency with Re. Notice discontinuities, and their coincidence with windows of chaos, as illustrated near A.

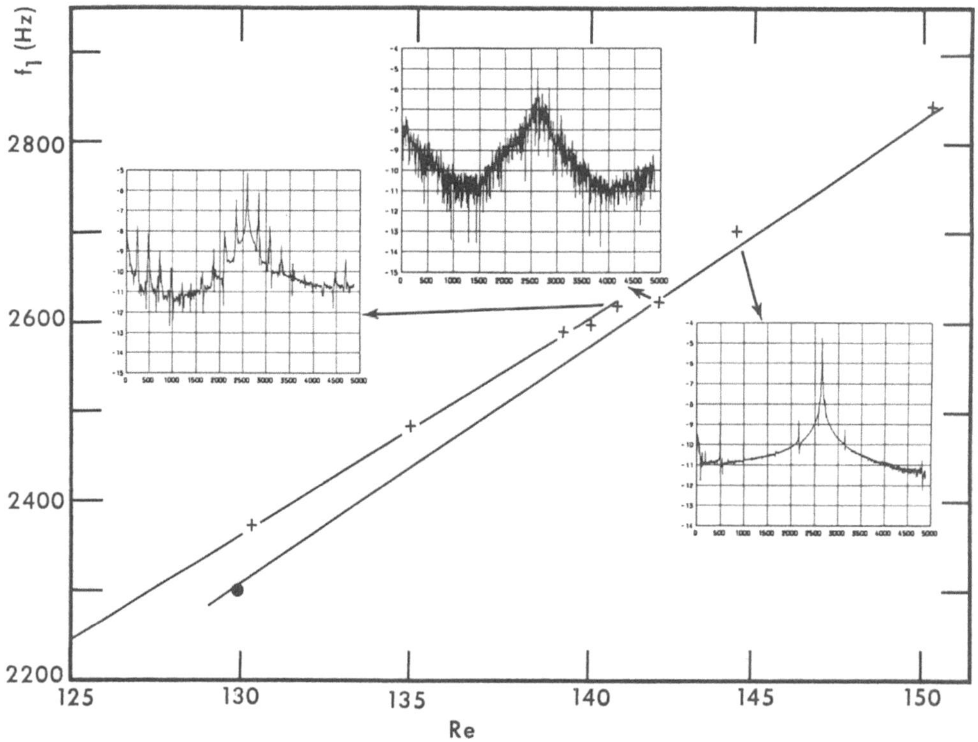

FIGURE 14: Expanded version of figure 13 near A. • shows a data point in another series of experiments where the window of chaos began at an Re ≈ 130.

Friehe varied the Reynolds number continuously at a small rate and obtained on an x-y plotter the frequency-Re variation directly. Although the appearance of the breaks has been disputed [28], our own data, presented here and elsewhere [10], support the conclusion that discontinuities do indeed appear.

Our interest here is in pointing out that the occurrence of these breaks coincides with the windows of chaos. To establish the connection better, we may consider in figure 14 the details of the break marked A in figure 13. Just upstream of the break, the spectral density is quite ordered (four-frequency quasiperiodicity) while it is broadband until the end of the break region coinciding with the upper end of the window of chaos; to the extent we can ascertain, the frequency spectrum shows a reordering immediately after the break.

The data shown by crosses in figures 13 and 14 were all obtained from one experimental run. In a repeat of the experiment the following day (for example) we found the same general features, except that chaos set in at different Reynolds numbers; the windows of chaos were also of different widths. The filled circle in figure 14 was obtained in a second series of experiments. It is seen that this point falls below the first set of data at the same Re, but it falls on the backward extrapolation of the line corresponding to the reordered state (Re $\geq$ 143) in the first set. It is hard to tell the differences between conditions in the two experiments without extensive documentation, but there are reasons to believe that the second experiment was conducted in a somewhat noisier environment. We thus speculate that the location as well as the widths of the windows of chaos are to some extent determined by noise characteristics — in a way that is not well understood at present.

It is interesting to note from figure 14 that the ratio f_2/f_1 (the so-called rotation number), where f_2 is the second largest independent frequency, changes its value abruptly across the narrow windows of chaos. Figure 15 is a plot of the rotation number with Re. It is seen that the number changes abruptly across all the windows of chaos, but only slowly within regions of order.

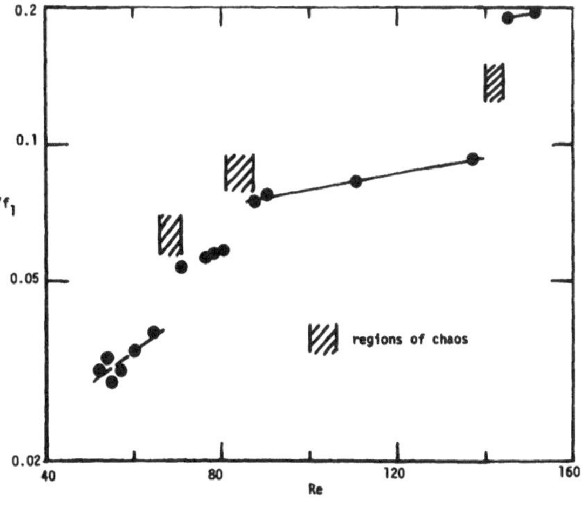

FIGURE 15: The variation of the rotation number with Reynolds number.

f. The amplitude of the vortex shedding mode and chaos

Since reordering is associated with the reemergence of stronger spectral peaks, it is natural to expect that there must be some relation between the amplitudes of the various modes and the occurrence of order and chaos. Figure 16 shows the amplitude of the vortex shedding mode (or the f_1 frequency) as a function of velocity. (The amplitude A_1 is expressed as a fraction of the freestream velocity U, but is given here to an arbitrary scale.) It is clear that O indicating order coincides with a local peak in A_1, C indicating the onset of chaos coincides with a local minimum, and, finally, RO indicating reordering coincides with the reappearance of a peak. Except for the first time that reordering occurs, every successive reordering is associated with a general lowering of the amplitude of the vortex shedding mode.

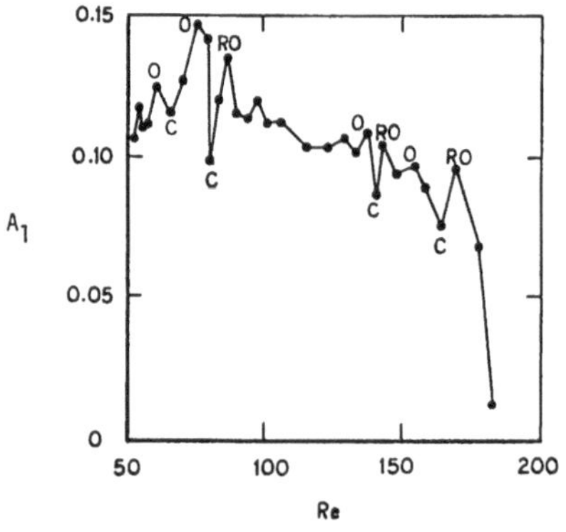

FIGURE 16: The amplitude of the vortex shedding mode as a function of Re. O is order, C chaos and RO is reordering; within a window of chaos, O and RO may in general indicate different states of order.

4. Results from the Dimension of the Attractor

a. The dimension

It is clearly worth inquiring whether there is any property of the attractor that successfully describes in some way the many subtle changes that occur in the frequency spectra and the related properties discussed in section 3. It appears that there indeed is such a quantity, namely the dimension of the attractor. Loosely speaking, the dimension of the attractor is related to the number of degrees of freedom — and hence its importance. The concept of the dimension is highlighted in studies of dynamical systems, and we may briefly digress here to discuss its meaning before presenting results from our measurements. It should be pointed out that, apart from our own earlier measurements of the dimension for turbulence attractors [9,10], such measurements have been made by others in the Taylor-Couette flow [5] and in the convection cell [29].

Let us consider an attractor (constructed as already discussed in section 3) from a measured temporal signal u(t) that is embedded in a (large) d-dimensional phase space. Let N(ε) be the number of d-dimensional cubes of linear dimension ε required to cover the attractor to an accuracy ε. Obviously, making ε smaller renders N larger, but if the limiting quantity

$$D = \lim_{\varepsilon \to o} \frac{\log N(\varepsilon)}{\log(\frac{1}{\varepsilon})} \tag{4.1}$$

exists, it will be called the dimension of the attractor. An important characteristic of a strange attractor is that D is small even though d is large. We should be interested in knowing whether transitional and turbulent signals have this property.

To see what the dimension means, let us write (4.1) as

$$N(\varepsilon) \sim \varepsilon^{-D}; \tag{4.2}$$

that is, if one specifies D and the accuracy ε to which we need to determine the attractor, we automatically know the number of cubes required to cover the attractor. The only missing information will now be the position of the cubes in the phase space. Thus, D can be considered as a measure of how much more information is required in order to specify the attractor completely; the larger the value of D, the larger is this missing information.

In general, the dimension D, as defined in (4.1), is fractional for strange attractors, and it has been called the fractal dimension by Mandelbrot [30] who has contributed a lot to our understanding of the quantity. As defined in (4.1), D is a geometric property of the attractor, and does not take into account the fact that a typical trajectory may visit some region of the phase space more frequently than others. Several measures, taking this probability into account, have been defined — and are believed to be closely related to the dynamical properties of the attractor. The most well-known among them are:

(a) the pointwise dimension

(b) the Grassberger-Proccacia dimension.

If the attractor is uniform, that is, every region in the phase space is as likely to be visited by the trajectory as every other, then the above two measures equal D defined by (4.1). Otherwise, they are generally smaller than D.

Let $S_\varepsilon(x)$ be a sphere of radius ε centered about a point x on the attractor, and let μ be the probability measure on the attractor. Then, the pointwise dimension is defined [31] as

$$d_p(x) = \lim_{\varepsilon \to o} \frac{\log \mu[S_\varepsilon(x)]}{\log \varepsilon} \tag{4.3}$$

or

$$\mu[S_\varepsilon(x)] \sim \varepsilon^{d_p} \tag{4.4}$$

Grassberger & Procaccia [32] have defined another measure ν which is related to the dimension of the attractor, as well as the entropy (see section 4d). The procedure for computing ν is as follows:

(i) Obtain the correlation sum $C(\varepsilon)$ from:

$$C(\varepsilon) = \lim_{N \to \infty} \frac{1}{N^2} \sum_{\substack{i=j=1 \\ i \neq j}}^{N} H[\varepsilon - |\underset{\sim}{u}_i - \underset{\sim}{u}_j|] \tag{4.5}$$

where H is the Heaviside step function and $\underset{\sim}{u}_i - \underset{\sim}{u}_j$ is difference in the two vector positions $\underset{\sim}{u}_i$ and $\underset{\sim}{u}_j$ on the phase space. Basically, what C does is to consider a window of size ε, and start a clock that ticks each time the difference $|\underset{\sim}{u}_i - \underset{\sim}{u}_j|$ lies within the box of size ε. Thus, one essentially has

$$C(\varepsilon) = \lim_{N\to\infty} \frac{1}{N^2} \{\text{number of pairs of points } (i,j) \text{ with } |\underset{\sim}{u}_i - \underset{\sim}{u}_j| < \varepsilon\}.$$

(ii) Obtain ν from the relation [32]

$$C(\varepsilon) \sim \varepsilon^{-\nu} \quad \text{as } \varepsilon\to o. \tag{4.6}$$

In practice, not all components of $\underset{\sim}{u}$ are known for constructing the phase space, but perhaps only one component, say u_m. As we discussed in section 3, one constructs a d-dimensional 'phase space' using delay coordinates

$$\{u_m(t_i),\ u_m(t_i+\tau),\ \ldots,\ u_m(t_i+(d-1)\tau)\},\ i = 1,\ \ldots,k,$$

where, again, τ is some interval which is neither too small nor too large and k is large (in principle, infinity!). Since one does not *a priori* know ν, one constructs several 'phase spaces' of increasingly large value of d and evaluates ν for each of them; ν will first increase with d and eventually asymptote to a constant independent of d. This asymptotic value of ν is of interest to us as a measure of the dimension of the strange attractor.

We have computed both d_p and ν as described above, using the streamwise velocity fluctuations u up to an Re of 10^4, and the delay coordinates. Our confidence in the numerical values of these measures of dimension is very good when they are less than about 5 or 6, but becomes increasingly shaky at higher values. However, we do believe that they are reasonable, judging from their repeatability and the several precautions we have taken (such as taking the proper limit as $\varepsilon\to o$ and using, in a couple of cases, double precision arithmetic in our computations). It would be interesting and useful to evaluate the dimension at high Reynolds numbers, but such calculations are likely to be of uncertain value (unless perhaps some carefully selected combination of experimental and computational conditions obtains): with increasing Re, the newly excited degrees of freedom can be expected to be of smaller and smaller scales, and to properly accommodate them in the dimension calculations requires that one must in practice look at increasingly smaller values of ε (see equation 4.6). Such efforts will very soon be frustrated by instrumentation noise and digitizer resolution problems.

b. Data for Re $\lesssim$ 100

It is convenient to consider first the data for Re $\lesssim$ 100 (figure 17). Concentrating on the data in the ordered states only, we may conclude the following. At Re = 36, where there is only one independent degree of freedom (corresponding to the periodic vortex shedding) — see figures 2 and 5 — the dimension of the attractor turns out to be about 1. When only two frequencies are present (figures 3 and 6) at Re =

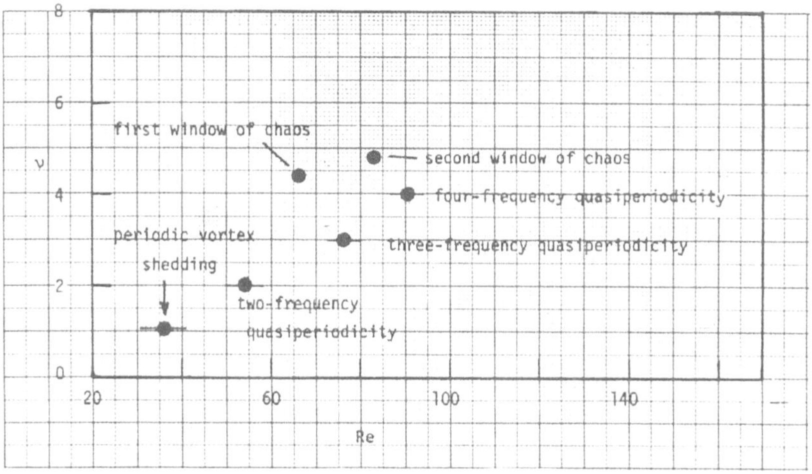

FIGURE 17: Variation of the dimension of the attractor with respect to Reynolds numbers. Note that the dimension is about 1 when there is only vortex shedding (Re = 36), about 2 when there are only 2 frequencies (Re = 54), about 3 when there are 3 frequencies (Re = 76), about 4 when there are 4 frequencies. The dimension jumps to higher noninteger values in the windows of chaos.

54, the dimension is about 2. At Re = 76 where there are three dominant frequencies (figure 10), the dimension is three to within experimental uncertainty. Lastly, at Re = 91 where there are four frequencies present, the calculated ν is very close to 4. Thus, to within computational uncertainties, it is seen that the dimension of the attractor is a reasonable representation of the number of degrees of freedom.

Now getting back to measurements in the windows of chaos, it is clear that the first appearance of chaos at Re = 66 is characterized by a jump in the dimension (to about 4.4 from 2 characteristic of the two-frequency quasiperiodicity), followed by a return to a value of 3 in the region of three-frequency quasiperiodicity. Similarly, the dimension of the attractor in the second chaotic window is about 4.8. As we discussed earlier, the dimension of the attractor in the chaotic windows is a fraction.

c. Higher Reynolds number data

Figure 18 shows the results of the dimension calculations up to an Re of about 10^4. Both ν and d_p increase to about 20 or so at an Re of 10^4, although the increase is not always monotonic. In fact, our calculations seem to suggest that the dimension settles down to about a value of 20!

If it is true that the dimension of the attractor retains, even at high Reynolds numbers, its meaning as an indicator of the number of dynamically significant degrees of freedom, common wisdom tells us that the dimension of the attractor should generally increase with Re. In contrast, the dimension does not increase continuously; further, its value is far lower than $Re^{9/4}$, which is the classical estimate (see Landau & Lifshitz [33]) for the number of degrees of freedom in a turbulent *flow*. It

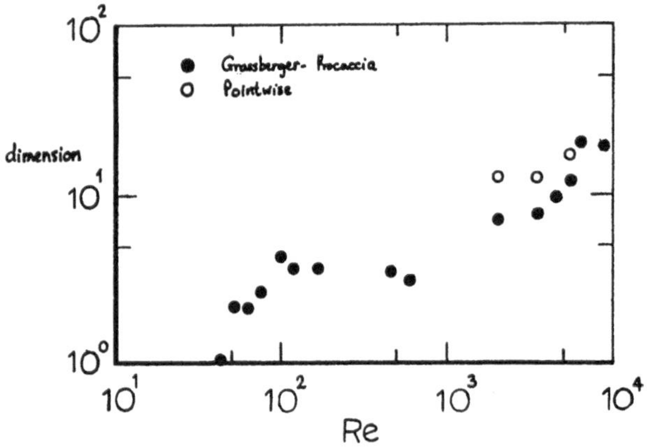

FIGURE 18: Dimension data for Re up to 10^4.

may be that the constancy of the dimension at higher Re is simply an artifact of resolution and computational problems, but if the result is genuine instead, it should provide an incentive for a suitable reformulation of 'turbulence problem'.

d. The Kolmogorov entropy

The Kolmogorov entropy has the property that it is positive for a chaotic signal, zero for ordered signals and infinite for a random signal with a space filling attractor. As already mentioned, there are conjectures that the entropy equals the sum of positive Lyapunov exponents, and hence, unlike the dimension D, is a dynamic measure of unpredictability of the motion.

Suppose the d-dimensional phase space housing the attractor is partitioned into boxes of size ε^d. Let $p(i_1, i_2, \ldots, i_d)$ be the joint probability of finding $\underset{\sim}{u}$ at time $t = \tau$ in box i_1, $\underset{\sim}{u}$ at time $t = 2\tau$ in box i_2, ..., $\underset{\sim}{u}$ at time $t = d\tau$ in box i_d. The Kolmogorov entropy is then defined [34] as

$$K = -\lim_{\varepsilon\to o}\ \lim_{\tau\to o}\ \lim_{d\to\infty} \frac{1}{d\tau} \sum_{i_1,\ldots i_d} p(i_1,\ldots i_d)\,\ell n\, p(i_1,\ldots i_d). \tag{4.7}$$

Grassberger & Procaccia [35] have defined a quantity K_2 which is close to K and further has the property that $K_2 > 0$ is a sufficient condition for chaos. Without going into too many details, we follow [35] and note that it can be computed by first obtaining $C(\varepsilon)$ as in Eq.(4.5) in section 4a for various d, and forming the ratio

$$K_{2,d}(\varepsilon) = \frac{1}{\tau}\,\ell n\,\frac{C_d(\varepsilon)}{C_{d+1}(\varepsilon)}\,, \tag{4.8}$$

where C_d indicates C for dimension d. In the limit,

$$\lim_{\substack{d\to\infty\\ \varepsilon\to o}} K_{2,d}(\varepsilon) \sim K_2.$$

Table 2 gives K_2 for Re = 66 and 81 within the first two windows of chaos. For comparison, the table also lists K_2 for the Hénon map from [35].

Table 2: The Kolmogorov entropy

Signal	K_2
u at Re = 66	≈ 0.22
u at Re = 81	≈ 0.24
The Hénon map	0.325 ± 0.02

5. Discussion of Results

We have shown that several features of transition to turbulence behind circular cylinders are in essential agreement with the behavior of low-dimensional dynamical systems. We emphasize that many details discussed above in the near-wake region hold also at around x/d ≃ 50, although less conspicuously.

One particularly important feature of this work is the discovery of windows of chaos interspersed between regions of order: these latter regions are three and four-frequency quasiperiodicities in the low Reynolds number range up to about 140 (possibly even higher!). Not all observations we have made can be understood within the present framework of chaos and dynamical systems, but we find it amazing that the dynamics of fluid motion which we believe are particularly governed by the NS equations should be at all represented by extremely simple systems. One aspect of this work is the fine resolution (in Reynolds number, frequency domain, as well as in the phase space) with which measurements have been made. It seems to us that even finer resolution, especially within the windows of chaos and regions bordering them, will perhaps disclose even more interesting aspects.

We have shown that, during early stages of transition, a strong connection (speculated previously, but never shown to be true conclusively) exists between the dimension of the attractor and the degrees of freedom as inferred from power spectral densities. Provided this interpretation is true also in windows of chaos and (moderately) high Reynolds number turbulence, our results suggest that the degrees of freedom are not too many even up to Reynolds number of the order of 10^4. Our numerical calculations based on Schewe's data lead us to expect that the dimension of the attractor, as computed according to (4.4) and (4.5), is not high even at higher Reynolds numbers corresponding to the fully turbulent state (Re ≈ 10^6). If the attractor is sufficiently low-dimensional, a clever projection of it can perhaps be used to our advantage. (If the attractor dimension is even as high as 20, however, no matter what projection one devises, it will perhaps look uniformly dark!) At this stage it is not clear how one could use this information, but, without entering into a detailed discussion, we may point out that it lends credence to concepts embodied in renormalization group theory, slaving principle, or, closer to home, large eddy simulation or orthogonal decomposition techniques.

We thus believe that there is much that we can learn about transition and turbulence from chaos theories. In the immediate future, these theories provide a strong motivation for looking into newer aspects of fluid flow phenomena; discoveries of close correspondence between fluid flows and low-dimensional chaotic dynamical systems will undoubtedly prove useful in the sense that the rich variety of results from dynamical systems can be brought to bear on fluid flow transition and, perhaps, even turbulence. In the long run, the hope is that they will help us in coming to grips with the eternal problem of turbulence, namely, the enormous amount of 'information' that seems to be available to us! Perhaps we can then model, even at high Reynolds numbers, at least local behaviors by low-dimensional dynamical systems.

Do we then conclude that the key to the understanding of transition and turbulence lies totally in low-dimensional dynamical systems? We think that such statements are optimistic at best and misguided at the worst. Apart from the fact that the spatial structure of turbulent flows, which is their single most important charactertistic, lies outside the scope of dynamical systems theories — at least as they stand today — there is a lot that they do not or, perhaps, cannot, tell: for example, they do not tell us anything about the origin and physical structure of the various bifurcations that can occur, or how the drag coefficient varies with Reynolds number. To answer these and similar questions of practical interest, we suspect that we have to revert to the NS equations!

One final comment should be made. It would be useful to make a concurrent flow visualization study and relate the various findings reported here to the spatial characteristics of the flow. It is unfortunate that we cannot use much the extensive flow visualization observations made by others (for example, Gerrard [36]) because the details from one experiment to another do not precisely match.

ACKNOWLEDGEMENTS

For their helpful comments during this work or on an earlier draft of the manuscript, it is my pleasant duty to thank Hassan Aref, Peter Bradshaw, B.-T. Chu, Rick Jensen, John Miles, Mark Morkovin, Turan Onat, David Ruelle, Paul Strykowski, Harry Swinney and Peter Wegener. I am indebted to Mike Francis who patiently accommodated this pursuit within an AFOSR grant I received for turbulence control work. Finally, presentation of these results in Evanston and preparation of this manuscript for this volume honoring Stan Corrsin has been a labor of love, and an expression of indebtedness I have for him.

Appendix

Let b_1, b_2, ..., b_n be the state variables of the system (1.1). In an n-dimensional space spanned by b_1, b_2, ..., b_n, each point determines the state of the system completely at a given time, t. As t evolves, we obtain a continuous sequence of points which form the trajectory of the system. As $t \to \infty$, the b_i's need not go to infinity, but may terminate (in two dimensions) either at a node or a focus or on a limit cycle or, in higher dimensions, on to a more complicated object. This object on which the trajectory terminates is called an attractor if all other trajectories starting near the said trajectory converge to the same object as $t \to \infty$. (That is, the attractor is the limit set of a representative point in phase space. Thus, an attractor attracts all nearby trajectories.)

If the system is stable and steady the attractor is a point — a node if the motion is critically damped (figure A1) or a focus if the motion is damped but oscillatory (figure A2). If the system executes a periodic motion, a limit cycle is observed in the phase plane (figure A3). Quasiperiodic motion with two incommensurate frequencies results in a two-torus (see figure A4), with the entire surface of the torus covered by the trajectory eventually. A projection of the torus on to a plane

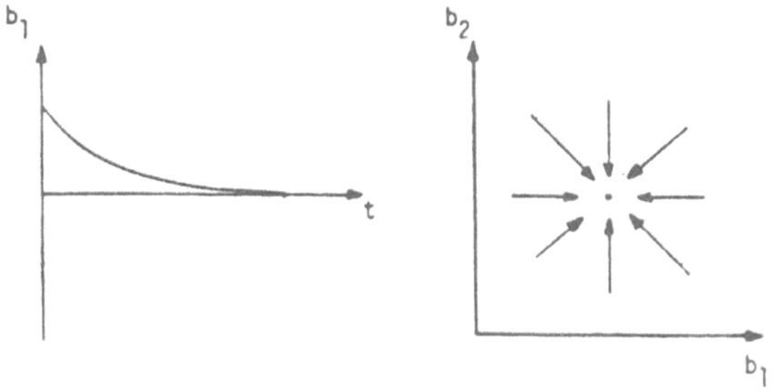

FIGURE A1: Stable node. (point attractor)

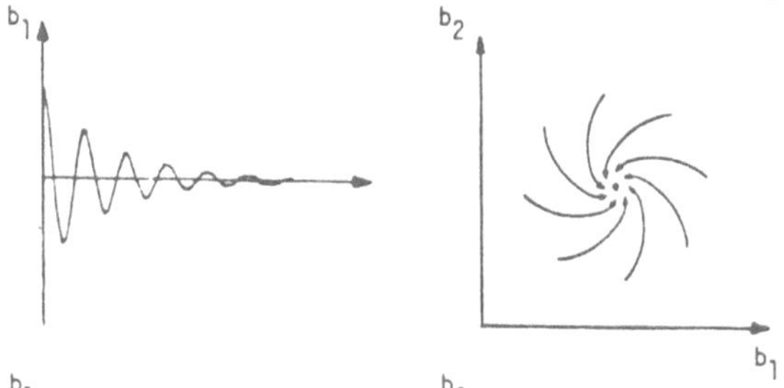

FIGURE A2: Stable focus. (point attractor)

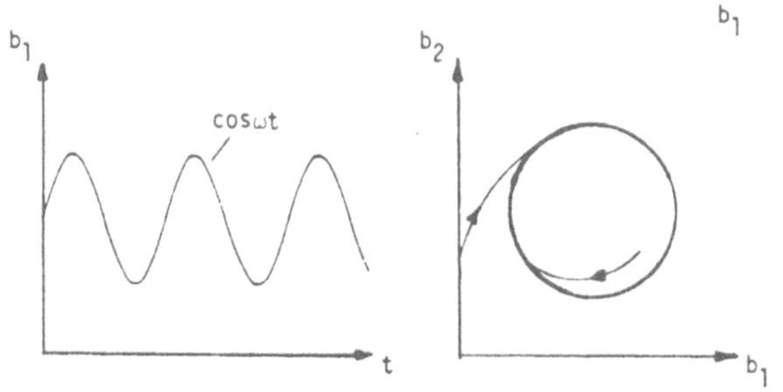

FIGURE A3: Limit cycle.

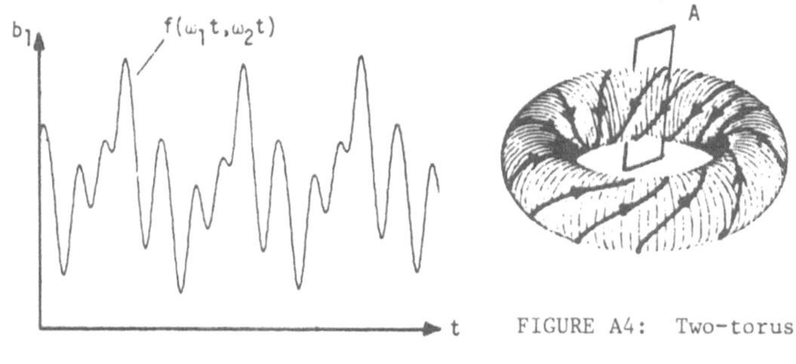

FIGURE A4: Two-torus. (perspective view)

may have different shapes depending on the orientation of the plane, but it is clear that a section of the torus, say, by the plane A in figure A4 (the Poincaré section) will yield a limit cycle. To obtain such a section in practice, one has to intercept the trajectory each time it crosses the plane (or 'sample' the system at the frequency f_1 and at fixed phase), and plot b_1 and b_2 (say) corresponding to these periodically sampled data. The phase portrait corresponding to the quasiperiodic motion with three frequencies is a three-torus, and so on.

The attractor has been called a 'strange attractor' if (roughly speaking) it is a complex surface repeatedly folded onto itself in such a manner that a line normal to the surface intersects it in a Cantor set. That is, if one successively magnifies regions of this intersection which appear, at some level of resolution, to be entirely 'filled', one sees regions of 'emptiness' interspersed between regions of 'occupation'. One cannot test this property of the strange attractor directly if it is constructed from experimental data (because of noise and the finite resolution of the instrumentation), and so, one uses several of its other properties to determine its occurrence. For example, any two neighboring trajectories on the strange attractor will diverge exponentially apart for small t (the so-called sensitivity to initial conditions, measured by positive Lyapunov exponents or the Kolmogorov entropy); the so-called dimension of the attractor (see section 4) is generally a non-integer; the spectral density of the temporal signal used to construct the attractor will have broadband components orders of magnitude above the instrumentation and other noise levels.

References

1. Ruelle, D. & Takens, F., Commun. Math. Phys. 20, 167 (1971).
2. Newhouse, S., Ruelle, D. & Takens, F., Commun. Math. Phys. 64, 35 (1978).
3. Gollub, J.P. & Swinney, H.L., Phys. Rev. Lett. 35, 927 (1975).
4. Fenstermacher, P.R., Swinney, H.L. & Gollub, J.P., J. Fluid Mech. 94, 103 (1979).
5. Brandstäter, A., Swift, J., Swinney, H.L. & Wolf, A., Phys. Rev. Lett. 51, 1442 (1983).
6. Libchaber, A. & Maurer, J., J. de Physique 41, C3-57 (1980).
7. Libchaber, A., In Chaos and Statistical Methods (ed. Y. Kuramoto, Springer 1984), p.221.
8. Coles, D., J. Fluid Mech. 21, 385 (1965).
9. Sreenivasan, K.R. & Strykowski, P.J., In Turbulence and Chaotic Phenomena in Fluids (ed. T. Tatsumi), to appear.
10. Sreenivasan, K.R., In Nonlinear Dynamics of Transcritical Flows (ed. H. Oertel), to appear.
11. Takens, F., In Lecture Notes in Mathematics 898 (eds. D.A. Rand and L.S. Young, Springer-Verlag 1981), p.366.
12. Chaitin, G.J., Scientific American, 246, 47 (May issue, 1982).
13. Thomas, R.M., J. Fluid Mech. 57, 549 (1973).
14. Gorman, M., Reith, L.A. & Swinney, H.L., Ann. N.Y. Acad. Sci. 357, 10 (1980).

15. Gollub, J.P., & Benson, S.V., J. Fluid Mech. 100, 449 (1980).

16. Greborgi, C., Ott, E. & Yorke, J.A., Phys. Rev. Lett. 51, 339 (1983).

17. Haken, H., Advanced Synergetics, Springer 1984.

18. Sparrow, C., The Lorenz Equations, Springer 1982.

19. Miles, J.W., Physica D (to appear), 1984.

20. Matsumoto, K. & Tsuda, I., J. Stat. Phys. 31, 87 (1983).

21. Roux, J.-C., Turner, J.C., McCormick, W.D., & Swinney, H.L., In Nonlinear Problems: Present and Future (eds. A.R. Bishop, D.K. Campbell, B. Nicolaenko, North-Holland Publishing Co., 1982), p.409.

22. Roshko, A., J. Fluid Mech. 10, 345 (1961).

23. Schewe, G., J. Fluid Mech. 133, 265 (1983).

24. Tritton, D.J., J. Fluid Mech. 6, 241 (1959).

25. Berger, E., Z. Flugwiss. 12, 41 (1964).

26. Tritton, D.J., J. Fluid Mech. 45, 749 (1971).

27. Friehe, C.A., J. Fluid Mech. 100, 237 (1980).

28. Gaster, M., J. Fluid Mech. 38, 565 (1969).

29. Malraison, B., Atten. P., Berge, P. & Dubois, M., J. Physique Lett. 44, L-897 (1983).

30. Mandelbrot, B., The Fractal Geometry of Nature, Freeman & Co., New York, 1983.

31. Farmer, J.D., Ott, E. & Yorke, J.A., Physica 7D, 153 (1983).

32. Grassberger, P. & Procaccia, I., Phys. Rev. Lett. 50, 346 (1983).

33. Landau, L.D. & Lifshitz, E.M., Fluid Mechanics (volume 6 of the Course of Theoretical Physics), Pergamon Press, 1982.

34. Barrow, J.D., Phys. Reports, 85, 1 (1982).

35. Grassberger, P. & Procaccia, I., Prepublication report (1983).

36. Gerrard, J.H., Phil. Trans. Roy. Soc. Lond. 288A, 29 (1978).

Some Contributions of Two-Point Closure to Turbulence

Jackson R. Herring
National Center for Atmospheric Research, P.O. Box 3000, Boulder, CO 80307, USA

Since their introduction over thirty years ago, two-point-moment closures have offered promise of providing practical and theoretical information concerning certain aspects of turbulent flows. More recently, both experiments and direct numerical simulations have suggested a high degree of organization (coherent structures) in even homogeneous flows, which contradicts--at least in its extreme form--the assumption of near Gaussianity central to closures. We examine here some successes and failures of closure for several problems including three-dimensional turbulence, and quasi-two-dimensional flows of the sort employed in geophysical context.

We begin with a brief discussion of the theoretical foundations of two-point closures, which are based upon modified pertubation theory or variational methods. We then consider what closures have contributed to the following problems: (1) the decay of homogeneous turbulence and scalar variance, (2) the return to isotropy of two- and three-dimensional flows, (3) two-dimensional flows at high Reynolds number, and (4) thermal convection between slip boundaries.

The results suggest that closures--overall--are sufficiently quantitative that in most cases the ease with which they may be performed (as compared to a full simulation) and the probability of their containing valid information make them a useful tool for considering turbulent flows.

1. Introduction

A central problem in the theory of turbulence is the prediction of the distribution of turbulent eddies that emerge from a simple initial state of Gaussian chaos. This is the problem posed--for example--in Batchelor's (1959) monograph, to which the early theories of Heisenberg (1948), Oboukhov (1941), and Kovasznay (1948) were addressed. At that time, there was perhaps hope that the detailed shapes of the eddies played only a secondary role, and it would be possible to determine their size distribution (in the sense of Fourier modes) without knowing very much about what constituted precisely the coherent structures or the flow. Such a characterization is reasonable for flows that depart weakly from Gaussianity. The approaches mentioned above were heuristic, relying upon dimensional as well as physical reasoning in fixing the functional forms that give energy transfer in terms of energy spectra, and to fix the empirical numbers entering the theories.

These heuristic theories pivot on the idea that the small scales act on the larger as an eddy viscosity. That such interactions can be so characterized is rooted in the same general concepts used in statistical mechanics to derive viscosity itself: the statistical independence of large and small scales. It remains a central concept in the more elaborate spectral theories whose results are the focus of the present paper. Their remaining ingredients are: (1) the conservation constraints of inviscid flows, and (2) the use of experiments to fix numerical constants.

As noted by Kraichnan and Speigel (1962), these early theories lacked an equipartitioning tendency, which--at least in a certain test-case--may be demonstrated as a rigorous consequence of the non-linear and pressure terms. The particular test-case consists of the Euler equations (zero viscosity) confined on a finite wave-number range. To fix ideas, consider the (complex) Fourier amplitudes of the velocity field $\underline{u}(\underline{k},t)$ on a wave number range $(0 < k < k_o)$. Then, (as first suggested by T. D. Lee [1950]) the real and imaginary components of $\underline{u}(\underline{k})$ constitute independent degrees of freedom of a conservative system, and in equilibrium (steady state) each degree of freedom shares equally the total available energy, E. In three dimensions this leads to an energy spectrum,

$$E(k) = 3\{E_v / k_o\} (k/k_o)^2 \tag{1.1}$$

Although our test-case exists only in a computer, it is nonetheless important to check that any proposal for E(k) equilibrates as (1.1) for $\nu=0$, and $(o < k < k_o)$. Isotropy is of course simply angular equipartitioning.

Equipartitioning and eddy viscosity concepts may be combined to give a plausible guess of how $\partial E(k,t)/\partial t$ behaves if k is much smaller than the energy containing range:

$$\partial E(k,t)/\partial t = T(k,t) = k^4 A\{E\} - (\nu + \nu_{eddy}) k^2 E(k,t) \tag{1.2}$$

Here A{E} and $\nu_{eddy}\{E\}$ are yet-to-be-determined functionals of E(k,t). (1.2) is the only (analytic) form of T(k,t) available as $k \to 0$ if we assert equipartitioning (for $\nu=T=0$) and eddy viscosity. As we show shortly, (1.2) alone suffices--in large measure--to determine the decay of total energy, $E_v(t)$;

$$E_V(t) = \int_0^\infty dk E(k,t) \qquad (1.3)$$

One of the earlier formal theories free of empiricism was the quasi-normal approximation proposed originally by Proudman and Reid (1954), and by Tatsumi (1955). The defects of this early theory are well known (see, e.g., Monin and Yaglom, 1975; Orszag, 1974, for summaries). However, later theories such as Kraichnan's (1959) direct interaction (DIA) approximation and its Lagrangian history variants (LHDI) are numerically only slightly more complicated, so that in practice a numerical code that will solve the quasi-normal approximation will--with only minor additions--also solve the DIA. The important point about these theories (the QN and DIA) is that they are free from empiricism, and directly generalizable to inhomogeneous problems such as shear flow and thermal convection. For completeness, we have recorded the DIA in the appendix for the special case of homogeneous and isotropic turbulence. We also have recorded there other more practical and simpler approximations, to be used in subsequent sections.

We now discuss how the formal closure is related to simpler heuristic procedures, and to equipartitioning. First, it is straightforward to demonstrate (1.2) from the DIA, by using suitable expansion of the right-hand side of (A.2) about k--> 0 (see, e.g., Kraichnan, 1976; Lesieur and Schertzer, 1978; Herring et al., 1982). The result for the functional $A\{E(k)\}$ so obtained has a simple interpretation in terms of how the theory represents Navier-Stokes dynamics. Thus, $A\{E\}k^4$ stems from the k-->0 expansion of the Fourier transform of

$$\int d\underline{x}''dt''\langle\{\underline{u}\cdot\nabla\underline{u}\}_S(\underline{x})\{\underline{u}\cdot\nabla\underline{u}\}_S(x'')\rangle_G G(\underline{x}'',t'';\underline{x}',t')$$

$$\longrightarrow (14/15)k^4\int_0^\infty dq/q^2\int_0^\infty ds E(p,t,s)E(q,t,s)G(k,t,s) \qquad (1.4)$$

with respect to $(\underline{x}-\underline{x}')$. In (1.4) $\{a\}_S$ is an incompressible part of a, and $\langle F\rangle_G$ evaluates the relevant moments as if $\underline{u}(\underline{x})$ were multi-variate Gaussian. The (time) integral of $G(\underline{x},\underline{x}';t,s)$ represents the time during which the force, $(\underline{u}\cdot\nabla\underline{u})$ acts to accelerate $\underline{u}(\underline{x},t)$. Eq. (1.4) has the form typical to Brownian motion theory: the acceleration $(\partial u/\partial t)$ is proportional to the mean-square random force $(\underline{u}\cdot\nabla\underline{u})$, integrated over a time typical for the force to act $(\int dsG)$. The eddy-viscuous term in (1.2) is necessary to restore energy conservation if the modeling of $(\underline{u}\cdot\nabla\underline{u})$ is via a random Gaussian field, $\underline{u}(\underline{x})$. Of course, $\underline{u}(\underline{x},t)$ is not Gaussian, but the above formulas emerge from a formal perturbation expansion in which the Gaussian part of $\underline{u}$ is assumed dominant.

Further insight into the closures is provided by examining its behavior at scales much smaller than the energy containing range, and making suitable expansions for large k. This is illustrated by considering a typical term that arises in the evaluation of E(k,t) (see Eq. (A.3) of the Appendix):

$$\int dpB(k,p,q)E(p)E(|k-p|) \ldots.$$

If $k >> k_o$ (k_o = the energy containing wave number), we may expand $E(|k-p|)$ about $\underline{p} = 0$, and retain the leading order term. Collecting all such terms (to order k^2) gives for the transfer function T(k):

$$T(k) \sim \partial/\partial k\{k^4 \partial/\partial k[k^{-2}E(k)/\tau(k)] \quad , \qquad (1.5)$$

where $\tau(k) \sim 1/\mu(k)$

and $\mu(k) \sim \{\int_o^k dpp^2 E(p)\}^{1/2}$.

The above transfer function is close to Leith's diffusion approximation (1968), except for the non-local nature of the diffusion coefficient. Eq. (1.5) seems a plausible model for E(k,t); it possesses a proper (k-5/3) inertial range, and a mechanism for equilibration of energy in k. However, a closer examination of its predictions shows it to be only order-of-magnitude accurate. Further, the large scale equipartitioning properties are wrong.

2. The Decay of Total Energy and Scalar Variance

We first turn to an old problem: the decay of total energy and--for a passive scalar--the variance. Much has been written about this simple problem, both theoretically and experimentally. Perhaps the best source of experimental information on this point is the paper of Corrsin and Compt-Bellot (1967). We recall that most recent experiments concerning the decay of kinetic energy suggest that $E(t) \sim t^{-n}$, with $n \sim (1-1.5)$ being the typical value. This aspect of turbulence decay has been explained by a number of theoretical ideas (see, e.g., Monin and Yaglom (1978) for a review). However, the concepts of equipartition, as discussed above, have some distinct predictions to make that are not offered by other approaches. To see how this happens, we first note from (1.2) that if the initial (t=0) spectrum behaves (as $k \to 0$) as $E(k,0) \to k^n$, then a short time later $E(k,t) \to k^{n'}$, $n' = 4$ for $n \geq 4$ and $n' = n$, $1 \leq n < 4$. Then to first order:

$$E(k) = Ck^{n'}, \quad k<k_o \qquad (2.1a)$$

$$= \varepsilon^{2/3} k^{5/3}, \quad k>k_o \qquad (2.1b)$$

where $\partial \int_o^{\infty} E(t)/\partial t = -\varepsilon = -2\nu \int_o^{\infty} dkk^2 E(k,t)$.

In (2.1a) (2.1b), k_o and C are related so that E(k) is continuous across k_o. To complete the story, observe that if $E(k) \to C(t)k^{n'}$ as $k \to 0$, then (1.2) implies C(t) is independent of t, provided n<4. If n=4, the argument breaks down, and more detailed analysis is needed to determine C(t). The problem is considered in Lesieur and Schertzer (1978). Briefly, if n>4, $\tilde{C}(t)$ is determined by (1.4). Dimensionally, this is ϵk_o^{-5}; the coefficient must then be determined by numerical integration of the closure. Their result (using the EDQNM) is:

$$E(t) \sim t^p, \; p=2(n+1)/(n+3), \; n<4, \; \sim t^{-2(5-\gamma)/7}, \; n \geq 4. \tag{2.2}$$

γ measures the magnitude of the input term (1.4). They estimate $\gamma \sim .16$, $p \approx 1.37$; however, the value of p(4) is a closure-dependent number. It is of interest to note that the actual inertial range slope used (the correct -5/3 or the incorrect -3/2 (DIA) is irrelevant to (2.2)). This point is discussed in more detail in the Appendix. The above calculation is similar to that given by Corrsin (1951), except that now there are some additional constraints implied by (1.3). We make two comments about these simple considerations. First, if the initial spectrum rises very sharply (n>4) at t=0, the (modified) Von-Karman $p \approx 1.37$ law takes on a quasi-universality; any spectrum sufficiently steep initially will have p=1.37 at later times. Secondly, if at t=0, n<4, (1.2) suggests such a power law remains--as a fossil--for all t.

As noted by Schertzer (1980), the above discussion uses only the rudiments of closure, as contained in (1.2). That the detailed spectral calculations confirm (2.2) is actually somewhat fortuitous. If we apply the above arguments to two-dimensional flow, the equivalent of (1,2) is Ak^3 instead of k^4A. Suppose, for example, that $E(k,0) \sim k$, and we assert, using the two-dimensional equivalent of (1.2), that C is constant. Then we may use the large k analog of (2.1b) (i.e., $E(k) \sim k^{-3}$), and the near constancy of mean-squared vorticity to infer that k_o is also a constant. But this is at variance with Batchelor's (1969) self-similarity solution $E(k) \sim V^2 t F(kVt)$, where V is the R.M.S. vorticity. Since Batchelor's theory uses only self-similarity and conservation of enstrophy for inviscid flows, and since these principles are also consistent with the closure, the k^3 range must intrude significantly toward $k \to 0$. A numerical study of the closure in two dimensions shows E(k,t) as in Fig. 1. We note that the dominant small k shape becomes k^3, consistent with Batchelor's theory. Thus, the two-dimensional spectrum possesses a higher degree of universality than the three-dimensional, because of the extra inviscid constraint. The two-dimensional problem is discussed in more detail in Lesieur and Herring (1984).

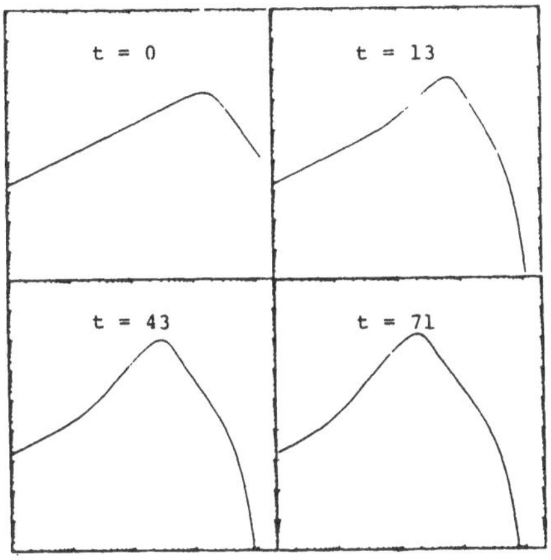

Fig. 1. Energy spectrum E(k,t) for two-dimensional turbulence for several t, measured in initial large-scale eddy turn-over times. ($E(k,0) = k/(1+k)^4$. Notice the evolution from k^1 to k^3 at small k. This is contrary to three-dimensional turbulence for which any spectrum beginning at small k with k^n, n<4 preserves its small k shape, according to Eq. (1.2). After Lesieur and Herring (1984).

Of perhaps more interest is the joint decay of turbulence and passive scalar. In this case, we may inject the scalar at quite different scales than that of the velocity field, and ask for dependence of the scalar spectrum on the energy. One interesting but simple aspect of this problem is to characterize the scalar decay rate, $\dot{E}_\Theta(t)/E_\Theta(t)$ in terms of the corresponding decay rate of the velocity field, and possibly other aspects of the velocity-scalar spectra. This problem has been studied experimentally (Warhaft and Lumley, 1987; Sreenivasan et al., 1980) and theoretically (Corrsin, 1951; Larcheveque et al., 1980; Kerr and Nelkin, 1979; and Herring et al., 1982). An interesting parameter is the decay ratio, r, defined by:

$$r = (\dot{E}_\Theta/E_\Theta)/(\dot{E}_V/E_V) \quad . \tag{2.3}$$

We recall Corrsin's proposal $r \sim \Gamma(L_V/L_\Theta)^{2/3}$, where the L's are integral length scales for the fields, and $\Gamma \sim 1$. We may derive this relation if E(k) has an inertial range, and if $L_V \gg L_\Theta$.

The closure may be used to make quantitative predictions for (2.3). To this end, we introduce profiles $(\tilde{E}_V(k), \tilde{E}_\Theta(k))$, which agree with $(E_V(k)$ and $E_\theta(k))$ at $k \ll k_{V,\theta}$ and also for k in the inertial range:

$$\begin{aligned} \tilde{E}_V(k) &= a_V\ (k/k_O)^n\ \varepsilon^{2/3}\ k_O^{-5/3}, \quad k<k_O \\ &= a_V\ \varepsilon^{2/3}\ k^{-5/3}, \qquad\qquad k>k_O \end{aligned}$$

with similar expressions for $\tilde{E}_\Theta(k)$. We may now compute r:

$$r \sim (a_V/a_\Theta)(k_\Theta/k_V)^{2/3} F_\Theta/F_V, \tag{2.4}$$

where

$$F_V = \int \tilde{E}_V(k)dk \,/ \int E_V(k)dk, \quad F_\Theta = \int \tilde{E}_\Theta(k)dk \,/ \int E_\Theta(k)dk \quad .$$

From (2.4) we note that the more rounded E_θ (k) is near k_θ, relative

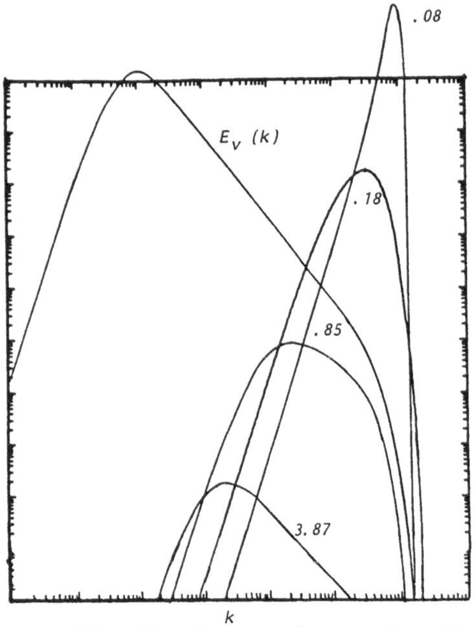

Fig. 2. Decay of a scalar field variance spectrum $2E_\theta\ (k,t)$ after injection into high Reynolds number(R_λ=2000) turbulence. The initial $E_\theta(k,0) = \delta(k-k_s)$, where k_s = the dissipation scale of the turbulence. Curves are labeled by run times t after initial injection. During the time shown, $E_V(k,t)$ changes insignificantly.

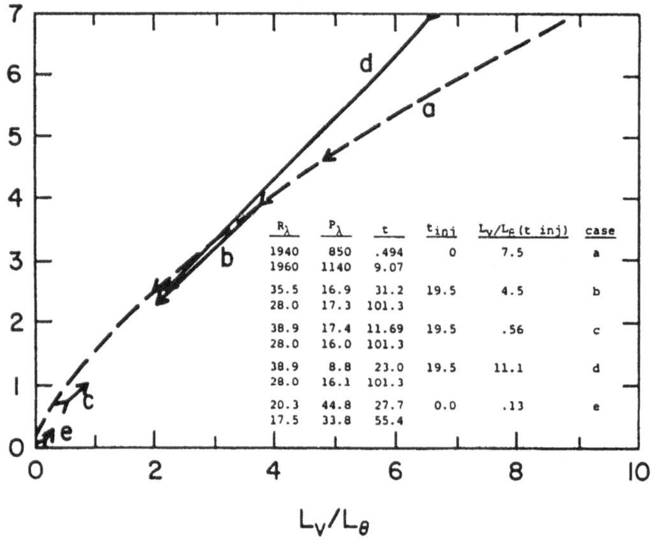

R_λ	P_λ	t	t_{inj}	L_V/L_θ (t inj)	case
1940	850	.494	0	7.5	a
1960	1140	9.07			
35.5	16.9	31.2	19.5	4.5	b
28.0	17.3	101.3			
38.9	17.4	11.69	19.5	.56	c
28.0	16.0	101.3			
38.9	8.8	23.0	19.5	11.1	d
28.0	16.1	101.3			
20.3	44.8	27.7	0.0	.13	e
17.5	33.8	55.4			

Fig. 3. Decay ratio r as a function of L_V/L_θ, where the L's are the integral scales of the velocity and scalar fields. Dashed line marks the trajectory of the run of Fig. 2. Other curves are for low R_λ (20-50). Note that the Corrsin prediction $(L_V/L_\theta)^{2/3}$ pertains only to the large R_λ run. The explanation seems to be that at low R_λ the scalar is subjected to the whole velocity field, and not to only a fraction $(L_\Theta/L_V)^{2\ 3}$ as at large R_λ as shown in Fig. 2. After Herring et al. (1982).

to $E_V(k)$, the larger Γ. Fig. 2 shows results of a TFM calculation in which the scalar is injected in the dissipation scale of $E_V(k)$. Note that as time proceeds, $E_\theta(k)$ is entrained towards smaller k by the equipartitioning mechanism discussed in Sec (2). The behavior of r(t) during this calculation is shown in Fig. 3 along with other lower Reynolds number runs. The latter apparently do not conform to (2.4), and are more nearly straight lines. The value of Γ from the large R_λ calculation is $\Gamma=2.1$, although other calculations for which the scalar spectrum is centered initially near the peak of $E_v(k)$ give $\Gamma=1.62$.

3. Return to Isotropy

We have already indicated that the two-point closures include a tendency towards isotropy as an aspect of their equipartitioning property. This has been examined via closure by Herring (1974), Schumann and Herring (1976), and Cambon et al. (1980). The first order of business is to examine the formal basis for the proposal of Rotta (1951) that an homogeneous flow relaxes toward isotropy linearly, if the initial departure from isotropy is small. Usually this proposal is stated as:

$$D\langle u_i u_j\rangle/Dt = - C\ (\varepsilon/E)\ \langle u_i u_j\rangle \quad , \tag{3.1}$$

where we have omitted the viscuous contribution for simplicity. Here D/Dt denotes the substantial derivative, and C is a constant of order unity (usually taken as C=1.4-1.1, by turbulence modelers). A derivation of (3.1) from more basic closures seems formidable. We may compare two-point closures with numerical predictions of C(t), as given by (3.1). Such a comparison (with DIA) for axi-symmetry is presented in Fig.4. The initial conditions are stated in the figure.

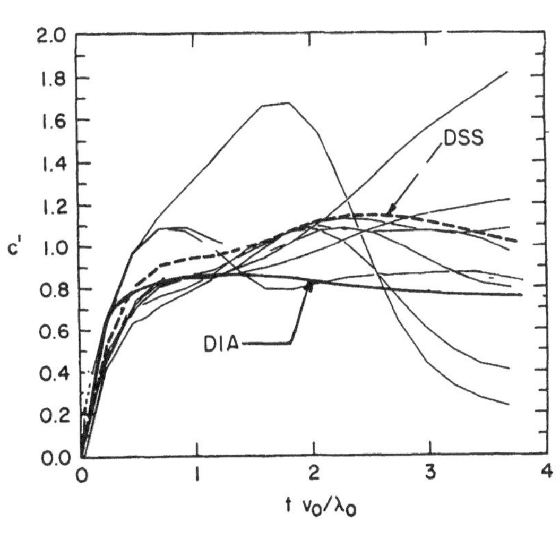

Fig. 4. Evolution of the "Rotta" constant C(t) (see equation (3.1) for direct numerical simulations (the light solid lines) and the DIA (the heavy solid line). Eight simulations were necessary to obtain an accurate representation of C(t) for the numerical simulation. Here $E(k,o) \sim k^4 \exp(-k^2)$, $R_\lambda(0)=30$.

The results are plausible, but a closer examination suggests that (3.1) is something of an oversimplification in several respects. First, the decay rate, C , depends on the symmetry of the initial field; axy-symmetric turbulence and shear flows clearly must be characterized by different C's. Secondly, (3.1) can only be expected if the turbulence length scales do not change significantly during the decay studied. The second point may be appreciated if we consider a stationary isotropic turbulence field into which a small anisotropy is introduced. Since the turbulence eddy turnover time decreases with increasing wave number, we expect the small scales to isotropize faster, leaving an isotropy at large scales only. Since there is no limit to the large scales, the decay will proceed progressively slower with time, leading to an algebraic (t^{-p}) decay of anisotropy instead of an exponential as implied by (3.1).

To illustrate some implications of the closure, we can make rough expansions of the wave number interactions at different scales, as indicated in Sec. I above (see Eq. (1.5)). For three dimensions, the formalism is somewhat obscure, so we state results here in two dimensions only, where the results are more transparent (Herring, 1975). The closures lead--roughly--to the following simple picture of the relaxation of the spectral Reynolds stress, $\Delta(k)$:

$$D\Delta/Dt = -(1/8)\{\partial(\kappa E(\kappa))/\partial\kappa\} \cdot \int_0^{\kappa} p^2 dp \Delta(p)\tau(p) - \mu(\kappa)\Delta(\kappa) \quad . \qquad (3.2)$$

Here μ is a local isotropization rate, and $\tau(k)$ represents the average time the large scale strain acts to distort the isotropic component ($E(k)$). Both are computable from the theory ($\tau \sim 1/\mu$, and $\mu \sim$ r.m.s. large scale strain). We note that (3.2) may be expected on heuristic grounds; anisotropy is produced by distorting isotropic κ-sized eddies by large scale anisotropy (the first term), while local interactions cause a dissipation of anisotropy. Further, if $E(\kappa)$ is steeper than $\sim\kappa^{-1}$ (its two-dimensional equipartition shape) the anisotropy at κ has the same sign as that at larger κ, and vice versa.

4. Two-Dimensional Turbulence

Beginning with the numerical studies of Regallo and Fornberg (1976), it has become increasingly clear that the decay of strictly two-dimensional turbulence may be very non-Gaussian, especially at small scales. The more recent studies of McWilliams (1984) have reinforced this impression, and have helped to delineate in some detail the characteristics of the coherent structures and their emergence from initial Gaussian chaos. The spectra of this highly intermittent flow are steeper than the closure-predicted $(\ln(k/k_1))^{-1/3}k^{-3}$, and it is not yet clear if it is strictly a power law or, indeed, if there exists a universal shape independent of initial conditions. On the other hand, randomly forced flows, with sufficient damping, may be quite close to Gaussian for a rather wide range of scales. It remains to be seen if such flows are sufficiently non-linear to be of interest.

These problems are of considerable significance in meterology, where the two-dimensional system serves as the simplest analog of large-scale planetary motions (Charney, 1972; Lilly, 1983; Gage 1979).

To discuss these issues in a quantitative sense, we may introduce an effective Reynolds number, R:

$$R = E(t)\ \sqrt{V(t)} / \int_0^\infty dk\, \nu(k) E(k) \quad . \tag{4.1}$$

Here V(t) is the r.m.s. vorticity, and $\upsilon(k) \sim \kappa^4$ is a "hyper viscosity," which allows for the dissipation of V a small scales. In many numerical experiments $\nu(k)$ is sharply peaked at high k, so as to provide for a long span of dissipationless large scales in the numerical study. For the earth's atmosphere, we estimate $R \sim 10^3$. Current numerical simulations (256×256 spectral resolution) reach only $R \sim 100$, even those that utilize a hyper viscosity to reduce the needed wave number span. Fig. 5 and Fig. 6 show a comparison of direct numerical simulations and the closure (TFM), for forced, dissipative flow (Fig. 5 for forced-

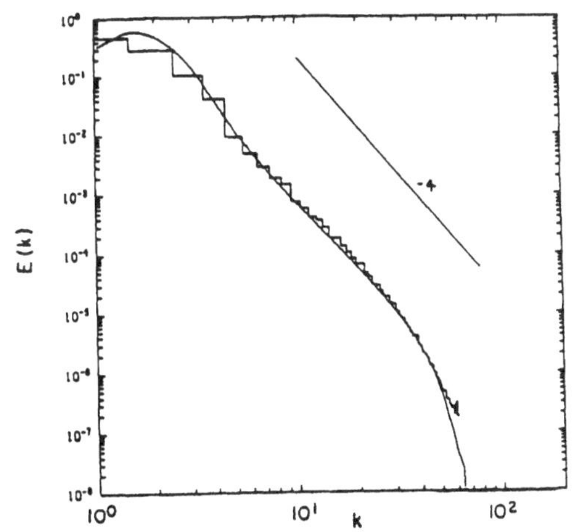

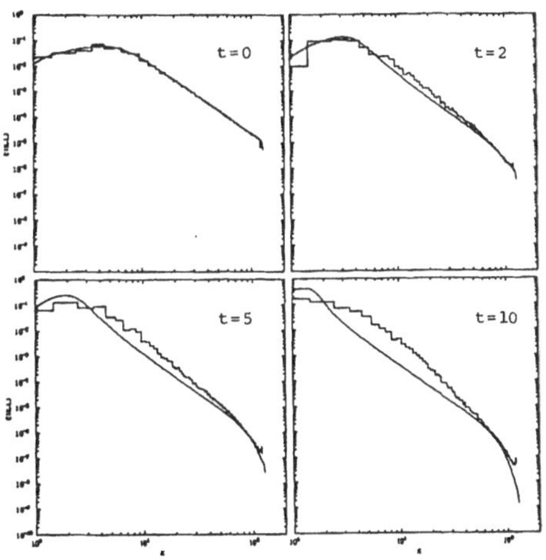

Fig. 5. Comparison of (256×256) numerical simulation for random forcing at k_0=20, and modified viscosity $\upsilon(k)=\gamma_0+\gamma_1 k^4$, compared to theory(TFM, described briefly in the Appendix). After Herring and McWilliams(1984).

Fig. 6. Comparison at late time of numerical simulation to the TFM for decay of two-dimensional flow. Resolution of simulation is 256x256, and modified viscosity is $\nu=2.\times10^{-8}k^4$. Enstrophy dissipation is maximal at t=4. After Herring and McWilliams (1984).

dissipative flow, and Fig. 6 for decaying flow). The agreement for the forced case is quite satisfactory. Here, R ≈ 10. Notice that the actual slope in the (enstrophy) inertial range is closer to k^{-4} than k^{-3}. The former is a particular prediction of Saffman (1968), and appears in all two-dimensional low R calculations. For this case, measures of non-Gaussianity are small (i.e., $\langle(\nabla x \underline{u})^4\rangle / \langle(\nabla x \underline{u})^2\rangle^2 \sim 3$.

This is not the case for the decay experiment for which $\langle(\nabla x \underline{u})^4\rangle / \langle(\nabla x u)^2\rangle^2 \sim 40$ at late times. We note in this case a strong disagreement between theory and simulations. A closer examination of the vorticity field shows that it consists of intense vortex regions, interspersed between quiet regions, where the vorticity is small. The decay (of enstrophy) process here consists of merging (and near merging) of these vortex regions, with vorticity dissipation at the time of merging. The spectral shape--at large k--is much steeper than that predicted by the theory. A closer examination of the decay experiment shows that theory and experiment are in good agreement for t less than the time at which the enstrophy dissipation achieves its maximum.

Both cases can be understood by arguing that the disruptive effects of random stirring prevents the formation of coherent structures in a considerable neighborhood of the stirring wave-number. It may be that in the atmospheric application, the energy input mechanism (baroclinic instability) simulates--roughly--the random stirring, which would restore to some extent the validity of theory.

5. Convection

Thermal convection is probably the simplest inhomogeneous flow to which to apply the theory, since the conditions of axisymmetry are more manageable. Moreover, if slip-boundary conditions are used on the confining plates, the flow becomes quickly turbulent just beyond the critical Rayleigh number, Rc=657. This is a rather surprising and recent result (see, e.g., Zippalius and Siggia, 1983).

Fig. 7 shows the vertical velocity field plan form for a numerical simulation at $Ra=7R_C, P_r=10$, and an aspect ratio of 4. The flow plan-form actually has considerable order. Fig. 8 shows a probe signal located near the mid-point of the flow. The flow field has a large component in kz=0, which is a two-dimensional turbulence generated by the "beating" interactions discussed in Sec. 2.
The maximally unstable component of the flow is destabilized by its interaction with the two-dimensional component, and hence the vigorous time dependence. Calculation of the Kurtosis of the various fields show only a small departure from Gaussianity.

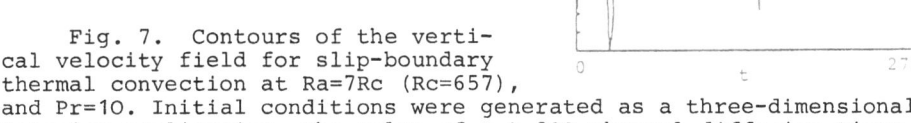

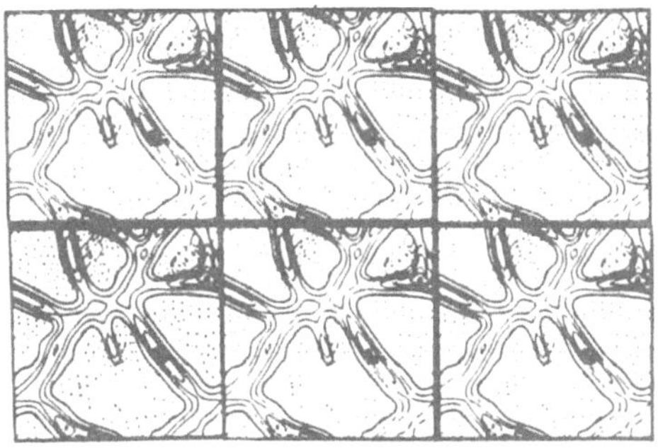

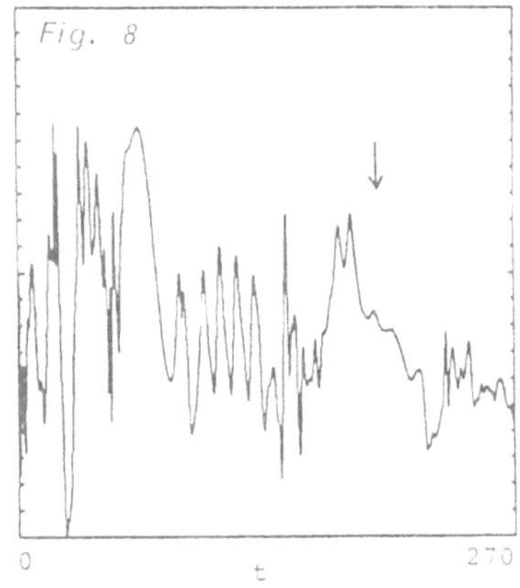

Fig. 7. Contours of the vertical velocity field for slip-boundary thermal convection at Ra=7Rc (Rc=657), and Pr=10. Initial conditions were generated as a three-dimensional, Gaussian realization. The value of t ~ 200 thermal diffusion times.

Fig. 8. Time series for the flow field shown in Fig. 7.

Calculations using DIA on this problem are currently being done by Dannevik (Thesis, U. St. Louis, 1984). For air (Pr≈.72), he reports good agreement between DIA and numerical simulations for both the Nusselt number as well as the heat and momentum flux budgets in the statistically steady state (<5.0%).
It may be that closures could be unexpectedly useful for this problem. As noted above, the cell plan-form of Fig. 7 is destabilized by the vertical-vorticity modes, which are essentially a two-dimensional turbulence. The spectrum of the latter is strongly peaked at the largest available scale suggesting that a much higher aspect ratio may be needed to accurately simulate the statistics of horizontally homogeneous flow. If an aspect ratio of 10×1 is needed, the simulation is prohibitive. On the other hand, closures have no problem with the very largest scales, and their effects can probably be represented analytically by an extension of the methods of Sec. 2.

The numerical task to solve the inhomogeneous DIA is prohibitive, at large Ra. The use of sophisticated algorithms may well help to overcome these difficulties, however. For example, Dannevik (1984) has introduced a double time scale analysis, and uses the Pade approximation to render the time-difference dependence of the problem manageable. Further simplifications of the numerical task is to use FFT's to evaluate the wave-number convolutions (Domaradzki and Orszag (1984), private communication).

6. Concluding Comments

Our results here suggest that two-point closures cannot be completely trusted to faithfully represent the physics of turbulent flows for all interesting cases. Rather, the best that can be offered is that such procedures are useful tools for certain flows, and that they may give some insight as to how certain (moment) aspects of the turbulence behave. We surmise that they have the best chance of succeeding for cases in which the flow is not far from Gaussian, a condition which can only be assessed a posteriori. From the examples surveyed here, it would seem that conditions of near two-dimensionality may be the most inhospitable for application; in this case--at least--the source of their failure may lie in the stability of intense vortex elements. We may thus also expect poor results for shear layers, and other quasi-two-dimensional flows. On the other hand, thermal convection and perhaps stratified flows may be a more hospitable regime--at least if the preliminary results of numerical simulations are a valid guide.

Are these methods useful in turbulence studies? The answer depends in part upon how difficult the (numerical) task of solving the closure equations is. As we have seen, a nontrivial amount of coding is needed to solve such equations. [A brief sketch of the numerical algorithms used here is given in Appendix B.] For homogeneous problems, once the investment is made, solutions are much easier to obtain than with the DNS. It is not so clear, however, for inhomogeneous problems--shear flow and convection--that closure competes well with DNS. However, we must bear in mind that new algorithms--e.g., the use of Pade tables (Dannevik, 1984) for time histories and FFT's for the convolution sums (Orszag and Domaradzki, 1984, private communication)--may soon reverse this.

Granted that new algorithms make closures competitive (and this is especially to be expected at sufficiently high Reynolds numbers), we must always stress that such methods furnish only limited information about turbulent flows. Their role is that of a computational tool, and as such their usefulness is mainly confined to quantitative information: relaxation rates, spectra, etc.

Among the problems surveyed here, we may consider the decay of homogeneous turbulence (bearing a passive scalar), the return to isotropy, and thermal convection at moderate Rayleigh numbers as cases in which closures succeed. The problems in two dimensions (M.H.D. and Navier-Stokes) represent more severe tests, and ones for which the results are mixed (see, e.g., Frisch et al., 1983). On the other hand, the question of singularities in non-dissipative flows clearly cannot be investigated along this avenue. Indeed, the question of the analytic character of such systems is as yet unsettled, and its resolution is a prerequisite for designing a closure to treat the statistics of the flow

Appendix A

Here we record the DIA equations, in their isotropic form. For the Navier-Stokes equations,

$$\partial \underline{u}(\underline{x},t)/\partial t = -\nabla p-\underline{u} \cdot \nabla \underline{u} + \nabla^2 \underline{u}, \ \nabla \cdot \underline{u} = 0 \quad . \qquad (A.1)$$

We form an ensemble mean,

$$\langle \underline{u}(x,t)\underline{u}(x',t')\rangle = \underline{\underline{U}}(x,x',t,t') \quad . \qquad (A.2)$$

The DI equations of motion for the Fourier transform of $\underline{\underline{U}}(\underline{x},\underline{x},'t,t')$ (which we simply denote as $U(k,t,t')$) is:

$$\partial U(k,t,,t')/\partial t = \int_\Delta B(k,p,q)dpdq[\int_0^{t'} U(p,t,s)U(q,t,s)G(k,t',s)ds$$

$$\int_0^{t} U(k,t',s)G(p,t,s)U(q,t,s)ds] \quad . \qquad (A.3)$$

Here, $B(k,p,q)$ is given by:

$$B(k,p,q) = (1/2)\sin^2(p,q)\{p^2-q^2)(k^2-q^2)+p^2k^2\}pq/k \quad . \qquad (A.4)$$

Equation (A.3) contains a Green's function, $G(k,t,t')$, whose equation of motion is needed to make a deterministic set with which to advance $U(k,t,t')$ forward in time. For the sake of brevity, we shall not record this equation, but instead shall simply indicate the physical characteristics of G. As originally proposed by Kraichnan (1959), G represented the response of $\underline{u}(\underline{k},t)$ if a small impulsive perturbation is applied to the right-hand side of (A.1) at time t'. The resulting equation for $G(k,t,t')$--as derived by a modified perturbation theory--resembled (A.3) (for $U(k,t,t')$), except that the first right-hand-side term is missing, and that the time integral on the second term is over $\{t',t\}$ instead of $\{0,t\}$ as in (A.3).

Most two-point closure calculations to date have used a much simpler set in place of (A.3). We refer here to the "Markovian" theories which involve only the simultaneous times (t,t), and use a simple approximate memory time to replace the effects of $G(k,t,t')$. The generic "Markovian" theory may be written simply as:

$$\{\partial/\partial t+2\nu\, k^2\}U(k,t) = \int_\Delta dpdq\tilde{B}(k,p,q)U(q,t)\{U(p,t)-U(k,t)\}, \qquad (A.5)$$

$$\tilde{B}(k,p,q) = B(k,p,q)\Theta(k,p,q) \ ,$$

$$\Theta(k,p,q) \sim [1 -\exp\{-\mu(k)+\mu(p)+\mu(q)\}t]/[\mu(k)+\mu(p)+\mu(q)]$$

Here, the triple moment memory time $\Theta(k,p,q)$ is simply characterized in terms of an amplitude dephasing rate $\mu(k)$. An approximate connection between (A.5) and the DIA is established by taking:

$$G(k,t,t')=\exp\{-\mu(k)(t-t')\} \ ; \ U(k,t,t')=U(k,t',t')G(k,t,t').$$

In this approximate characterization, the DIA is nearly the same as Edwards (1962) theory. The expression for $\mu(k)$ in that case is

$$\mu(k) = \int dpdqB(k,p,q)U(q)/[\mu(k)+\mu(p)+\mu(q)] \tag{A.6}$$

$$\sim k \{\int_0^k dpE(p)\}^{1/2} . \tag{A.7}$$

(A.7) is a rough approximation to (A.6) in which we have assumed $U(q)$ to vary more rapidly than $\mu(q)$ in order approximately to evaluate the integral equation. The above approximations (A.3) and (A.5) do not behave properly under random Galilean transformations largely because of the lack of invariance of (A.6). This error means that the inertial range computed on the basis of (A.6) (A.7) will not be correct. Several methods have been proposed to build perturbation theories that more properly incorporate Galilean invariance. The gist of most of these proposals is to replace the $\mu(k)$, Equation (A.7), by the large-scale (r.m.s) strain rate:

$$\mu^2(k) = \gamma\{\int_0^k dpp^2E(p)\} \tag{A.8}$$

The value of γ (or its generalization) is generally fixed by the theory, or--as in the Test Field Model (TFM) (Kraichnan, 1971)--by comparing the approximate procedure for some simple problem to a more exact method. The closure utilizing (A.8) is called the Eddy Damped Quasinormal Markovian (EDQNM) approximation. The test field model has a more elaborate determination of a generalized μ, but effectively reduces to (A.8) at large k. At small k it has a term $\sim k\int_k^\infty dpE(p)$, in addition to (A.8). See Herring et al. (1982) for details.

Generally, the larger the μ (as in DIA), the smaller the transfer. The sensivity of decay to variations in μ is presented in Fig. 9, which shows $E_V(t)$ for Edwards theory (μ given by (A.6)), the TFM, and the Markovian Quasinormal approximation (Tatsumi et al., 1978). For the

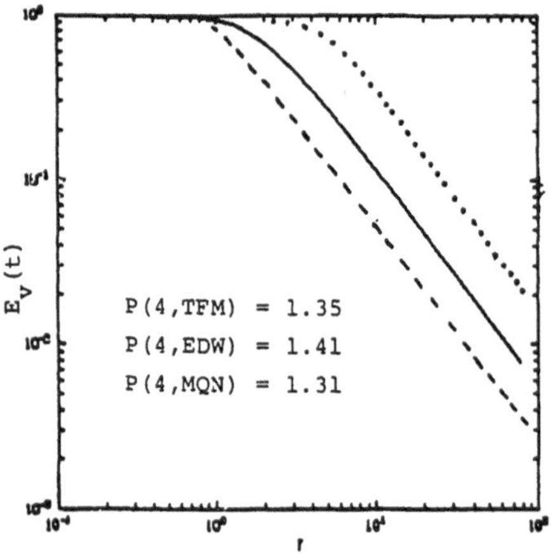

Fig. 9. $E_V(t)$ vs. t (units of initial large-eddy turn-over time) for three theories: TFM (solid line), Edwards (dotted line), and the Markovian Quasinormal (dashed). $R_\lambda(t)$ ranges from 600 at t=0 to 50 at t$\hat{=}$100. The value of P for the Quasinormal theory may not yet be converged at t~100.

latter, μ=0. Notice that the decay exponent p(4) (see eq. (2.2)) is quite similar (1.35, 1.41, 1.31) for TFM, Edwards, and MQN). However, the values of $E_V(t)$ differ markedly. This variability may be understood by the analysis of Sec. 2.

Appendix B

In this section, we sketch some numerical techniques we deem appropriate for the closure equations (essentially (A.3) or (A.5)). The two problems to be addressed are: (1) the wave-number dependence (including the evaluations of convolutions ($\int_\Delta dpdq$)), and (2) time stepping. After considerable experimentation, we have found that the wave number representation is economically written in terms of collocation utilizing (cubic) splines. To explain this, we abbreviate (A.5) as:

$$\partial U/\partial t = F(U) \quad . \tag{B.1}$$

Now approximate U(k,t) in terms of a discrete set $\{U_i\}$ (i=1,...,N):

$$\bar{U}(k) = \sum_{i=1}^{N} U_i s_i(k) \quad . \tag{B.2}$$

Here $\{s_i(k)\}$ is an as yet unspecified set of linearly independent functions. We now determine $\{U_i\}$ by the integral relation

$$\int x_i(k)dk\{\partial\bar{U}(k)/\partial t - F(\bar{U})\}, \ (i=1,2,\ldots,N) \quad . \tag{B.3}$$

In (B.3), $x_i(k)$ (i=1,...N) is a set of arbitrary test functions. Note that (B.3) may be converted to a set of N equations for dU_i/dt.

Our choice for the $s_i(k)$ are B-splines (see DeBoor, 1977). Briefly, these are (C-2) unit-height pulses centered at $\hat{k}_i$ (i=1, N-2) vanishing (cubically) outside the interval ($\hat{k}_{i-2} < k < \hat{k}_{i+2}$), for 4 < i < N-3. They are related to the standard (cardinal) splines, $\bar{s}_i(k)[\bar{s}_i(\hat{k}_j) = \delta_{ij}]$ by a linear transformation. For $i \leq 4$ $s_i(k)$ vanish at $\hat{k}_{i+1}$ and are unity at $\hat{k}_1$. The $s_i(k)$, i > N-3 mirror this behavior at $\hat{k}_N$. We choose the set of computational points:

$$[k_1, k_2, \ldots k_N] \quad , \tag{B.4}$$

at which the spectrum is specified as identical to $\{\hat{k}_i\}$ except for k_2 and k_{N-1}. Their distribution on the computational domain $[k_B, k_T]$ will be discussed presently. The B-splines have the advantage over the cardinal splines, $\bar{s}_i(k)$, of greatly economizing the (dpdq)-wave number convolution integrals in (A.5). In addition, the splines seem optimally suited to representing accurately spectral shapes which may vary by several orders of magnitude over the computational domain. Furthermore, they yield considerably more accurate evaluation of the (p-q) integrals than, for example, linear interpolation.

If $x_i(k) = s_i(k)$, (B.3) is a Galerkin procedure, while if $x_i(k) = \delta(k-k_i)$ $(i=1,\ldots N)$, it is collocation. For the former, exact energy conservation is incorporated, whereas for the latter, it must be added as an additional constraint. However, preserving exact conservation laws is not necessarily a part of the best numerical procedure, and for the present problem, we have found that collocation is actually superior. Conservation properties may then be used as tests of numerical accuracy.

In approximating the right-hand side of (A.5) with (B.2) and (B.3), we first rewrite it in terms of an appropriate (ds) integral of $U(k,t,s)$, where

$$U(k,t,s) = \exp(-\mu(k)(t-s))U(k,t) \quad ,$$

and apply spline interpolation to $U(k,t,s)$. This makes the approximation procedure parallel to that for the DIA, and more importantly obviates an awkward functional dependence of $\theta(k,p,q)$ on $s(k)$ which otherwise would occur. This amounts to applying the rule,

$$F(p,q) \tilde{=} \sum_{i,j} F(p_i,q_j)\bar{s}_i(p)\bar{s}_j(q) \quad ,$$

where needed. Here we recall that $\bar{s}_i(k)$ are the cardinal splines.

The final approximation equations to be solved are but a discretized version of (A.5), with different coefficients B:

$$(\partial/\partial t + 2\nu k_n^2)U(n,t) =$$

$$\sum_{m,\ell} B(n,m,\ell)\Theta(n,m,\ell)U(\ell,t)\{U(m,t)-U(n,t)\} \quad , \qquad (B.5)$$

where,

$$U(n) = U(k_n) \quad ,$$

$$B(n,m,\ell) = \sum_{m'}^{N}\sum_{\ell'}^{N} A_{mm'}A_{\ell\ell'}\int B(k_n,p,q)dpdq\,s_{m'}(p)s_{\ell'}(q) \quad ,$$

and,

$$A_{nm} = [s_n(k_m)]^{-1} \; .$$

The (p-q) integrals are here effected by the appropriate (Legendre or Tschebycheff) method, depending on the analytic character of $B(k,p,q)$.

There remains to discuss the distribution of computational points (B.4). Other computations (beginning with those of Kraichnan, 1964) utilize an exponential distribution $k_n = \exp(an)$. This distribution has the advantage of rendering linear interpolation exact for power law spectra. In addition, it concentrates points near the spectral peak, where energy transfer changes rapidly, and has relatively few points in the large k dissipation range, where transfer is small and smooth. The approach used here is dictated by the following considerations: (1) the $\{k_i\}$-distribution should be such as to optimize the satisfaction of the existing conservation constraints; (2) cusps in $U(k,t)$ near k_B or k_T may develop in certain studies, and hence $\{k_i\}$ should have high density at end points to preserve accuracy; (3) if a wave number diffusion approximation to the closure like (1.5) is an approximate guide, a (piece-wise) Gaussian collocation $\{k_i\}$ may be optimal (Printer 1975).

Guided by these considerations, we settled on the following stretched Gaussian points as fairly optimal:

$$k_i = B(\exp(Ay_i - 1) + k_B \quad . \tag{B.6}$$

Here y_i are Gaussian points mapped onto $\{0,N-1\}$. The two numbers $[A,B]$ are chosen so that: (1) (B.6) yields $k(N)=k_T$, and (2) energy conservation by non-linear interactions is optimized for an anticipated test $U(k)$. The calculations reported here all use $N=32$. This appears also adequate even for quite large Reynolds numbers.

The time integral of (B.5) may be most economically effected by a quasi-linearization of the right-hand side (Herring and Kraichnan, 1979). To this end, we rewrite it as

$$\partial U_n/\partial t = \sum_{m-1}^{N} A_{nm}\,(t)U_m \quad , \text{ where,} \tag{B.7}$$

$$A_{nm} = -2\nu k_n^2 \delta_{nm} + (1/2)[\sum_{\ell}\{(\bar{B}(n,m,\ell)+\bar{B}(n,\ell,m))]U(\ell)$$

$$- \sum_{r} [\bar{B}(n,r,m)U(n) + \delta_{nm}\bar{B}(n,r,\ell)U(\ell)]\} \quad , \tag{B.8}$$

$$\bar{B}(n,m,\ell) = B(n,m,\ell)\theta(n,m,\ell,t) \quad .$$

The basic strategy in solving (B.7) on an interval (t_1,t_2) is to solve first for $U(t_2)$ using $A(t_1)$, and then iterate (B.8) using the predicted value for $U(t_2)$. In practice, it suffices to simply use the average $(1/2)(A(t_1)+A(t_2))$ on the corrector phase. This procedure is able to take steps ~100 times the time step in an Adams-Bashforth or Euler time step scheme (for the same accuracy). Hence, the time spent in eigen mode computations is more than compensated in increasing the allowed time step. [We note that a typical time for eigen-made analysis of (B.7) is about twice the time needed to evaluate the right-hand side of (B.5) for N=32.] The physical reason for this economizing can be appreciated if we recall that the spectrum moves coherently to larger scale, without much distortion. The main feature is then characterized by the value of the spectral peak, $k_o(t)$. The maximum quasi-linear time time step is a significant fraction of $k_o(dt/dk_o)$, whereas the more standard methods may take only a fraction of the small-scale eddy turnover time before diverging.

References

Batchelor, G. K., 1959: The Theory of Homogeneous Turbulence. Cambridge at the University Press. 197 pp.

_____, 1969: Computation of the energy spectrum in homogeneous two-dimensional turbulence. Phys. Fluids Suppl., 12, II, 233.

Boor, C. de, 1977: Package for calculating with B-splines. SIAM. J. Numer. Anal., 14, 441.

Cambon, C., D. Jeandel, and J. Mathieu, 1980: Spectral modelling of homogeneous nonisotropic turbulence. J. Fluid Mech., 104, 247-262.

Charney, J. G., 1971: Geostropic turbulence, J. Atmos. Sci., 28, 1087-1095.

Comte-Bellot, G. and S. Corrsin, 1966: The use of a contraction to improve the isotropy of grid-generated turbulence. J. Fluid Mech., 24, 657-682.

Corrsin, S., 1951: The decay of isotropic temperature fluctuations in an isotropic turbulence. J. Atmos. Sci., 18, 1951.

_____, 1964: The isotropic turbulent mixer. II. Arbitrary Schmidt Number. A. I. Ch. E. J., 10, 870.

Dannevik, W., 1984: Two-point closure study of covariance budgets for turbulent Rayleigh-Benard convection. Ph.D. Thesis, St. Louis University, St. Louis, MO.

Edwards, S. F., 1964: The statistical dynamics of homogeneous turbulence. J. Fluid Mech., 18, 239-273.

Fornberg, B., 1977: A numerical study of 2-D turbulence. J. Comp. Phys., 25, 1.

Frisch, U., A. Pouquet, P. L. Sulem, and M. Meneguzzi, 1983: The dynamics of two-dimensional ideal magnetohydrodynamics. J. Mécanique Théor. Appl. Suppl., R. Moreau, Ed., 191-216.

Gage, K. S., 1979: Evidence for a $k^{-5/3}$ law inertial range in mesoscale two dimensional turbulence. J. Atmos. Sci., 36, 1950-1954.

Heisenberg, W., 1948: Zur statistichen Theorie der Turbulenz. Z. Physik, 124, 1628-657.

Herring, J. R., 1975: Theory of two-dimensional anisotropic turbulence. J. Atmos. Sci., 32, 2254-2271.

_____, and R., M. Larcheveque, J.-P. Chollet, M. Lesieur, and G. R. Newman, 1982: A comparative assessment of spectral closures as applied to passive scalar diffusion. J. Fluid Mech., 124, 411-437.

_____, and J. C. McWilliams, 1984: Comparison of direct numerical simulation of two-dimensional turbulence with two-point closure. To appear in J. Fluid Mech.

_____, and R. H. Kraichnan, 1978: A numerical comparison of velocity-based and strain-based Lagrangian-history turbulence approximations. J. Fluid Mech., 91, 581-597.

Kovasznay, L. S. G., 1948: Spectrum of locally isotropic turbulence. J. Aeronaut. Sci., 15, 657-674.

Kraichnan, R. H., 1959: The structure of isotropic turbulence at very high Reynolds numbers. J. Fluid Mech., 5, 497-543.

_____, and E. A. Speigel, 1962: Model for energy transfer in isotropic turbulence. Phys. Fluids, 5, 583-588.

_____, 1964: Decay of isotropic turbulence in the Direct Interaction Approximation. Phys. Fluids, 7, 1030.

_____, 1971: An almost-Markovian Galilean-invariant turbulent model. J. Fluid Mech., 47, 513-524.

_____, 1976: Eddy viscosity in two and three dimensions. J. Atmos. Sci., 33, 1521.

_____, and J. R. Herring, 1978: A strained based Lagrangian-history turbulence theory. J. Fluid Mech., 88, 355-367.

Lee, T. D., 1950: Note on the coefficient of eddy viscosity in isotropic hydromagnetic turbulence in an incompressible fluid. Ann. Phys.,321, 292-321.

Leith, C. E., 1968: Diffusion approximation for turbulent scalar fields. Phys. Fluids, 11, 1612-1617.

Lesieur, M. and D. Schertzer, 1978: Amortissement auto similarité d'une turbulence à grand nombre de Reynolds. J. de Mécanique, 17, 609-646.

Lilly, D. K., 1983: Stratified turbulence and the mesoscale variability of the atmosphere. J. Atmos. Sci., 40, 749-761.

McWilliams, J. C., 1984: The emergence of isolated, coherent vortices in a turbulent flow. To appear in J. Fluid Mech.

Monin, A. S. and A. M. Yaglom, 1975: Statistical Fluid Mechanics: Mechanics of Turbulence, Vol. 2. The M.I.T. Press, Cambridge, MA, and London.

Oboukhov, A. M., 1941: Spectral energy distribution in a turbulent flow. Akad. Nauk. USSR, 32, 22-24.

Orszag, S. A., 1974: Statistical Theory of Turbulence: Les Houches Summer School on Physics. Gordon and Breach, 216 pp.

Printer, P. M., 1975: Splines and variational methods. John Wiley & Sons, New York. 321 pp.

Proudman, I. and W. H. Reid, 1954: On the decay of normally distributed and homogeneous turbulent velocity field. Phil. Trans. Roy. Soc., A247, 163-189.

Saffman, P. G., 1971: On the spectrum and decay of random two-dimensional vorticity distribution of large Reynolds numbers. Studies in Applied Math., 50, 377.

Schertzer, D., 1980: Comportements auto-similaires en turbulence homogene isotrope. C. R. Acad. Sci., Paris, 280, 277.

Schumann, U., and J. R. Herring, 1976: Axisymmetric homogeneous turbulence: A comparison of direct spectral simulations with the direct interaction approximation. J. Fluid Mech., 76, 755-782.

Sreenivasaan, K. R., S. Tavoularis, R. Henry, and S. Corrsin, 1980: Temperature fluctuations and scales in grid-generated turbulence. J. Fluid Mech., 100, 597-621.

Tatsumi, T., 1955: Theory of isotropic turbulence with the normal joint-probability distribution of velocity. Proceedings 4th Japan Nat. Congr., Appl. Mech., Tokyo, 307-311.

_____, S. Kida, and J. Mizushima, 1978: The multiple scale cumulant expansion for isotropic turbulence. J. Fluid Mech., 85, 97-142.

Warhaft, Z., and J. L. Lumley, 1978: An experimental study of the decay of temperature fluctuations in grid-generated turbulence. J. Fluid Mech., 88, 659.

Zippelius, A., and E. D. Siggia, 1982: Disappearance of stable convection between free-slip boundaries. Phys. Rev., A26, 1788-1790.

Intermittent Turbulent Flow

W. Kollmann
Department of Mechanical Engineering, University of California, Davis, CA 95616, USA

1. Introduction

Corrsin [1], [2] established several decades ago the existence of a sharp interface between turbulent and nonturbulent fluid for a variety of shear flows. Since then a large number of papers appeared reporting experimental results on conditioned variables. For the flat plate boundary layer measurements of the intermittency factor and conditional mean velocity components [2]-[11], including flows with non-zero pressure gradients [3] and normal stress components, point statistical expectations [4]-[11] are now available. The turbulent zone shear stress was reported in [6], and fundamental questions concerning discrimination and the dependence of conditional moments on threshold levels received attention [7], [8]. Measurements in plane jets including the initial region with potential core and point statistical moments were reported by several authors [12]-[16]. For round jets measured intermittency factor and some conditional moments are available [2], [17]-[21]. The comparison of free and ducted round jets [17] shows that the radial intermittency factor profiles deviates significantly from the common Gaussian shape due to the presence of ducts. For mixing layers and wakes less data for conditional quantities are available [22]-[24]. Complex flows such as interacting mixing layers [25]-[28] wall jets [29], transitional flows [30] and reacting flows [31]-[33] are receiving increasing attention. The existence of large scale structures lead to new views on entrainment [34]-[35] and averaging [38], [39]. The theoretical treatment of intermittently turbulent flows was initiated by Corrsin [2], [40] and continued by Corrsin and Phillips [41], [42], and Gibson [43], [44] and Lumley [45], [46]. The statistics of multi-valued random variables was introduced [45] to deal with the interface separating turbulent from non-turbulent fluid. Closure models were first suggested by Libby [47], [48] and further developed [49], [50] to allow calculation of intermittency factor and conditional moments [51]. The advantage of conditional moments over unconditional quantities is exemplified in the measured pdf of two velocity components in a v-shaped premixed flame in fig. 8 of ref. [32]. The pdf has two distinct maxima and very small probability for velocity values between the maxima. In this case corresponds the unconditional mean to values which have, thus, small probability of actually occurring in the flow, whereas conditioning produces the mean values near each of the maxima of the pdf. Thus it can be argued that conditioning leads to less complicated statistics and explicit transfer processes between different zones.

The present paper investigates conditioning and the corresponding intermittency factors. Representations for the sources of intermittency are established. Reacting flows are considered as an application and the closure problem for the intermittency source is related to an existing closure for the pdf-equation. Finally, some results obtained with a conditional second order closure model for incompressible flows are presented.

2. Conditional Events and Their Description

The experimental evidence indicates that turbulent shear flows have random and non-random properties. Thus, an important question arises for the theoretical description of turbulent flows: how to incorporate chaotic/random and non-random features. First, it should be noted that the distinction between random and non-random properties is by no means an easy matter, because non-random solutions to equations describing mechanical systems can have the complicated appearance of randomness without being random. A second point to consider is the fact that the experimental observation of a non-random structure in a turbulent shear flow is usually incomplete and the non-random features can only be extracted by pattern recognition techniques involving some short time or spatial near-field averaging processes [34] [52]. The probabilistic analysis of turbulent shear flows can be based on the following considerations. The variables describing the flow (velocity, pressure, etc.) are decomposed into contributions pertaining to different properties of the flow. Statistical operations are then applied observing certain conditions. Both decomposition and conditioning can be achieved in many ways and their choice will determine the usefulness of the resulting equations.

2.1 Decomposition

The decomposition of the flow variables will be classified in terms of the decomposition of the physical flow domain $\mathcal{D}$ (with boundary $\partial\mathcal{D}$) it induces. Denoting the flow variable by $\psi(\underline{x},t)$ (which can be a scalar or vector quantity), the decomposition

$$\psi(\underline{x},t) = \psi^{(0)}(\underline{x},t) + \psi^{(1)}(\underline{x},t) + \psi^{(2)}(\underline{x},t) + \ldots$$

is either disjoint, such that for a point $\underline{x} \in \mathcal{D}$ only one member $\psi^{(i)}$ is non-zero, or it is overlapping where for any $\underline{x} \in \mathcal{D}$ at least two members of the decomposition $\psi^{(i)} \neq \psi^{(j)}$ are non-zero. The tagging of material points in a subdomain of the flow field $\mathcal{D}$ at some reference time with a non-diffusive property is an example for a disjoint decomposition: at an Eulerian point $\underline{x}$ either a tagged or a non-tagged material is present but never both. An important example for an overlapping decomposition is obtained by means of Fourier transformation. A disjoint decomposition of the image variables in Fourier space produces members in physical space (wave packets) which are non-zero nearly everywhere in $\mathcal{D}$ [53]. Vice versa is a disjoint decomposition overlapping in Fourier space. A particular overlapping decomposition was suggested by Hussain and Reynolds [38] which uses conditioned and unconditioned variables in the decomposition.

2.2 Conditioning

The decomposition of the flow variables is achieved by prescribing conditions for which a member $\psi^{(i)}$ can assume non-zero values. The conditions defining the members in a decomposition of the flow variables can be classified (referring to physical space) as local, where information at a single point $\underline{x}$ in $\mathcal{D}$ and a single instant in time is required for the evaluation of the conditions, and non-local, where at least two points in time or space are required. In the following it will be assumed that the conditions can always be expressed in terms of a finite set of discriminating variables defined in $\mathcal{D}$ or a transformed space (i.e., Fourier-space). Then it is always possible to form a finite set of indicator functions

$$\chi_i(\underline{x},t) = \begin{cases} 1 & \text{condition (i) is satisfied at } (\underline{x},t) \\ 0 & \text{otherwise} \end{cases} \tag{1}$$

defined in physical space, or

$$\chi_i(\underline{k},t) = \begin{cases} 1 & \text{condition (i) is satisfied at } (\underline{k},t) \\ 0 & \text{otherwise} \end{cases} \tag{2}$$

defined in the transformed space. The members of the decomposition are then defined as

$$\psi^{(i)}(\underline{x},t) = \chi_i(\underline{x},t) \cdot \psi(\underline{x},t) \tag{3}$$

and analogously in the transformed space. In order to show the flexibility of this approach two examples will be considered. First, consider a single non-negative scalar $\phi(\underline{x},t)$ as discriminator in physical space such that the condition is expressed as inequality

$$\phi(\underline{x},t) \geq d > 0 \tag{4}$$

where d is a chosen threshold value. The indicator function χ marks all points of the flow field where this inequality is satisfied. For local discrimination the inequality is evaluated at the same point in space and time as the flow variable . If the discriminator ϕ and the threshold d are such that the condition amounts to distinguishing turbulent from non-turbulent flow, the standard intermittency factor is obtained as expectation of the indicator function. As an example of non-local conditioning, consider then

$$\phi(\underline{x},\tau) \geq d \quad \text{for} \quad \tau < t \quad \text{and} \quad \phi(\underline{x},t) \geq d \tag{5}$$

which amounts to requiring that the scalar is above the threshold at the present time and at some earlier time irrespective of the events in between. A more restrictive condition would be

$$\phi(\underline{x},\tau) \geq d \quad \text{for} \quad t - T \leq \tau \leq t \tag{6}$$

involving a finite time interval T. Comparison of the time history of the discriminating scalar with a chosen function $S(\tau)$

$$S(\tau) - d \leq \phi\,(\underline{x},\tau) \leq S(\tau) + d \quad \text{for } t - T \leq \tau \leq t \tag{7}$$

allows conditioning with chosen patterns $S(\tau)$ within an interval 2d, which itself could be time dependent. In each case the indicator function $\mathcal{X}$ can be formed and its expectation

$$\gamma(\underline{x},t/T,S,d) \equiv \langle \mathcal{X}(\underline{x},t/T,S,d) \rangle \tag{8}$$

(notation for comparison with $S(\tau)$ in time interval T) is called intermittency factor. It measures at a given point $\underline{x}$ of the flow field the statistical frequency of the event that the condition is satisfied. Conditioning non-local in space an be done in analogous fashion. If the flow variable ψ is taken at $\underline{x}$ and

$$\phi(\underline{x}_i,t) \geq d \quad , \quad i = 1, \ldots n, \; n > 1$$

is required, various subsets of the flow domain $\mathcal{D}$ are obtained depending how $\underline{x}$ and/or the $\underline{x}_i$ are varied and whether $\underline{x}$ is identical with one of the $\underline{x}_i$ or not. If $\underline{x}$ is kept fixed and $\underline{x} \neq \underline{x}_i$ for a i = 1, . . . n and varied arbitrarily

$$\mathcal{X}(\underline{x},t) = 1 \quad \text{for} \quad \phi \geq d \text{ somewhere in } \mathcal{D}$$

and no condition is imposed on the sampling of $\psi(\underline{x},t)$ if d is small enough.

If $\underline{x} = \underline{x}_j$ then

$$\mathcal{X}(\underline{x},t) = \begin{cases} 1 & \text{for} \quad \phi(\underline{x},t) \geq d \\ 0 & \text{otherwise} \end{cases}$$

because all $\underline{x}_i = \underline{x}$ is a possible configuration and local conditioning results. If the configuration of the $\underline{x}_i$ (relative distances and enclosed angles) is kept fixed and the configuration is allowed to translate and rotate genuine non-local conditioning is obtained. As for non-local conditioning in time conditions analogous to (7) on subsets of $\mathcal{D}$ with non-zero volume can be constructed in order to single out chosen spatial structures in terms of the scalar variable ϕ.

As second example conditioning in Fourier-space is considered. Let $\psi(\underline{x},t)$ and $\phi(\underline{x},t)$ be realizations of the flow field at a given instant t, then

$$\hat{\psi}(\underline{k},t) = \frac{1}{(2\pi)^{3/2}} \iiint_{\mathcal{D}} dx\,\psi(\underline{x},t)\, e^{i\underline{k}\cdot\underline{x}}$$

and

$$\hat{\phi}(\underline{k},t) = \frac{1}{(2\pi)^{3/2}} \iiint dx\ \phi(\underline{x},t)\ e^{i\underline{k}\cdot\underline{x}}$$

Now let the condition be

$$|\phi(\underline{k},t)| \geq d$$

at the same wave-number as $\hat{\psi}$ for local conditioning. The conditioned variable is

$$\psi^*(\underline{k},t) = \chi(\underline{k},t)\,\hat{\psi}(\underline{k},t) \tag{9}$$

and gives transformed back into physical space

$$\psi_c(\underline{x},t) = \frac{1}{(2\pi)^{3/2}} \iiint d\underline{k}\,\chi(\underline{k},t)\,\hat{\psi}(\underline{k},t)\ e^{-i\underline{k}\cdot\underline{x}} \tag{10}$$

The transformation acts only on those wave-numbers $\underline{k}$ for which $|\hat{\phi}(\underline{k},t)| \geq d$. If ϕ is proportional to a positive power of $|\underline{k}|$ the high (or low) wave-number range of $\hat{\psi}$ can be singled out. Furthermore a combination of conditions in physical and Fourier spaces is possible, where the condition in Fourier-space acts as a filter and the condition in physical space acts on the filtered realization.

Both examples show that for non-local conditioning the geometric relations of the conditioning and sampling point together with the conditions must be given. Furthermore, it is possible to deal with non-random structures if some of their properties are known a priori. Non-local conditioning in space or time can be set up with appropriate discriminating variables to recognize them and, therefore, sample in phase with their appearance and state. The problem of existence and generation of non-random structures is presently excluded, however.

3. Conditional Moments

Expectations of conditional variables and pdf's form the basis of the present investigation. The methods of conditioning discussed in the previous chapter will be applied to obtain the intermittency factor and conditioned moments and their exact transport equations.

3.1 Intermittency factor: local conditions

First local conditioning in physical space is considered. This case was first analyzed by Corrsin [1] and Corrsin and Kistler [2] for turbulent-nonturbulent discrimination in several shear flows. An equation for the intermittency factor appeared subsequently in Libby's papers [47], [48] and received exact interpretation by Dopazo and O'Brien [49], [50]. The analysis is based on a non-negative scalar $\phi(\underline{x},t)$ which is transported by convection and diffusion and may be created or destroyed by various sources or sinks. Thus,

$$\frac{\partial \phi}{\partial t} + v_\alpha \frac{\partial \phi}{\partial x_\alpha} = \frac{\partial}{\partial x_\alpha} (\Gamma \frac{\partial \phi}{\partial x_\alpha}) + R(\phi) \tag{11}$$

holds instantaneously in the flow field $\mathcal{D}$. The local condition is expressed as

$$\chi(\underline{x},t) = \begin{cases} 1 & \text{for } \phi(\underline{x},t) \geq d \\ 0 & \text{otherwise} \end{cases}$$

or

$$\chi(\underline{x},t) = \begin{cases} 1 & \text{for } \phi(\underline{x},t) \geq d \text{ or } \phi(\underline{x},t) \leq 1\text{-}d \\ 0 & \text{otherwise} \end{cases}$$

for a scalar bounded from above and below and normalized. The threshold level $d > 0$ can be considered a function of the Reynolds-number in particular if $\phi \geq d$ corresponds to turbulent flow at $\underline{x},t$ and $\phi < d$ to non-turbulent flow. For sufficiently smooth boundary conditions and source R, the solution of (11) is twice differentiable and then defines for any instant t

$$S(\underline{x},t) \equiv \phi(\underline{x},t) - d = 0 \tag{12}$$

a subset of the flow domain, which has a bounding surface determined by the limit points of sequences y_i with $S(y_i,t) \leq 0$. This surface is called an interface between the zones with $\phi \geq d$ and $\phi < d$. A point on this interface moves with velocity V in the direction normal to the interface relative to the fluid. Thus

$$n_\alpha V = v_\alpha - v_\alpha^S \quad , \quad \underline{n} = \frac{\nabla S}{|\nabla S|} \quad \text{for} \quad |\nabla S| > 0 \tag{13}$$

where v_α^S denotes the interface velocity. If the scalar variable $\phi(\underline{x},t)$ is sufficiently smooth in $\underline{x}$ the normal vector n_α will exist nearly everywhere. For the indicator function follows then ([54] Ch. III)

$$\frac{\partial \chi}{\partial t} + v_\alpha \frac{\partial \chi}{\partial x_\alpha} = \frac{DS}{Dt} \delta(S) \tag{14}$$

where $\delta(S)$ emerges as a derivative of χ with respect to its argument S (D/Dt denotes the substantial derivative). Ensemble averaging results in the exact transport equation for the intermittency factor

$$\frac{\partial \gamma}{\partial t} + \frac{\partial}{\partial x_\alpha} \langle v_\alpha \chi \rangle = \dot{Q} \tag{15}$$

where $\dot{Q}$ denotes the intermittency source defined by

$$\dot{Q} \equiv \langle DS/Dt \; \delta(S) \rangle \tag{16}$$

The source term of the intermittency equation can be expressed in terms of the relative progression velocity V of the interface $S = 0$. Thus, the first representation of $\dot{Q}$ is obtained

$$\dot{Q} = \langle V |\nabla S| \; \delta(S) \rangle \tag{17}$$

which follows from the fact that the left hand side of (14) is zero for points moving with v_α^S. A second representation for the intermittency source $\dot{Q}$ can be given by applying (11) to (16)

$$\dot{Q} = \left\langle \left\{ \frac{\partial}{\partial x_\alpha} \left(\Gamma \frac{\partial \phi}{\partial x_\alpha} \right) + R(\phi) \right\} \delta(S) \right\rangle \tag{18}$$

linking $\dot{Q}$ to the dynamics of the discriminating scalar variable . These representations allow one to draw several conclusions. Considering (17) we note that this representation is equivalent to Dopazo's form of the intermittency source [49]

$$\dot{Q}(\underline{x}, t) = \lim_{\nu(U) \to 0} \left\langle \frac{1}{\nu(U)} \iint_{U_S} dAV \right\rangle \tag{19}$$

as shown in the appendix. U denotes here a particular sequence of neighborhoods of the point $\underline{x}$ and U_S is the intersection of U with the surface S = 0 (see appendix for details).

Dopazo's representation (19) can be interpreted [69] as a correlation of the relative progression velocity V with a local measure for the amount of interface area per unit volume. The relative progression velocity is determined by

$$V\,\delta(S) = \frac{1}{|\nabla \phi|} \left\{ \frac{\partial}{\partial x_\alpha} \left(\Gamma \frac{\partial \phi}{\partial x_\alpha} \right) + R(\phi) \right\} \delta(S) \tag{20}$$

as shown by Corrsin [55] and Gibson [43]. V can grow over all bounds if points in the flow field appear where $\nabla\phi = 0$ and either $R(\phi) \neq 0$ or molecular diffusion has non-zero effect (extremal points and lines). If a fluid is considered which is initially at rest and a scalar ϕ such that the sources are zero initially, then ϕ is distributed in $\mathcal{D}$ with extremal values at the boundary. Two agents are then able to produce extremal points (maxima or minima) or extremal lines (saddle lines) in $\mathcal{D}$ as time evolves: convection as shown by Gibson [43] and sources. If the scalar ϕ is non-negative and the threshold level d in (12) is sufficiently small the gradient $|\nabla\phi|$ will be small and the second derivatives will be positive with high probability thus making

$$\frac{1}{|\nabla \phi|} \frac{\partial}{\partial x_\alpha} \left(\Gamma \frac{\partial \phi}{\partial x_\alpha} \right) \geq 0 \quad \text{for} \quad \underline{x}\colon \phi(\underline{x}, t) = d > 0$$

Molecular diffusion will therefore most likely produce a positive progression of the interface in the direction of the negative normal i.e., into the zone with $S < 0$. Little can be said about the source term unless a particular scalar ϕ is chosen except

$$\lim_{\phi \to 0} R(\phi) \geq 0 \tag{21}$$

because ϕ must remain non-negative. The influence of $R(\phi)$ for small threshold levels is however worth investigating because it represents the growth of the zone

$S \geq 0$ not linked to transport of mass. For this purpose the scalar

$$\phi(\underline{x},t) \equiv \omega'_\alpha \omega'_\alpha \tag{22}$$

is considered as discrimination between turbulent ($S \geq 0$) and nonturbulent zones. The source term follows from the vorticity transport equation as

$$R = -2\, v'_\beta \omega'_\alpha \frac{\partial \langle \omega_\alpha \rangle}{\partial x_\beta} + 2\, \omega'_\alpha \frac{\partial}{\partial x_\beta} (\langle v'_\beta \omega'_\alpha \rangle - \langle v'_\alpha \omega'_\beta \rangle) +$$

$$+ \langle \omega_\beta \rangle\, \omega'_\alpha S'_{\alpha\beta} + \omega'_\alpha \omega'_\beta \langle S_{\alpha\beta} \rangle + \omega'_\alpha \omega'_\beta S'_{\alpha\beta} - 2\nu \frac{\partial \omega'_\alpha}{\partial x_\beta} \frac{\partial \omega'_\alpha}{\partial x_\beta} \tag{23}$$

where $S_{\alpha\beta} \equiv \frac{1}{2} (\frac{\partial v_\alpha}{\partial x_\beta} + \frac{\partial v_\beta}{\partial x_\alpha})$ denotes the rate of strain. In this case, therefore, $\lim_{\phi \to 0} R(\phi) = 0$ because all contributions to R are proportional to the fluctuating vorticity. Consider now the free boundary of a turbulent shear flow where the spatial variation of the mean quantities is small and outside the zone $S < 0$ the motion is nearly inviscid. Then for points $(\underline{x},t)$ with $S < 0$ the ϕ-source is dominated by the vortex-stretching term

$$R \cong \omega'_\alpha \omega'_\beta S_{\alpha\beta}$$

If vorticity fluctuations are present in this zone at a level below the threshold d, they will be amplified by the strain rate stretching them most efficiently if they are aligned with the dominant direction of stretching. Conversely, if the low level vorticity is homogeneously distributed little effect on the scalar ϕ results, because the strain tensor has zero trace. Thus, only in the case that low level vorticity in the zone $S < 0$ is nearly aligned with the stretching direction created by the large scale motion in the turbulent zone, will the interface defined by (22) move faster than the interface defined by a conserved scalar with the same diffusivity.

3.2 Intermittency factor: non-local conditions

Conditioning at more than one point in space and/or time leads to moments or pdf's that contain a certain amount of scale information. The range of possibilities for non-local conditioning is vast and for this reason the simplest case of two points is discussed first. A single scalar $\phi(\underline{x},t)$ is considered as discriminating variable and the conditions are given at two points $\underline{x},\underline{y}$ at the same time t

$$\chi_2(\underline{x},t/\underline{y},t) = \begin{cases} 1 & \text{If } \phi(\underline{x},t) \geq d \text{ and } \phi(\underline{y},t) \geq d^* \\ 0 & \text{otherwise} \end{cases} \tag{24}$$

where d, d* are positive threshold levels. Since the condition at $(\underline{x},t)$ and $(\underline{y},t)$ is the logical product of the conditions at each point, the two-point indicator function $\mathcal{X}_2(\underline{x},t/\underline{y},t)$ can be considered as a product of single-point indicator functions

$$\mathcal{X}_2(\underline{x},t/\underline{y},t) = \mathcal{X}(\underline{x},t)\,\mathcal{X}^*(\underline{y},t) \tag{25}$$

where the asterisk denotes the level d*. Now the two surfaces, S = 0 and S* = 0 (in the three-dimensional flow domain $\mathcal{D}$), are defined by

$$S \equiv \phi(\underline{x},t) - d \quad \text{and} \quad S^* \equiv \phi(\underline{y},t) - d^*$$

determine the value of $\mathcal{X}_2$. They can be viewed as a single surface in the six-dimensional space of point $(\underline{x},\underline{y})$ in $\mathcal{D} \times \mathcal{D} \subseteq R^3 \times R^3$. The equation for the rate of change of the indicator function follows for (24) using (25) as in the case of local conditioning

$$\frac{\partial \mathcal{X}_2}{\partial t} + v_\alpha(\underline{x},t)\,\frac{\partial \mathcal{X}_2}{\partial x_\alpha} + v_\alpha(\underline{y},t)\,\frac{\partial \mathcal{X}_2}{\partial y_\alpha} = \mathcal{X}^*(\underline{y},t)\;\delta(S)\;\frac{D\phi}{Dt}(\underline{x},t) + \mathcal{X}(\underline{x},t)\;\delta(S^*)\;\frac{D\phi}{Dt}(\underline{y},t) \tag{26}$$

Ensemble averaging leads to the exact transport equation for the two-point intermittency factor

$$\gamma_2(\underline{x},t/\underline{y},t) \equiv \langle \mathcal{X}_2(\underline{x},t/\underline{y},t) \rangle \tag{27}$$

given by

$$\frac{\partial \gamma_2}{\partial t} + \frac{\partial}{\partial x_\alpha} \langle v_\alpha(\underline{x},t)\,\mathcal{X}_2 \rangle + \frac{\partial}{\partial y_\alpha} \langle v_\alpha(\underline{y},t)\,\mathcal{X}_2 \rangle = \dot{Q}_2(\underline{x},t/\underline{y},t) \tag{28}$$

where $\dot{Q}_2$ denotes the intermittency source defined by

$$\dot{Q}_2 \equiv \langle \mathcal{X}^*(\underline{y},t)\;\delta(S)\;\frac{D\phi}{Dt}(\underline{x},t) \rangle + \langle \mathcal{X}(\underline{x},t)\;\delta(S^*)\;\frac{D\phi}{Dt}(\underline{y},t) \rangle \tag{29}$$

Representations (17) to (19) carry over to (29) without modification. If $\dot{q}(\underline{x},t)$ denotes the instantaneous value of $\dot{Q}$ in (16), the structure of $\dot{Q}_2$ becomes evident as

$$\dot{Q}_2 = \langle \mathcal{X}^*(\underline{y},t)\;\dot{q}(\underline{x},t) \rangle + \langle \mathcal{X}(\underline{x},t)\;\dot{q}(\underline{y},t) \rangle$$

where we have the contributions of the single-point intermittency source at each point $\underline{x}$, $\underline{y}$ provided that the condition is met at the other point. A two-point intermittency factor is able to capture topological properties of the interface between two zones that the corresponding single-point variable does not see. In order to show this a fixed configuration of points $\underline{x},\underline{y}$ is considered where $\underline{x}$ varies in the flow field and $\underline{y} = \underline{x} + \underline{r}$ with fixed $\underline{r} \neq 0$. The equation for the intermittency factor $\gamma_2(\underline{x},\underline{r},t)$ can be recast then in the form

$$\frac{\partial \gamma_2}{\partial t} + \frac{\partial}{\partial x_\alpha} \langle v_\alpha(\underline{x},t)\,\mathcal{X}_2 \rangle = \dot{Q}_2 - \langle (v_\alpha(\underline{x}+\underline{r},t) - v_\alpha(\underline{x},t))\,\frac{\partial \mathcal{X}_2}{\partial r_\alpha} \rangle \tag{30}$$

analogous to the single-point equation (15).

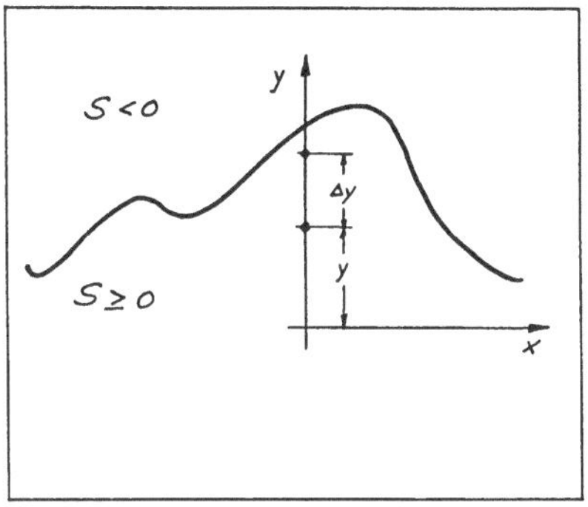

Fig. 1 Interface without folding in y-direction

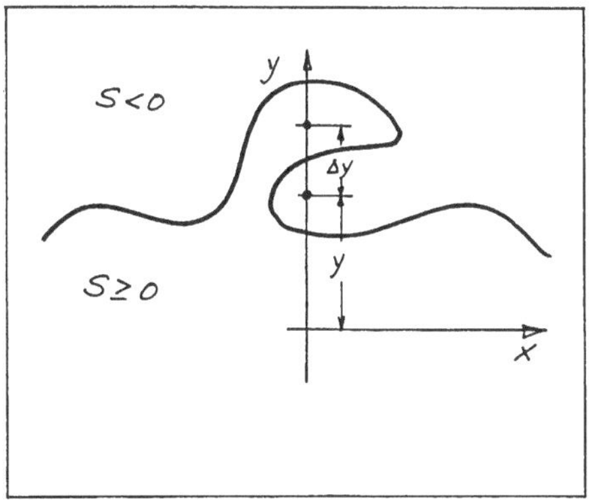

Fig. 2 Interface with folding in y-direction

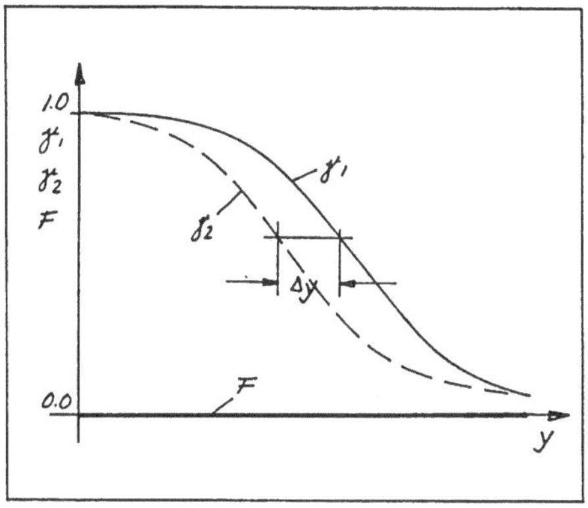

Fig. 3 Intermittency factor $\gamma 1$ and $\gamma 2$ and function F defined in (31) for fig. 1

Now consider the free boundary of a turbulent flow with dominant direction of the mean flow along the x-axis as sketched in fig. 1 and fig. 2. Two cases are considered: in fig. 1 the fluctuations of the interface does not produce folds with respect to the y-axis (crossflow direction), whereas in fig. 2 the same fluctuations of the interface are assumed with folds added to the interface with positive probability. The shift $\underline{r}$ in the two-point intermittency factor is taken as $\underline{r} = (0,\Delta y,0)$ with $\Delta y > 0$ and of the order of the spatial macro-scale of the discriminating scalar ϕ. Let the single-point intermittency factor $\gamma(x,y)$ be known, indicated as the full line in fig. 3 and fig. 4. Then we note that $\gamma_2(\underline{x},\underline{r},t)$ in case A (fig. 1) turns out to be

$$\gamma_2(x,y) = \gamma_1(x,y+\Delta y)$$

for x held constant, shown as the broken line in fig. 3. This follows immediately from the assumption that in this case no folds with respect to the y-axis appear, because then the event that the condition is satisfied at $y + \Delta y$ always implies that the condition is satisfied at y. Thus $\mathcal{X}_2(x,y/x,y+\Delta y) = \mathcal{X}(x,y)\mathcal{X}(x,y+\Delta y) = \mathcal{X}(x,y+\Delta y)$ and ensemble averaging yields the relation above and no new information is contained in γ_2 compared to γ_1 in the case of no folds of the interface.

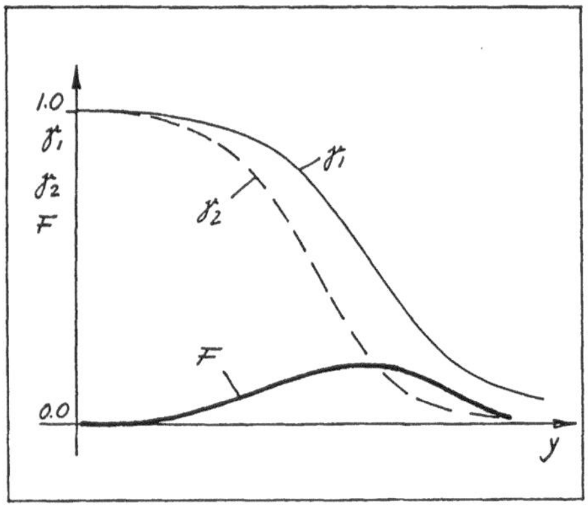

Fig. 4 Intermittency factor γ1 and γ2 and function F defined in (31) for fig. 2

Now considering the case with folds added to the interface as sketched in fig. 2 we note that always

$$\gamma_2(x,y) \leq \gamma_1(x,y) \quad , \quad \gamma_2(x,y) \leq \gamma_1(x,y+\Delta y)$$

must hold, because the condition in γ_2 is more restrictive than the condition in γ_1 at either point (x,y) or $(x,y+\Delta y)$. The presence of folds, however, will for given $(x,y+\Delta y)$, such that the condition is satisfied at this point, reduce the probability (which is unity without folds), that the condition is satisfied at (x,y) too. Hence

$$\gamma_2(x,y) < \gamma_1(x,y+\Delta y)$$

for a shift Δy of the order of the fold size. If we now define a foldedness measure $F(\underline{x},\underline{r},t)$

$$F(\underline{x},\underline{r},t) \equiv \gamma_1(\underline{x},\underline{r},t) - \gamma_2(\underline{x},t/\underline{x}+\underline{r},t) \tag{31}$$

we note that this function has several interesting properties:

1) $F(\underline{x},\underline{r},t) = 0$ if no folds appear in the direction of the shift $\underline{r}$.
2) $F(\underline{x},\underline{r},t) > 0$ if folding appears with respect to $\underline{r}$.
3) F is composed of statistical moments for which exact transport equations can be derived.

Furthermore we note that F depends, as two-point correlation, on six independent spatial coordinates. If, however, a reasonable estimate for the fold size is known (i.e., macro-scale) then $\underline{r}$ can be set to a constant as in the example above and the transport equation for F can be included in single-point closure schemes to provide, for instance, improved treatment of the entrainment mechanism at free boundaries of turbulent shear flows. In this case the form (30) of the transport equation for γ_2 is appropriate and closure assumptions for the right hand side must be developed. A growing body of experimental information [34] exists on development and properties of large scale structures, which can be expected to reveal conditions for folding of iso-scalar surfaces and the transport processes connected with it. Furthermore, quantities such as γ_2 or F are directly measurable and, therefore, the closures of equations (15) and (30) is not an impossible task.

Finally it should be noted that condition (24) has the particular structure given by (25), which implies that the two-point intermittency factor γ_2 is equal to the two-point correlation of indicator functions $\mathcal{X}(\underline{x},t)$ and $\mathcal{X}^*(\underline{y},t)$. This relation (25) would not hold if the condition in (24) involves a function f of the values of ϕ at both points.

$$\mathcal{X}_2(\underline{x},t/\underline{y},t) = \begin{cases} 1 & \text{for } f[\phi(\underline{x},t),\phi(\underline{y},t)] \geq d \\ 0 & \text{otherwise} \end{cases}$$

which defines an interface

$$f(\underline{x},\underline{y},t) - d = 0$$

in $\mathcal{D} \times \mathcal{D}$, but the projection of this interface into $\mathcal{D}$ is not necessarily a surface and may have non-zero volume.

The more complex case of conditions involving more than two points in space (and/or time) can be handled in terms of the n-point probability density function $P_n(\lambda^1, \ldots, \lambda^n, \underline{\boldsymbol{v}}^1, \ldots, \underline{\boldsymbol{v}}^n, \underline{x}^1, \ldots, \underline{x}^n,t)$. The equation for P_n can be established either using Lundgren's [56] or Monin's [57] method. Choosing the former it can be given in divergence form (λ^i denotes the value of $\phi(\underline{x}^i,t)$)

$$\frac{\partial P_n}{\partial t} + \sum_{i=1}^{n} \frac{\partial}{\partial \lambda^i} \langle \frac{\partial \phi}{\partial t} (\underline{x}^i,t)\hat{P}_n \rangle + \sum_{i=1}^{n} \sum_{\alpha=1}^{3} \frac{\partial}{\partial \boldsymbol{v}_\alpha^i} \langle \frac{\partial v_\alpha}{\partial t} (\underline{x}^i,t)\hat{P}_n \rangle = 0 \qquad (32)$$

where

$$\hat{P}_n \equiv \prod_{i=1}^{n} \delta\,(\phi(\underline{x}^i,t) - \lambda^i) \prod_{\alpha=1}^{3} \delta\,(v_\alpha(\underline{x}^i,t) - \boldsymbol{v}_\alpha^i)$$

The explicit form of (32) follows by replacing the time derivatives of the scalar ϕ and the velocity v_α with the scalar and momentum balances. The non-local conditions are now defined in terms of the scalar ϕ at the n points $\underline{x}^i$.

$$\mathcal{X}_n(\underline{x}^1,t/\underline{x}^2, \ldots, \underline{x}^n) = \begin{cases} 1 & \text{for } f_k(\phi(\underline{x}^1,t), \ldots, \phi(\underline{x}^n,t)) \geq d_k,\ k=1, \ldots, L \\ 0 & \text{otherwise} \end{cases}$$

The n-point intermittency factor γ_n is the expectation of $\mathcal{X}_n$ expressed in terms of P_n by

$$\gamma_n(\underline{x}^1,t/\underline{x}^2, \ldots, \underline{x}^n) = \underbrace{\int d\lambda^1 \ldots \int d\lambda^n}_{f_k(\lambda^1, \ldots, \lambda^n) \geq d_k} \int_{-\infty}^{\infty} d\underline{v}^1 \ldots \int_{-\infty}^{\infty} d\underline{v}^n\, P_n$$

Thus follows the transport equation for γ_n by intergration from (32) without recourse to the notion of an interface. From this brief outline can be concluded that non-local intermittency factors can be constructed to suit particular needs in terms of the function f_k and the exact transport equation for γ_n can be established.

3.3 Conditional moments

The indicator function discussed above allows definition of conditional moments: let $A(\underline{x},t)$ be a fluctuating variable (such as velocity, pressure), for $0 < \langle \chi \rangle < 1$

$$\bar{\tilde{A}} \equiv \frac{\langle \chi A \rangle}{\langle \chi \rangle} \quad , \quad A^* \equiv A - \bar{\tilde{A}} \tag{34}$$

the expectation in the zone with $S \geq 0$

$$\tilde{\tilde{A}} \equiv \frac{\langle (1-\chi)\, A \rangle}{1 - \langle \chi \rangle} \quad , \quad A^0 \equiv A - \tilde{\tilde{A}} \tag{35}$$

conversely the expectation in the zone $S < 0$ and A^*, A^0 are the corresponding fluctuations. Considering local conditioning according to (11) and (12), the Navier-Stokes system can be averaged conditionally and yield for constant density the equations

$$\frac{\partial}{\partial x_\alpha} (\gamma\, \bar{\tilde{v}}_\alpha) = \langle v_\alpha\, n_\alpha |\nabla S| \delta(S) \rangle \tag{36}$$

as a consequence of mass conservation and

$$\frac{\partial \bar{\tilde{v}}_\alpha}{\partial t} + \bar{\tilde{v}}_\beta \frac{\partial \bar{\tilde{v}}_\alpha}{\partial x_\beta} = - \frac{1}{\gamma} \frac{\partial}{\partial x_\beta} (\gamma\, \widetilde{v^*_\alpha v^*_\beta}) - \frac{1}{\rho} \frac{\partial \bar{\tilde{p}}}{\partial x_\alpha} + \frac{\partial}{\partial x_\beta} (\nu \frac{\partial \bar{\tilde{v}}_\alpha}{\partial x_\beta}) + \bar{\tilde{S}}_\alpha \tag{37}$$

and

$$\frac{\partial \tilde{\tilde{v}}_\alpha}{\partial t} + \tilde{\tilde{v}}_\beta \frac{\partial \tilde{\tilde{v}}_\alpha}{\partial x_\beta} = - \frac{1}{1-\gamma} \frac{\partial}{\partial x_\beta} ((1-\gamma)\, \widetilde{v^0_\alpha v^0_\beta}) - \frac{1}{\rho} \frac{\partial \tilde{\tilde{p}}}{\partial x_\alpha} + \frac{\partial}{\partial x_\beta} (\nu \frac{\partial \tilde{\tilde{v}}_\alpha}{\partial x_\beta}) + \tilde{\tilde{S}}_\alpha \tag{38}$$

follow from the momentum balance. The terms $\bar{\tilde{S}}_\alpha$ and $\tilde{\tilde{S}}_\alpha$ reflect the transport of momentum through the interface and can be viewed as force per unit mass exerted by the zone $\chi = 1$ on the zone $\chi = 0$ and vice versa [49]. These sources can be represented in terms of point-statistical moments [49] as follows. Introducing [58]

$$F_\alpha \equiv \langle (v_\alpha V + \frac{p}{\rho} n_\alpha - \nu \frac{\partial v_\alpha}{\partial x_\beta} n_\beta) |\nabla S| \delta(S) \rangle - \frac{\partial}{\partial x_\beta} \langle \nu\, v_\alpha n_\beta |\nabla S| \delta(S) \rangle \tag{39}$$

we get

$$\bar{\tilde{S}}_\alpha = \frac{1}{\gamma} F^* \qquad \text{and} \qquad \tilde{\tilde{S}}_\alpha = - \frac{1}{1-\gamma} F^0 \tag{40}$$

where the superscript at F applies to velocity and presure in (39). The momentum sources $\bar{\tilde{S}}_\alpha$ and $\tilde{\tilde{S}}_\alpha$ have no counterpart in unconditionally averaged balance equation. They contain new unknown correlations and for this reason it is important to investigate their properties. It will be shown that $\bar{\tilde{S}}_\alpha$ and $\tilde{\tilde{S}}_\alpha$ are not independent, but connected with the intermittency source Q via a local relation. This follows from the fact that the unconditional mean velocity is given by

$$\langle v_\alpha \rangle = \gamma\, \bar{\tilde{v}}_\alpha + (1 - \gamma)\, \tilde{\tilde{v}}_\alpha \tag{41}$$

and the unconditional stress tensor by

$$\langle v'_\alpha v'_\beta \rangle = \gamma\, \widetilde{\widetilde{v^*_\alpha v^*_\beta}} + (1 - \gamma)\, \overline{\widetilde{v^o_\alpha v^o_\beta}} + \gamma\,(1 - \gamma)\, (\bar{\tilde{v}}_\alpha - \tilde{\tilde{v}}_\alpha)\, (\bar{\tilde{v}}_\beta - \tilde{\tilde{v}}_\beta) \tag{42}$$

locally. Applying these relations to (37), (38), and combining to obtain the equation for the unconditional mean yields the result

$$\gamma\, \bar{\tilde{S}}_\alpha + (1 - \gamma)\, \tilde{\tilde{S}}_\alpha = (\tilde{\tilde{v}}_\alpha - \bar{\tilde{v}}_\alpha)\, \dot{Q} + \frac{1}{\rho}\, (\tilde{\tilde{p}} - \bar{\tilde{p}})\, \frac{\partial \gamma}{\partial x_\alpha} \tag{43}$$

as claimed. This relation has important implications for the closure of conditional moment equations, because it essentially removes one of the momentum sources from the set of unknowns. This type of local consistency requirements carry over to higher order moments such as Reynolds stress, but become increasingly complex.

Non-local conditioning offers too many possibilities for an exhaustive discussion. We note, however, that for the case of two points in space the decomposition

$$v_\alpha = \mathcal{X}_2\, v_\alpha + (\mathcal{X}_1 - \mathcal{X}_2)\, v_\alpha + (1 - \mathcal{X}_1)\, v_\alpha \tag{44}$$

with $\mathcal{X}_1 = \mathcal{X}_1(\underline{x},t)$ and $\mathcal{X}_2(\underline{x}-\underline{r},t/\underline{x},t)$ as defined in (24) leads to three contributions representing the velocity in the core of the zone $S > 0$, the velocity in the zone $S > 0$ near folds and the velocity outside this zone. The shift $\underline{r}$ is considered constant. Averaging and appropriate definition of zonal means shows that the second part in (44) is proportional to the foldedness measured introduced in (31).

4. Turbulent Flows with Chemical Reactions

Turbulent flows with chemical reactions can be separated in zones with distinct properties by the extent of reaction that takes place. If significant amounts of heat are released by the reactions, the density will be variable. In this chapter two cases of turbulent flows with chemical reactions in the gas phase will be discussed: non-premixed and premixed combustion. The chemistry of combustion in the gas phase involves complex systems of reaction steps with numerous components. In order to keep the problem tractable, only a greatly simplified and global description of chemistry will be employed. In both cases particular values of scalar variables arise, which lend themselves to conditioning and the corresoponding sources for the intermittency factor will be analyzed.

4.1 Non-premixed combustion: diffusion flames

The simplest model for the chemical reactions in diffusion flames consists of a single infinitely fast and global step, in which fuel reacts with oxidizer to product. The instantaneous description of the local thermo-chemical state can then be reduced [59], [60] to a single conserved but non-passive scalar variable $\phi(\underline{x},t)$ with values in the interval [0,1]. The thermodynamic variables density, temperature and composition are local functions [60] of this scalar $\phi(\underline{x},t)$ which is usually called mixture fraction [61]. For the combustion of hydrogen and hydrocarbons these local relations are strongly non-linear and information on the pdf is requird in order to calculate the expectations of thermodynamic variables. The composition of fuel and oxidizer streams determine the stoichiometric value λ_{st} of ϕ, at which the infitely thin flame sheet is located. The conserved and non-negative scalar mixture fraction is now taken as discriminator and the stoichiometric value λ_{st} is considered threshold value. Hence the fluid consisting of fresh oxidizer and product (plus inert components) is the zone where $S \equiv \phi - \lambda_{st} < 0$ and the fluid consisting of fuel and product (plus eventual inert components) is the zone $S \geq 0$. The surface separating the two zones is identical with the infinitely thin flame sheet. Furthermore, both zones are turbulent but show different levels of density fluctuations: the zone $S < 0$ exhibits strong fluctuations whereas the zone $S \geq 0$ shows weak fluctuations (for fuels like H_2 they are negligible). Following the suggestion of Janicka [62] the indicator function is weighted with the density and the intermittency factor follows then as

$$\gamma \equiv \langle w\mathcal{X}\rangle \quad , \quad \mathcal{X}(\underline{x},t) = \begin{cases} 1 & \text{for } \phi(\underline{x},t) \geq \lambda_{st} \\ 0 & \text{otherwise} \end{cases} \tag{45}$$

where $w = \rho/\langle\rho\rangle$ denotes the weight function. The conditional moments analogous to (34) and (35) are now

$$\bar{\tilde{A}} \equiv \frac{\langle w\ A\rangle}{\gamma} \quad , \quad A^* \equiv A - \bar{\tilde{A}} \tag{46}$$

and

$$\tilde{\tilde{A}} \equiv \frac{\langle w\ (1-\ )A\rangle}{1-\gamma} \quad , \quad A^o \equiv A - \tilde{\tilde{A}} \tag{47}$$

Then the dynamics of the weighted intermittency factor are determined by

$$\frac{\partial}{\partial t}(\langle\rho\rangle\ \gamma) + \frac{\partial}{\partial x_\alpha}(\langle\rho\rangle\ \bar{\tilde{v}}_\alpha\ \gamma) = \dot{Q} \tag{47}$$

with the source

$$\dot{Q} \equiv \langle\rho\ \frac{DS}{Dt}\ \delta(S)\rangle \tag{49}$$

This source is closely connected with the pdf of the scalar ϕ. This connection can be analyzed by considering a second indicator function $\mathcal{X}_1$.

$$\mathcal{X}_1(\underline{x},t) = \begin{cases} 1 & \text{for } \phi(\underline{x},t) \leq \lambda_{st} + \Delta\lambda \\ 0 & \text{otherwise} \end{cases} \tag{50}$$

with $\Delta\lambda > 0$. Then follows that the difference of the corresponding intermittency factor is

$$\gamma - \gamma_1 = \int_{\lambda_{st}}^{\lambda_{st}+\Delta\lambda} d\lambda \; P(\lambda) \frac{\rho(\lambda)}{<\rho>} = \Delta\lambda \; \tilde{P}(\lambda_{st}) + O(\Delta\lambda^2)$$

where $P(\lambda,\underline{x},t)$ is the pdf of $\phi(\underline{x},t)$ and $\tilde{P} \equiv wP$. From (48) follows

$$\frac{\partial}{\partial t} (<\rho> (\gamma-\gamma_1)) + \frac{\partial}{\partial x} (<\rho><w \; v_\alpha \; (\chi-\chi_1)> = \Delta\lambda \; [\frac{\partial}{\partial t} (<\rho> \tilde{P}) + \frac{\partial}{\partial x_\alpha} <\rho> \tilde{\tilde{v}}_\alpha \tilde{P})] + O(\Delta\lambda^2)$$

and

$$\frac{\partial}{\partial t} (<\rho> \tilde{P}) + \frac{\partial}{\partial x_\alpha} (<\rho> \tilde{\tilde{v}}_\alpha \; \tilde{P}) = \frac{1}{\Delta\lambda} (\dot{Q}(\lambda_{st}) - \dot{Q}(\lambda_{st} + \Delta\lambda)) + O(\Delta\lambda)$$

where $\tilde{\tilde{v}}_\alpha$ is given by (46). Going to the limit $\Delta\lambda \to 0$ and using the scalar pdf-equation for high Re_t numbers

$$\frac{\partial}{\partial t} (<\rho> \tilde{P}) + \frac{\partial}{\partial x_\alpha} (<\rho> \tilde{\tilde{v}}_\alpha \; \tilde{P}) = - \frac{\partial}{\partial \lambda} (<\rho> R\tilde{P}) - \frac{\partial^2}{\partial \lambda^2} <\rho \Gamma \nabla\phi . \nabla\phi\hat{P}>$$

the relation of the intermittency source $\dot{Q}$ to the pdf is obtained (for $\lambda = \lambda_{st}$)

$$\frac{\partial \dot{Q}}{\partial \lambda} = \frac{\partial}{\partial \lambda} (<\rho> R\tilde{P}) + \frac{\partial^2}{\partial \lambda^2} <\rho \Gamma \nabla\phi . \nabla\phi\hat{P}> \tag{51}$$

where $\hat{P} \equiv \delta(\phi - \lambda_{st})$ [56]. Integration yields finally

$$\dot{Q} = <\rho> R\tilde{P} + \frac{\partial}{\partial \lambda} <\rho \Gamma \nabla\phi . \nabla\phi P> \tag{52}$$

For diffusion flames the scalar ϕ is conserved and thus

$$\dot{Q} = \frac{\partial}{\partial \lambda} <\rho \; \varepsilon_\phi \; \delta(S)> \tag{53}$$

where $\varepsilon_\phi \equiv \Gamma\nabla\phi . \nabla\phi$ denotes the instantaneous scalar dissipation. This representation of the intermittency source shows several properties. The derivation indicates clearly that conditioning (50) introduces a new independent variable λ and separation of the flow field in several zones is closely related to the pdf $\tilde{P}(\lambda)$ of the scalar via (51). For diffusion flames with infinitely fast single step chemistry the intermittency source (53) can be viewed as divergence of a flux due to scalar dissipation along the scalar axis taken at the value of the threshold. For small threshold values (53) is positive. It must change sign at least once with respect to λ and therefore $\dot{Q}$ is negative in the range of large threshold levels, because (53) integrates to zero over all vaules of ϕ. Finally, it is worth noting, that closure models for (53) are already available in the context of pdf-transport closures [63], [64] hence no new effort is required.

4.2 Premixed combustion

The thermodynamic state in premixed combustion can be described locally in terms of a single progress variable (Bray-Libby-Moss model [65], [66]) for simplified finite rate chemistry. This scalar variable $\phi(\underline{x},t)$ is, however, not conserved and its transport equation contains a nonlinear source/sink term. A small threshold value for ϕ is introduced in the BML model and used twice to define three zones: unburnt mixtures for $\phi < d$, reaction zone for $d \leq \phi \leq 1-d$, burnt mixture for $\phi > 1-d$. In the BML model a powerful assumption is introduced, namely that the probability of finding the reaction zone at a given location is very small compared to burnt and unburnt zones. A different point of view can be taken by taking the value ϕ_{ig} of the progress variable that corresponds to the ignition temperature as threshold level. Then follows for the intermittency source from (52)

$$\dot{Q} = \langle\rho\rangle R\tilde{P} + \frac{\partial}{\partial\lambda} \langle\rho\ \varepsilon_{\phi}\ \delta(S)\rangle \tag{54}$$

From this relation it becomes apparent, that the progress of the zone, where $\phi \geq \phi_{ig}$ (main reaction zone and burnt mixture), depends on the reaction kinetics as well as the turbulence dynamics. Since the threshold level is not necessarily small enough, such that the turbulence contribution is positive, conditions can be inferred from (54) for $\dot{Q} \leq 0$. If the turbulence level in the fresh unburnt fluid is high enough, then $\langle\rho\ \varepsilon_{\phi}\ \delta(S)\rangle$ can decrease with λ and result in negative production of the intermittency factor. This implies that the zone containing reacting and burnt mixture ceases to grow at the expense of the fresh mixture.

5. Applications

The approach discussed in the previous chapters can be illustrated in many ways. Three cases were selected from the areas of numerial simulation, experiment, closure model. First, the numerical simulation done by W. T. Ashurst of Sandia, Livermore of a plane v-shaped flames in a premixed fluid is considered. The dynamics of turbulence is represented by a large number of vortex lines in the two zones of burnt and unburnt mixture, each having different density. The reaction zone appears as a thin sheet as shown by the full line in fig. 5 for a particular instance in time. The indicator functions

$$\mathcal{X}(\underline{x},t) = \begin{cases} 1 & \text{burnt fluid at } (\underline{x},t) \\ 0 & \text{otherwise} \end{cases}$$

and

$$\mathcal{X}_2(\underline{x},t/\underline{x}+\underline{r},t) = \begin{cases} 1 & \text{burnt fluid at } (\underline{x},t) \text{ and } (\underline{x}+\underline{r},t) \\ 0 & \text{otherwise} \end{cases}$$

are defined as suggested in Chapter 4, where $\underline{x}$ and $\underline{x}+\underline{r}$ are on the slanted straight

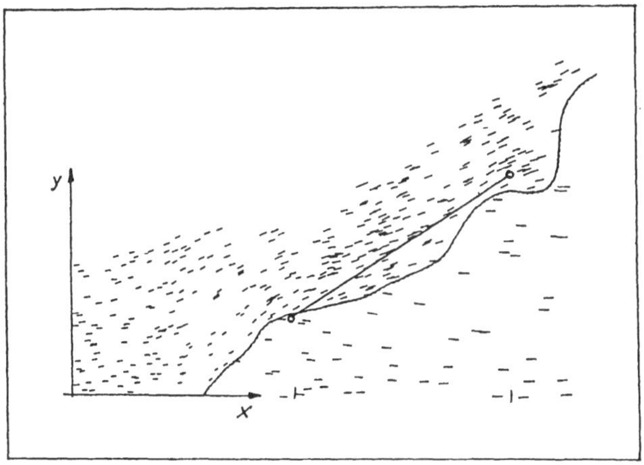

Fig. 5 Numerical simulation of a v-shaped premeixed flame (courtesy of W. T. Ashurst, Sandia Livermore). Full line: Interface burnt and unburnt mixture.

line shown in fig. 5. Taking a sufficient number of realizations the intermittency factors $\gamma \equiv \langle \chi \rangle$ and $\gamma_2 \equiv \langle \chi_2 \rangle$ can be calculated as a function of $\underline{x}$ and γ_2 as a function of $\underline{r}$ for $\underline{x}$ fixed. The result is shown in fig. 6 where the open symbol represents γ and the full symbol γ_2. The single point intermittency factor γ is nearly constant, whereas γ_2 falls off with increasing relative distance. This implies that the foldedness measure defined in (31) is non-zero, hence considerable folding takes place in the direction of the slanted line (fig. 5). If the same calculation is done normal to the slanted line in fig. 5, γ and γ_2 turn out to have the same profile shifted by $\underline{r}$, hence the folding measure (31) is zero.

The experiments of Dibble and Schefer [67] in a turbulent H_2-air diffusion flames provide instantaneous density records at fixed x/D = 50 and y/D = 5.1.

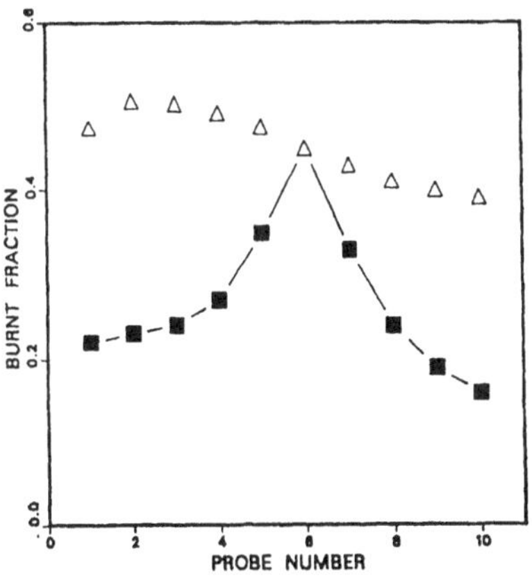

Fig. 6 Numerical simulation of a v-shaped premixed flame (courtesy of W. T. Ashurst, Sandia Livermore). Intermittency factor $\gamma 1$ and $\gamma 2$ along slanted straight line in fig. 5.

These records can be used to establish two-point intermittency factors $\gamma_2 = < {}_2 >$

$$\chi_2(\underline{x}, t/\underline{x}, t+\tau) = \begin{cases} 1 & \text{for } \rho(\underline{x}, t) \leq K_1 \rho_{max} \text{ and } \rho(\underline{x}, t+\tau) \geq K_2 \rho_{max} \\ 0 & \text{otherwise} \end{cases}$$

with $K_1 = 0.9$ and $K_2 = 0.9/0.6$ as shown in fig. 7. The intermittency factor γ_2 is in this case equal to the two-point correlation of the single-point indicator functions. The curves in fig. 7 show the typical form of two-point correlations remaining, however, always positive.

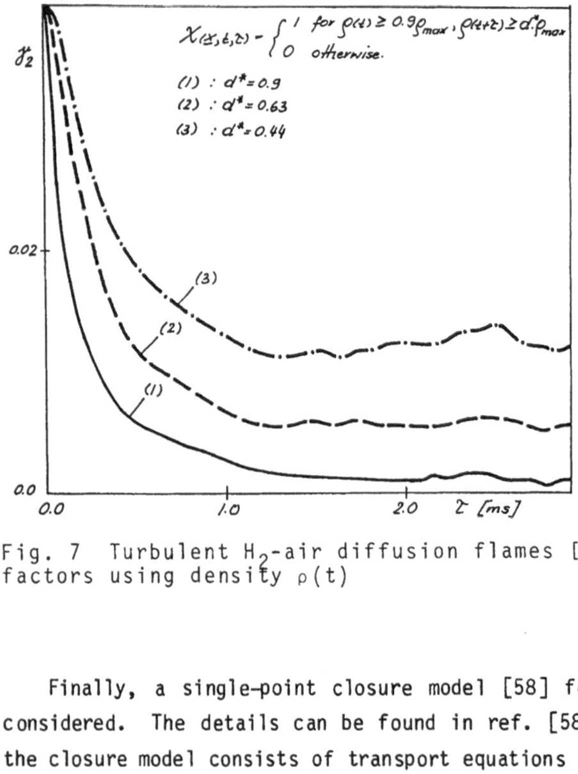

Fig. 7 Turbulent H_2-air diffusion flames [68]: Two-point intermittency factors using density $\rho(t)$

Finally, a single-point closure model [58] for intermittent shear flows in considered. The details can be found in ref. [58], but it should be noted that the closure model consists of transport equations for the intermittency factor γ, the turbulent zone and non-turbulent zone mean velocities, the turbulent zone stress tensor and the dissipation rate. The stress tensor in the non-turbulent zone is related locally to the turbulent zone tensor using a closed form of the Corrsin-Kistler [49] equation. The closure model is applied to the plane jet and compared to the measurements of Gutmark and Wygnanski [15] in fig. 8 to fig. 10. The intermittency factor γ in fig. 8 is determined with a closure for the source, $\dot{Q}$, which includes the effects of the mean strain rate in the turbulent zone and the spatial variation of γ itself. The mean velocities in the two zones in fig. 9 show that the non-turbulent zone mean decays significantly slower than the turbulent zone mean, since the shear stress in the former zone is much smaller.

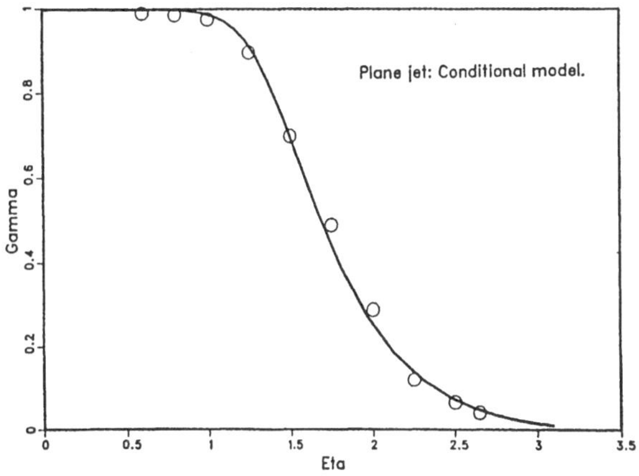

Fig. 8 Conditional closure model [58]: Intermittency factor in plane jet (symbols [15])

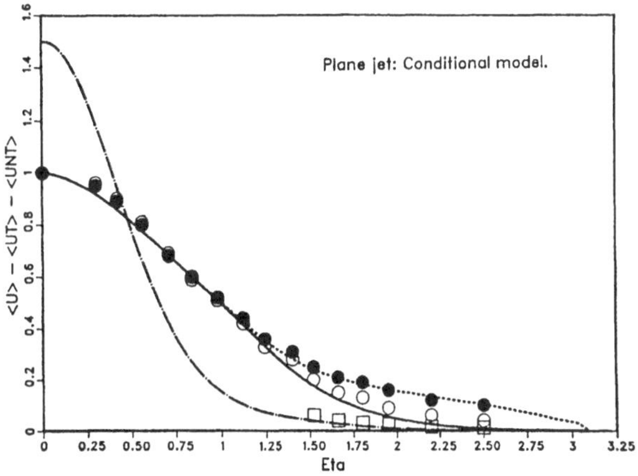

Fig. 9 Conditional closure model [58]:Conditional and unconditional mean velocities (symbols [15])

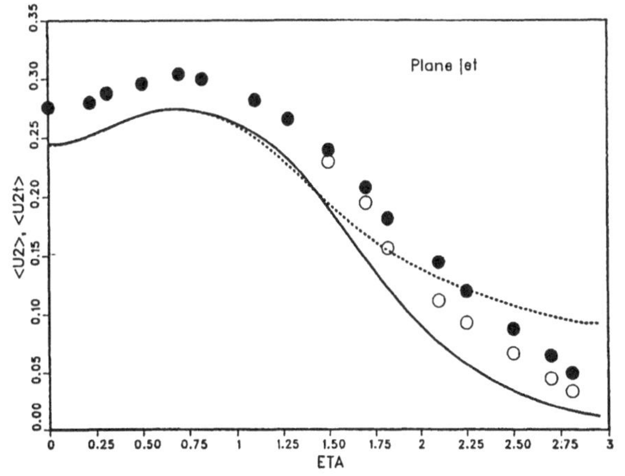

Fig. 10 Conditional closure model [58]: Conditional and unconditional normal stress (symbols [15])

The normal stress component $<u'^2>$ for both zones and the unconditional case in fig. 10 combine nearly linearly because the contribution from the relative movement of the two zones is small. This is not always the case it is possible that the unconditional fluctuations are larger than both zonal fluctuations (for density fluctuations in ref. [68]).

6. Conclusions

The idea of conditioning provides a concept for the classification of methods of decomposition of flow variables describing turbulence. The classification in local and non-local conditions was discussed and a variety of examples was given. In particular, allows non-local conditioning the construction of expectations in phase with recognizable structures provided some of their properties are known. Conditions, which can be expressed in terms of scalar variables called discriminators, allow the construction of indicator functions and intermittency factors as their expectation. The source term in the equation for the intermittency factor and for local conditioning can be represented in several ways leading to the following conclusions. The source is the expectation of the relative progression velocity of the interface defined by the local condition times the absolute value of the gradient of the discriminator. This product has well behaved statistics and is determined by molecular transport and the rate of formation or destruction of the discriminator at the preset threshold level. Hence, for example, the discrimination between turbulent and non-turbulent fluid with conserved and non-conserved discriminators is not equivalent unless the formation rate of the discriminator at the threshold level is sufficiently small. The simple case of two-point conditioning as example for the non-local case allows construction of a measure for the amount of folding of the interface between two zones without recourse to statistics of multi-valued random functions [45]. Application of the idea of conditioning to turbulent reacting flows reveals the close connection to the pdf of the discriminating scalar, which is chosen as mixture fraction for diffusion flames and as progress variables for premixed flames. Existing closure models for the scalar dissipation in pdf equations can be applied to the intermittency source in diffusion flames. The flame sheet can be the interface if the threshold value for the discriminator is set to the stoichiometric value of the mixture fraction. Several applications show that conditional information can be extracted from experiment and numerical simulation and closure models for turbulent shear flows can benefit from these results.

Acknowledgement

This work was supported by DOE-AS03-76SF00034 and Sandia National Laboratories, Livermore. The author is particularly grateful to W. T. Ashurst of CRF, Sandia Livermore for making some of his unpublished results (fig. 5 and fig. 6) available.

References

1. S. Corrsin, NACA Wartime Reports No. 94 (1943).
2. S. Corrsin and A. L. Kistler, NACA TN-3133 (1954).
3. H. E. Fiedler and M. R. Head, JFM 25 (1966), pp. 719.

4. R. E. Kaplan and J. Laufer, Proc. 12th Congress Appl. Mech. (1968), pp. 236.
5. L. S. G. Kovasznay, V. Kibens, and R. F. Blackwelder, JFM 41 (1970), pp. 283.
6. R. F. Blackwelder and L. S. G. Kovasznay, Phys. Fluids 15 (1972), pp. 1545.
7. T. B. Hedley and J. F. Keffer, JFM 64 (1974), pp. 625.
8. C.-H. P. Chen and R. F. Blackwelder, JFM 89 (1978), pp. 1.
9. C. Charnay, J. P. Schon, E. Alcaraz, and J. Mathieu, Turbulent Shear Flows I, Springer Verlag (1979), pp. 104.
10. M. R. Raupach, JFM 108 (1981), pp. 363.
11. J. Murlis, H. M. Tsai, and P. Bradshaw, JFM 122 (1982), pp. 13.
12. G. Heskestad, Trans. ASME Ser. E 32 (1965), pp. 721.
13. L. J. S. Bradbury, JFM 23 (1965), pp. 31.
14. M. Sunyach and J. Mathieu, Int. J. Heat Mass Transfer 12 (1969), pp. 31.
15. E. Gutmark and I. Wygnanski, JFM 73 (1976), pp. 465.
16. J. W. Oler, P. E. Jenkins, and V. W. Goldschmidt, Phys. Fluids 24 (1981), pp. 1235.
17. H. A. Becker, H. C. Hottel, and G. C. Williams, Tenth Symnp. Comb. (1965), pp. 1253.
18. I. Wygnanski and H. E. Fiedler, JFM 38 (1969), pp. 577.
19. R. A. Antonia, A. Prabhu, and S. E. Stephenson, JFM 72 (1975), pp. 455.
20. R. Chevray and N. K. Tutu, JFM 88 (1978), pp. 133.
21. R. Chevray and N. K. Tutu, Lecture Notes in Physics 76, Springer Verlag (1978), pp. 73.
22. I. Wygnanski and H. E. Fiedler, JFM 41 (1970), pp. 327.
23. C. Beguier, L. Fulachier, J. F. Keffer, JFM 89 (1978), pp. 561.
24. I. S. Gartshore, JFM 30 (1967), pp. 547.
25. A. D. Weir and P. Bradshaw, JFM 78 (1976), pp. 641.
26. R. B. Dean, D. H. Wood, and P. Bradshaw, JFM 107 (1981), pp. 237.
27. C. Chandrsuda and P. Bradshaw, JFM 110 (1981), pp. 171.
28. D. H. Wood and P. Bradshaw, JFM 122 (1982), pp. 57.
29. S. T. Paizis and W. H. Schwarz, JFM 63 (1974), pp. 315.
30. R. E. Meyer (ed.), Transition and Turbulence, Pub. no. 46, Math. Res. Center, Madison, WI (1980), Academic Press.
31. R. W. Pitz and M. C. Drake, AIAA-84-0197, AIAA 22nd Aerosp. Sci. Meet, Reno (1984).
32. R. K. Cheng, sub. to Comb. Sci. Technol.
33. R. K. Cheng, L. Talbot, and F. Robben, to be published.
34. A. K. M. F. Hussain, Phys. Fluids 26 (1983), pp. 2816.
35. B. Cantwell and D. Coles, JFM 136 (1983), pp. 321.
36. P. E. Dimotakis, R. C. Miake-Lye, and D. A. Papantoniou, XV Int. Symp. Fluid Dyn., Jachranka, Poland (1981).
37. P. E. Dimotakis, AIAA-84-0368, AIAA 22nd Aerosp. Sci. Meet, Reno (1984).
38. W. C. Reynolds and A. K. M. F. Hussain, JFM 54 (1972), pp. 263.

39. D. D. Knight, B. T. Murray, Lect. Notes in Phys. 136, Springer V (1981), pp. 62.
40. S. Corrsin, Quart. Appl. Math. 12 (1955), pp. 404.
41. S. Corrsin and O. M Phillips, J. SIAM 9 (1961), pp. 395.
42. O. M. Phillips, JFM 51 (1972), pp. 97.
43. C. H. Gibson, Phys. Fluids 11 (1968), pp. 2305.
44. C. H. Gibson, Phys. Fluids 11 (1968), pp. 2316.
45. J. L. Lumley, J. Math. Phys. 5 (1961), pp. 1198.
46. J. L. Lumley, Prediction Methods for Turbulent Flows, Hemisphere (1981), pp. 1.
47. P. A. Libby, JFM 68 (1975), pp. 273.
48. P. A. Libby, Phys. Fluids 19 (1976), pp. 494.
49. C. Dopazo, JFM 81 (1977), pp. 433.
50. C. Dopazo, E. E. O'Brien, in Turbulent Shear Flows I, Springer Verlag (1979), pp. 6.
51. S. Byggstoyl and W. Kollmann, Int. J. Heat Mass Transfer 24 (1981), pp. 1811.
52. J. Jimenez, M. Cogollos, and L. P. Bernal, Fourth Symp. Turb. Shear Flows, Karlsruhe (1983), pp. 16.7.
53. C. M. Tchen, Tellus. 26 (1974), pp. 565.
54. I. M. Gelfand and G. E. Shilov, Generalized Functions, vol. I, Academic Press (1964).
55. S. Corrsin, Notes on Fluid Mechanics, to be published.
56. T. S. Lundgren, Phys. Fluids 10 (1967), pp. 969.
57. A. S. Monin, Prikl. Mat. Mekh. 31 (1967), pp. 1057.
58. J. Janicka and W. Kollmann, Fourth Symp. Turb. Shear Flows, Karlsruhe (1983), pp. 14.13.
59. R. W. Bilger, Progr. Energy Comb. Sci. 1 (1976), pp. 87.
60. R. W. Bilger, in Turbulent Reactions Flows, Springer Verlag (1980), p. 65.
61. D. B. Spalding, Chem. Eng. Sci. 26 (1971), pp. 95.
62. J. Janicka, Habilitation Thesis, RWTH, Aachen (1983).
63. S. B. Pope, to appear in Progr. Energy Comb. Sci.
64. W. Kollmann and J. Janicka, Phys. Fluids 25 (1982), pp. 1755.
65. K. N. C. Bray, in Turbulent Reacting Flows, Springer Verlag (1980), pp. 115.
66. K. N. C. Bray, P. A. Libby, and J. B. Mass, to be publ. in Comb. Flame.
67. J. F. Driscoll, R. W. Schefer, and R. W. Dibble, 19th Symp. Comb. (1982), pp. 477.
68. R. W. Schefer and R. W. Dibble, AIAA-83-0401, 21st Aerosp. Sci. Meet, Reno (1983).
69. C. L. Dancey, Ph.D. Thesis, Cornell University (1983).
70. W. Feller, An Introduction to Probability Theory, Vol. II, Wiley (1971), second edition.

Appendix

Consider the representation (17) of the intermittency source $\dot{Q}$

$$\dot{Q}(\underline{x},t) = \langle V \, |\nabla S| \, \delta(S) \rangle \tag{A1}$$

and Dopazo's [49] representation

$$\dot{Q}^*(\underline{x},t) = \lim_{\mathcal{V}(U)\to 0} \langle \frac{1}{\mathcal{V}(U)} \iint_{U_S} dA \; V \rangle \tag{A2}$$

where U_S denotes the intersection of the neighborhood U of $\underline{x}$ with the interface given implicitly by $S(\underline{x},t) = 0$ and $\mathcal{V}(U)$ is the volume of U. The surfaces S = const are assumed to be orientable and sufficiently smooth (at least once continuity differentiable) such that a unique normal vector n_α (positive into zone $S > 0$) exists nearly everywhere. Both the sequence of neighborhoods U of $\underline{x}$, and the statistics of V must satisfy certain restrictions to afford the representation (A2) as explained below. Then holds

$$\dot{Q} = \dot{Q}^* \tag{A3}$$

To show (A3) we consider a realization for S and V and the fundamental sequence for the Dirac function

$$\delta_n(S) = \begin{cases} 1/\varepsilon_n & \text{for } -\varepsilon_n/2 \le S \le \varepsilon_n/2 \\ 0 & \text{otherwise} \end{cases}$$

with $\varepsilon_n > 0$ and $\varepsilon_n \to 0$ as $n \to \infty$. Then the set

$$K_n(t) \equiv \left\{ \underline{y} \in \mathcal{D} : -\frac{\varepsilon}{2} \le S(\underline{y},t) \le \frac{\varepsilon}{2} \right\}$$

is formed ($\mathcal{D}$ denotes the three-dimensional flow domain). Now consider a point $\underline{x} \in \mathcal{D}$, then either $S(\underline{x},t) = 0$ or $S(\underline{x},t) \neq 0$ because $S = 0$ is a smooth surface. If $S \neq 0$ an index n^* can be found such that

$$\delta_{n}^*(S) = 0$$

and thus

$$|\nabla S| \, \delta_n(S) = 0 \quad \text{for} \quad n \ge n^* \; .$$

If $S = 0$ then (with probability one) the normal vector n_α through $\underline{x}$ on $S = 0$ exists and the intersecting points $\underline{x}^+$, $\underline{x}^-$ of n_α with the surfaces $S = \varepsilon/2$ and $S = -\varepsilon/2$ can be found for $n \ge n^*$ for some n^*, because the surfaces are sufficiently smooth. Thus, for $\underline{x} \in K_n(t)$

$$|\nabla S| = \frac{\varepsilon}{|\underline{x}^+ - \underline{x}^-|} + 0\,(|\underline{x}^+ - \underline{x}^-|)$$

and ($V < \infty$ can be assumed, because for V unbounded $|\dot{Q}|$ and $|\dot{Q}^*| \to \infty$)

$$V|\nabla S|\delta_n(S) = \begin{cases} \dfrac{V(\underline{x},t)}{|x^+ - x^-|} + V.O(|x^+ - x^-|) & \text{for } \underline{x} \in K_n(t) \\ 0 & \text{otherwise} \end{cases} \tag{A4}$$

for $n > n^*$. Now construct the neighborhood U_n of $\underline{x}$ as cube with side-length $|x^+ - x^-|$ centered at $\underline{x}$ and a side aligned with the normal n_α, then is

$$\frac{1}{\mathcal{V}(U_n)} \iint_{U_S} dAV = \begin{cases} \dfrac{V(\underline{x},t)}{|x^+ - x^-|} + V.O(|x^+ - x^-|) & \text{for } U_S \neq \phi \\ 0 & \text{otherwise} \end{cases} \tag{A5}$$

where $\mathcal{V}(U_n) = |\underline{x}^+ - x^-|^3$ is the volume of U_n and U_S is the intersection of U_n with the surface $S = 0$. Comparison with (A4) yields the result

$$V|\nabla S|\,\delta_n(S) = \frac{1}{\mathcal{V}(U_n)} \iint_{U_S} dAV + O(|x^+ - x^-|)$$

and letting $n \to \infty$

$$V|\nabla S|\,\delta(S) = \lim_{n\to} \frac{1}{\mathcal{V}(U_n)} \iint_{U_S} dAV$$

Ensemble averaging leads to

$$<V|\nabla S|\,\delta(S)> = <\lim_{n\to\infty} \frac{1}{\mathcal{V}(U_n)} \iint_{U_S} dAV>$$

The limit on the right hand side commutes with the expectation if the random variables X_n

$$X_n \equiv \frac{1}{\mathcal{V}(U_n)} \iint_{U_S} dAV$$

have distribution functions $F_n(x)$ such that the expectations

$$\int_{-\infty}^{\infty} A(x)\, F_n(dx)$$

converge to a finite limit for any continuous $A(x)$ (see theorem 2 of [70] Ch. VIII.1). Thus under these restrictions on V and the sequence U_n constructecd above, follows

$$<V|\nabla S|\,\delta(S)> = \lim_{n\to\infty} <\frac{1}{\mathcal{V}(U_n)} \iint_{U_S} dAV> \tag{A6}$$

The Spectra of Single Reactants in Homogeneous Turbulence

Edward E. O'Brien

Department of Mechanical Engineering, State University of New York at Stony Brook
Stony Brook, NY 11794, USA

A closure is proposed which is appropriate for a numerical study of the spectrum of a single, chemically-reactive species in turbulence. The objective is to predict the effect of moderate rate reaction kinetics on scalar spectra in situations in which the fluid density is not significantly affected by the reaction. The method of approach is to use the Fourier transform of the two-point scalar probability density function as the primary independent variable whose evolution from an initial state is computed. Two closure approximations are needed, one for turbulent transport and one for molecular diffusion. We propose an EDQNM-style spectral closure to represent the former and a cluster expansion closure due to Lundgren for the latter. These two closures are combined to preserve almost all realizability constraints of the system except the coincidence property of two-point density functions. Preliminary calculations have been made in the no-reaction limit to reproduce previous spectral results. Computations of linear reaction kinetic spectra are under way with the expectation that the results can be compared with previous flux cascade spectral predictions. More general reaction kinetic situations will also be investigated.

1. Introduction

The earliest theoretical study of chemical reaction in statistically isotropic velocity and concentration fields is due to Corrsin[1]. It followed on the heels of his pioneering work on the properties of temperature fluctuations[2,3] in turbulence which, in tandem with the Obukhoff[4] paper, laid the ground work for subsequent investigations in that important field. The problem of chemical reaction in turbulence is also of significant practical importance, so important in practice that much of the work of the last two decades has concentrated on modeling inhomogeneous flows and reactions fields which have immediate applicability to a laboratory or industrial reacting flows.

Study of the spectra of reactants in turbulence has not been as wide spread as the study of temperature fluctuations. Corrsin[5] contributed one of the few papers on this aspect of concentration fluctuations with his study of the spectrum of a first order reaction. He used on Onsager-like cascade process applied to the statistically stationary portion of the spectrum in the equilibrium range of wave numbers. His student, Y. H. Pao[6], extended these ideas to turbulent mixing of a multicomponent mixture of reactants, all of which are first order. The second order, single species reaction was considered in Consin's earliest paper[1] and its asymptotic behavior in the final period of decay was subsequently deduced[7]. Other studies pursued a numerical treatment of the species conservation equation using a modified version[8] of the Direct Interaction[9] approximation for turbulent advection and an inequality preserving moment closure approximation for the reactive terms[10].

Over the last two decades turbulent transport of reacting scalars has received considerable attention from two fresh viewpoints. One stream of research has focussed on the probability density function (pdf) of scalar concentrations as the primary quantity whose evolution is to be modeled and computed[11,12,13], while others, especially from the perspective of combustion, have emphasized the relationship between turbulence structure and the reaction process[14,15,16]. There have been some efforts to combine both approaches[17,18]. In this paper we pursue a continuation of the former method which has certain important advantages such as a close attachment to the conservation laws and the ability to incorporate in it various chemical kinetic schemes without requiring a new closure with each.

It also has disadvantages, the chief of which is the high dimensionality of the defining equations. In practice, with only a few exceptions to be mentioned later, this has meant limiting consideration to the 1-point pdf[19]. From a theoretical point of view the 1-point pdf is on a par with the moment closure theories of turbulence and turbulent scalar diffusion in the sense that the evolution of length scale, or any other parameter requiring two-point information such as dissipation rate, must be brought into the description in an ad-hoc manner. The recent work of Pope[20] demonstrates this; he has brought the 1-point pdf description of reactive scalars into close agreement with second order moment closure practice for the transport of non-reacting scalars in certain turbulent shear flows. A fundamental flaw of one-point descriptions is their inability to bring in the basic physics associated with a spectrum of eddy sizes.

What seems to be missing from the literature on reactive flows is a genre of spectral closures such as exists for turbulent non-reactive scalar transport and which have been used in that field to gain insight into the degree of universality of certain parameters of second-order modeling. They have also been used to deduce the proper forms of sub grid-scale eddy parameters in large eddy simulations. The spectral closures for turbulence which have been most developed are the eddy-damped quasinormal Markovian[21] (EDQNM) approximation, the test field model[22] (TFM), the direct interaction approximation[23] (DIA) and its cousin the Lagrangian history direct interaction[24] (LHDIA) approximation. A recent paper[25] elaborates on the successes of EDQNM and TFM with regard to their influence on second order modeling and their relationship to the more complicated DIA and LHDIA methods.

In this paper we develop, in the 2-point pdf format, a description of the spatial (spectral) effects of turbulent transport at a level equivalent to the EDQNM and TFM approximation; while simultaneously representing, by a suitable closure, composition space transport due to reaction and diffusion. The choice of the EDQNM and TFM methods to approximate turbulent effects is based on the practical consideration that both DIA and LHDIA in their own right are very demanding on computational resources. The EDQNM and TFM methods have been shown to preserve energy positivity and Kolmogoroff scaling at inertial-range wavelengths. Their more subtle properties have been reviewed recently by Herring et al[25]; enough is known about them to make either of them plausible representations of the major spectral features of scalar mixing in homogeneous turbulence. In this paper we arbitrarily choose EDQNM over TFM, a decision which can be reversed if necessary.

There is no closure for composition space transport which is as well developed as the EDQNM closure is for turbulent spectral transport. The paper by Kuo and O'Brien[26], which developed the 2-point Ievlev[27] approximation for a stochastic system undergoing molecular diffusion and reaction, is one of the few to have looked in detail at two-point pdf equations; another is the seminal work of Lundgren[28] who described the equilibrium range of isotropic turbulence starting from a two-point velocity pdf description and a pdf closure hypothesis. The Ievlev closure mentioned above is a formal one which, at the two-point level, has been shown to satisfy all of the realizability conditions of reduction, coincidence, separation and normalization and to give a physically reasonable representation of the evolution of a reacting scalar field in composition space[26].

The Ievlev closure can also be used to represent turbulent convective transport in Fourier space[27], but a previous study[29] showed that it was not feasible to carry out a numerical solution of the resulting closed equations. Even for transport in composition space the computer memory requirements of Ievlev's closure are likely to be excessive when it is applied in conjunction with an EDQNM representation of turbulent spectral transport. In the following pages we outline the Ievlev closure with a view toward incorporating it into the study when methods are found to make its solution a reasonable numerical pursuit. However, we initiate the study by adapting two other closure proposals to the problem of transport of the two-point pdf in composition space. One is due to Lundgren[28] and amounts to an expansion about the jointly normal distribution, the other due to Kuo[30] is based on a linear mean square estimate closure. Both are simpler than Ievlev's formalism but both represent the coincidence property only approximately. Lundgren's closure is the more attractive because connection to a standard statistical state is more apparent, as we show later.

2. Turbulent Mixing Approximation

A scalar concentration $\phi(\underset{\sim}{x},t)$ satisfying the rate equation

$$\frac{\partial\phi}{\partial t} + \underset{\sim}{u}.\nabla\,\phi = D\,\nabla^2\,\phi + \dot{\omega}(\phi) \tag{1}$$

has a corresponding single-point fine grained density

$$\rho = \delta(\hat{\phi} - \phi(\underset{\sim}{x},t))$$

which satisfies[1]

$$\frac{\partial \rho}{\partial t} + \underset{\sim}{u}.\nabla \rho - D \nabla^2 \rho = - \frac{\partial}{\partial \hat{\phi}}[\dot{\omega}(\hat{\phi})\rho] - D\frac{\partial^2}{\partial \hat{\phi}^2}[(\nabla\phi\cdot\nabla\phi)\rho] \tag{2}$$

where $\underset{\sim}{u}$ is the turbulent velocity, $\hat{\phi}$ designates a (non-random) value of $\phi(\underset{\sim}{x},t)$ and $\dot{\omega}$ is the chemical production rate term supposed, for our purposes, to depend explicitly only on concentration.

In (2) the right hand side is concerned with composition space transport, the left hand side with transport in physical space.

If, for the moment, we ignore the composition space terms, ρ obeys an equation analogous to the conservation equation for a non-reacting scalar ϕ

$$\frac{\partial \phi}{\partial t} + \underset{\sim}{u}.\nabla \phi = D \nabla^2 \phi \tag{3}$$

Equation (3) is the starting point for the derivation of the EDQNM or TFM spectral closure methods[25]. For the purpose of this study we will discuss only EDQNM; both it and TFM are readily found in the literature. By an analogous series of arguments one can obtain a closed EDQNM equation for the physical space transport of the two-point pdf $P_2(\hat{\phi}_1,\hat{\phi}_2,r,t) = \langle\rho(\hat{\phi}_1,\underset{\sim}{x}_1,t)\ \rho(\hat{\phi}_2,\underset{\sim}{x}_2,t)\rangle$ where the values of $\hat{\phi}$ at the two points $(\underset{\sim}{x}_1,\underset{\sim}{x}_2)$ are labelled $(\hat{\phi}_1,\hat{\phi}_2)$, $r = |\underset{\sim}{x}_1-\underset{\sim}{x}_2|$ and the angle brackets denote ensemble averages. A form of the result can be written in terms of the one dimensional Fourier transform of P_2, say $H(\hat{\phi}_1,\hat{\phi}_2,k)$, where

$$P_2(\hat{\phi}_1,\hat{\phi}_2,r) = \int_0^\infty H(\hat{\phi}_1,\hat{\phi}_2,k)\ \frac{\sin k\,r}{k\,r}\ d\,k.$$

We find

$$\frac{\partial H}{\partial t} = \iint_\Delta \Theta^\rho_{kpq}(x\,y + z)\{k^2\ p\ H(p)\ E(q) - p^3\ H(k)\ E(q)\}\ \frac{dpdq}{pq} \tag{4}$$

$E(q)$ is the energy spectral density of the turbulence which can be described by a similar well-known EDQNM expression[25]; q and k are wave numbers; the integral $\iint_\Delta dpdq$ is over all (p,q) for which (k, p, q) can form a triangle and (x,y,z) are cosines of the interior angles opposite (k,p,q). The quantity Θ^ρ_{kpq} is an inverse relaxation time whose prescription is a fundamental parameter of the EDQNM method. Its form has been developed for the mixing of non-reacting scalars so as to guarantee realiability properties of the resulting spectrum. It may need to be reconsidered carefully in this context, but we begin with the presumption that it is identical to the analogous expression in the EDQNM equation for $G(k,t)$, where G is the scalar spectrum.

The two terms on the right hand side of (2) carry information about pdf transport in composition space due to chemical reaction and turbulent diffusive smearing of concentration fluctuations. The latter term requires a closure approximation in composition space[12]. Analogously there are residual composition space terms in the equation for $P_2(\hat{\phi}_1,\hat{\phi}_2,r,t)$ which have to be subjected to an appropriate closure.

3. Transport in Composition Space Due to Molecular Diffusion and Reaction

Equation (4) is in closed form; it must be supplemented by an expression for the time rate of change of H due to composition space transport whenever $\dot{\omega}(\phi)$ is non-linear[31]. If the reaction is linear in ϕ, $\dot{\omega}(\phi)$ can be eliminated from (1) by a simple transformation of the time coordinate which leaves the remaining terms in (1) unchanged; there is then no need to use composition space. One way to treat non-linear reactions is to use the Ievlev approximation technique[27] applied to the diffusive-reactive terms in the equation for $P_2(\hat{\phi}_1,\hat{\phi}_2,r,t)$. The two-point pdf for a stochastic diffusive-reactive system has been explored previously[26] and yields the following equation for $P_2(\hat{\phi}_1,\hat{\phi}_2,r,t)$:

$$\frac{\partial P_2}{\partial t} + \frac{\partial}{\partial \hat{\phi}_1}[\dot{\omega}(\hat{\phi}_1)P_2] + \frac{\partial}{\partial \hat{\phi}_2}[\dot{\omega}(\hat{\phi}_2)P_2] = \frac{\partial}{\partial \hat{\phi}_1}[c^{(1)}P_2] + \frac{\partial}{\partial \hat{\phi}_2}[c^{(2)}P_2] \tag{5}$$

where $c^{(1)} = \lim\limits_{\underset{\sim}{x}_3 \to \underset{\sim}{x}_1} D\,\nabla^2_{\underset{\sim}{x}_3}\, E(\hat{\phi}_3|\hat{\phi}_1,\hat{\phi}_2)$

and $c^{(2)} = \lim\limits_{\underset{\sim}{x}_3 \to \underset{\sim}{x}_2} D\,\nabla^2_{\underset{\sim}{x}_3}\, E(\hat{\phi}_3|\hat{\phi}_1,\hat{\phi}_2)$

In the above expressions the subscripts on $\hat{\phi}$ and $\underset{\sim}{x}$ refer to a particular spatial location. For example, $\hat{\phi}_2$ is shorthand for $\hat{\phi}(\underset{\sim}{x}_2,t)$, $E(\hat{\phi}_3|\hat{\phi}_1,\hat{\phi}_2)$ represents a conditional expected value and $\underset{\sim}{x}_3$ is a third spatial point whose existence clearly demonstrates that (5) is not closed but must be approximated. Ievlev provided a framework for doing this; the details of its application to this particular problem are in the literature and cannot be repeated here. The end result can be written in the format of equation (5) but with the following constraints on $c^{(1)}$ which yields a closed equation for P_2, satisfying both realizability conditions and uniqueness.

$$\int c^{(1)} P_2\, d\hat{\phi}_1 = \nabla^2_r \int \hat{\phi}_2 P_2\, d\hat{\phi}_1$$

$$\int c^{(1)} P_2\, d\hat{\phi}_2 = \lim_{r\to 0} \nabla^2_r \int \hat{\phi}_2 P_2\, d\hat{\phi}_2 \tag{6}$$

The realizability conditions guaranteed by the set of equations (5) and (6) are[26]

Reduction $\int P_2(\hat{\phi}_2,\hat{\phi}_1)\, d\hat{\phi}_2 = P_1(\hat{\phi}_1)$

Coincidence $\lim\limits_{\underset{\sim}{x}_2 \to \underset{\sim}{x}_1} P_2(\hat{\phi}_1,\hat{\phi}_2) = P_1(\hat{\phi}_1)\delta(\hat{\phi}_1-\hat{\phi}_2)$

Separation $\lim\limits_{|\underset{\sim}{x}_2 - \underset{\sim}{x}_1|\to\infty} P_2(\hat{\phi}_1,\hat{\phi}_2) = P_1(\hat{\phi}_1)P_1(\hat{\phi}_2)$

Normalization $\int P_2(\hat{\phi}_1,\hat{\phi}_2)\, d\hat{\phi}_1\, d\hat{\phi}_2 = 1$

This system has been examined numerically and has been shown to give a physically reasonable representation of the evolution of P_2 in composition space under diffusion and reaction[26]. It has also been demonstrated that singular (atomic) portions of a two-point pdf are mixed correctly by this approximation.

It is easily shown that the Ievlev closure can be rewritten straight forwardly in Fourier space as follows:

$$\frac{\partial H}{\partial t}(\hat{\phi}_1,\hat{\phi}_2,k,t) = \frac{1}{2}\frac{\partial}{\partial\hat{\phi}_1}\iint_{\Delta}\frac{k}{pq}\,[g_1(\hat{\phi}_1,q) + h_1(\hat{\phi}_2,q)]\,H(\hat{\phi}_1,\hat{\phi}_2,p,t)\;d\,p\;d\,q$$

$$+ \frac{1}{2}\frac{\partial}{\partial\hat{\phi}_2}\iint_{\Delta}\frac{k}{pq}\,[g_2(\hat{\phi}_1,q) + h_2(\hat{\phi}_2,q)]\,H(\hat{\phi}_1,\hat{\phi}_2,p,t)\;d\,p\;d\,q \qquad (7)$$

where g_1, g_2, h_1 and h_2 are weighting functions uniquely determined by the constraints on P_2 and its symmetry properties. The symbol Δ is shorthand for the constraint $\underset{\sim}{k} + \underset{\sim}{p} + \underset{\sim}{q} = 0$

The realizability conditions listed following Eq. (6) can also be expressed in Fourier space as:

Reduction	$\int \hat{P}_2\; d\,\hat{\phi}_2 = P_1(\hat{\phi}_1)\;\delta(\underset{\sim}{k})$	
Coincidence	$\int P_2(\hat{\phi}_1,\hat{\phi}_2,\underset{\sim}{k})\; d\,\underset{\sim}{k} = P_1(\hat{\phi}_1)\;\delta(\hat{\phi}_1-\hat{\phi}_2)$	
Separation	$P_2(\hat{\phi}_1,\hat{\phi}_2,\underset{\sim}{k}) = P_1(\hat{\phi}_1)\;P_1(\hat{\phi}_2)\;\delta(k) + B(\hat{\phi}_1,\hat{\phi}_2,\underset{\sim}{k})$	(8)
Normalization	$\iint P_2(\hat{\phi}_1,\hat{\phi}_2,\underset{\sim}{k})\; d\,\hat{\phi}_1\; d\,\hat{\phi}_2 = \delta(\underset{\sim}{k})$	

Equations (6) and (7) can be solved simultaneously by numerical computation given an initial state, and it can be shown that the solution will satisfy (8) if the initial state does.

In principle the Ievlev closure for the composition space evolution of $H(\hat{\phi}_1,\hat{\phi}_2,k,t)$ can be combined with the EDQNM closure for simultaneous evolution of H in wave number space. The combination of the two closures can be shown to be formally adequate in the sense of satisfying constraints. We have found it to be unrealistically demanding of computational resources at present because of the multiple integral format of the Ievlev equation which exacerbates the difficulty of high dimensionality presented by $H(\hat{\phi}_1,\hat{\phi}_2,k,t)$. We now invoke two simpler closed representations of composition space transport due to diffusion as potential replacements for the Ievlev approximation.

The first is a linear mean square estimate (LMSE) closure suggested by Kuo[20]. It satisfies all the realizability conditions (8) except coincidence, which it satisfies exactly only when the process under consideration is jointly normal. Using LMSE approximations, one can write

$$E(\hat{\phi}_3|\hat{\phi}_1,\hat{\phi}_2) = a_1\hat{\phi}_1 + a_2\hat{\phi}_2$$

where $a_1 = \dfrac{\rho_{31} - \rho_{32}\rho_{12}}{1 - \rho_{12}^2}$ and $a_2 = \dfrac{\rho_{32} - \rho_{31}\rho_{12}}{1 - \rho_{12}^2}$

$\rho_{ij}(r,t)$ is the correlation coefficient of ϕ_i,ϕ_j.

Since only small structure is likely to be well represented by the homogeneity assumption it is consistent to approximate ρ_{ij} by the form

$$\rho \sim \exp[-r^2 \lambda_s^{-2}(t)]$$

where λ_s is the scalar microscale which will be obtained from the two-point pdf, at each time step in the solution process and ρ_{ij} is replaced by a single scalar correlation function $\rho(r,t)$.

The connection between ρ and H can be derived

$$\rho(r,t) = \overline{\phi^2}^{-1} \iiint \hat{\phi}_1 \hat{\phi}_2 \, H(\hat{\phi}_1,\hat{\phi}_2,k,t) \frac{\sin k r}{k r} \, d\,\hat{\phi}_1 \, d\,\hat{\phi}_2 \, d\,k$$

We have found it to be most convenient to represent the effect of molecular diffusion in composition space by using spatial coordinates as independent variables and subsequently converting to a wave number description by a fast Fourier transform after each time step of the evolution.

In composition space using the LMSE approximation for diffusion alone (reaction can be treated exactly) we find

$$\frac{\partial P_2}{\partial t}(\hat{\phi}_1,\hat{\phi}_2,r,t) = D \frac{\partial}{\partial \hat{\phi}_1} \left\{ -\left(6 + \frac{4r^2\rho}{1-\rho^2}\right) \hat{\phi}_1 \, P_2(\hat{\phi}_1,\hat{\phi}_2) + \frac{4r^2\rho^2}{1-\rho^2} \hat{\phi}_2 \, P_2(\hat{\phi}_1,\hat{\phi}_2) \right.$$

$$\left. + D \frac{\partial}{\partial \hat{\phi}_2} \left\{ \frac{4r^2\rho^2}{1-\rho^2} \hat{\phi}_1 \, P_2(\hat{\phi}_1,\hat{\phi}_2) - \left(6 + \frac{4r^2\rho}{1-\rho^2}\right) \hat{\phi}_2 P_2(\hat{\phi}_1,\hat{\phi}_2) \right\} \right. \quad (9)$$

This approximate representation of composition space transport, combined with the role of reaction in composition space can then be added to the EDQNM approximation for spectral transport[4] to produce a closed evolution equation appropriate to describe non-linear reactants.

A second closure approximation which might usefully replace the Ievlev scheme is a two-point pdf closure used by Lundgren[18] in a study of turbulence dynamics. It can be represented as a closure for the three-point pdf which is embedded in conditional expectations such as $E(\hat{\phi}_3|\hat{\phi}_1,\hat{\phi}_2)$. For example

$$E(\hat{\phi}_3|\hat{\phi}_1,\hat{\phi}_2) = \iint \hat{\phi}_3 \, P_3(\hat{\phi}_3|\hat{\phi}_1,\hat{\phi}_2) \, d\,\hat{\phi}_1 \, d\,\hat{\phi}_2$$

$$= \frac{1}{P(\hat{\phi}_1,\hat{\phi}_2)} \iint \hat{\phi}_3 \, P_3(\hat{\phi}_3,\hat{\phi}_1,\hat{\phi}_2) \, d\,\hat{\phi}_1 \, d\,\hat{\phi}_2 \quad (10)$$

By analogy with Lundgren's work an approximation for $P_3(\hat{\phi}_1,\hat{\phi}_2,\hat{\phi}_3)$ is invoked which is an expansion about the multivariate Gaussian state. Writing $P(\hat{\phi}_1,\hat{\phi}_2,\hat{\phi}_3)$ as $P_3(1,2,3)$ and analogously for lower order pdf's the approximation is:

$$P_3(1,2,3) = P_1(1)P_2(2,3) + P_1(2)\,P_2(1,3) + P_1(3)\,P_2(1,2) - 2P_1(1)\,P_1(2)\,P_1(3)$$

$$+ P_3^G(1,2,3) - P_1^G(1)\,P_2^G(2,3) - P_1^G(2)\,P_2^G(1,3) - P_1^G(3)\,P_2^G(1,2)$$

$$+ 2\,P_1^G(1)\,P_1^G(2)\,P_1^G(3), \quad (11)$$

where P_n^G, for n = 1,2 or 3 is the jointly Gaussian density with the same mean and correlation values as P_n.

This expansion satisfies all realizability constraints except coincidence; and it can be combined with EDQNM and composition space transport due to reaction to produce a closed evolution equation for $H(\hat{\phi}_1,\hat{\phi}_2,k,t)$. The relationship between Lundgren's expansion (11) and Kuo's closure (9) can be examined by computing $E(\hat{\phi}_3|\hat{\phi}_1,\hat{\phi}_2)$ from (10) using (11) and inserting the result into the diffusive terms of (5). It turns out, for example, that

$$\begin{aligned} c^{(1)} = \text{LMSE} &+ D \frac{\partial}{\partial\hat{\phi}_1}\left\{\nabla_r^2\, E(\hat{\phi}_1|\hat{\phi}_2)\, P_1(\hat{\phi}_1)\, P_1(\hat{\phi}_2)\right\} \\ &+ D \frac{\partial}{\partial\hat{\phi}_1}\left\{\nabla_{r\to 0}^2\, E(\hat{\phi}_2|\hat{\phi}_1)\, P_1(\hat{\phi}_1)\, P_1(\hat{\phi}_2)\right\} \\ &- D \frac{\partial}{\partial\hat{\phi}_1}\left\{\nabla_r^2\, \rho(r)(\hat{\phi}_2-\bar{\phi})\, P_1(\hat{\phi}_1)\, P_1(\hat{\phi}_2)\right\} \\ &- D \frac{\partial}{\partial\hat{\phi}_1}\left\{\nabla_{r\to 0}^2\, \rho(r)(\hat{\phi}_1-\bar{\phi})\, P_1(\hat{\phi}_1)\, P_1(\hat{\phi}_2)\right\} \end{aligned}$$

where the first term, written LMSE, signifies the terms obtained using Kuo's approximation. It is apparent that the two approximations coincide when the process is a normal one but, in general, Lundgren's closure expresses better the departure from normality.

4. Combined Closures for Turbulent Mixing of a Reactive Scalar

For freely decaying isotropic fluctuation fields the case of turbulent transport of a linear reaction is amenable to a mathematical transformation of the species conservation equation which reduces it to an equation of transport of a non-reacting scalar[31]. All of the closures of the previous section can be shown to preserve this property.

For two-species reacting systems which have unity Lewis number it is possible to obtain the statistics of a progress variable related to species concentrations by employing the equation of transport of a non-reacting scalar[17]. When the species are unpremixed and the reaction rate is also extremely rapid the statistics of the concentration fluctuations can be computed directly from that of the progress variable[32]. All the closures of the previous section also preserve this feature. More generally for very rapid reaction of dynamically passive reactants which are not premixed the reaction effects are diffusion limited and modify only the small scale segment of the scalar spectrum. In this case one can reasonably expect to modify the known statistics of scalars in the equilibrium ranges of high Reynolds number turbulence to describe the behavior of reactants which obey specified kinetic laws, without having to invoke the full closure apparatus described in this paper. We do not know that this has yet been done.

A more difficult situation arises with finite rate reactions which may only be treated by direct numerical computation of spectral decay. Since the task is formidable it makes sense to follow Corrsin's lead and begin by investigating a dynamically passive, single-species, dilute reaction to establish the feasability of formulating and solving a closure equation which is adequate in both composition and spectral spaces.

The strategy is to obtain a closed equation for $H(\phi_1,\phi_2,k,t)$ whose evolution has terms incorporating turbulent mixing (EDQNM) and transport in composition space due to reaction and molecular diffusion (Ievlev, Kuo or Lundgren). Ideally, the closure should be such as to satisfy (4) in the absence of reaction so that standard non-reacting scalar spectral results obtain, and should satisfy (5) and (6) when turbulence is absent (stochastically distributed molecular diffusion and reaction).

The four realizability conditions on $P(\phi_1,\phi_2,k,t)$, (8), should also be satisfied by the closure. This suggests combining (4) and (7) by simply summing the right hand sides of each equation and adding the closed chemical reaction term to it to obtain an expression for $\frac{\partial H}{\partial t}(\phi_1,\phi_2,k,t)$, which will automatically satisfy the constraints mentioned above. A time splitting technique[33] can be used to compute alternately the evolution of H in each of the two spaces, Fourier space and composition space. Fourier space transport has been successfully computed for the evolution of the spectrum of a single passive scalar[34] using a technique due to Leith[35] to solve the time-consuming integrals over all triangles $\underset{\sim}{k} + \underset{\sim}{p} + \underset{\sim}{q} = 0$ which occur in the EDQNM approximation. We have successfully repeated those calculations for (4).

Preliminary calculations of (7) and the constraints associated with it, have shown that it is excessively demanding of computer resources to compute (4) and (7) simultaneously unless novel technique can be found to simplify the numerical solution. An alternative is to replace the Ievlev closure (7) by either the Kuo closure (9) or the Lundgren closure (11). We are focussing on the latter because it appears to be better suited for the treatment of non-normal random variables, which a non-negative variable such as concentration must always be. Since (11) is clearly an expansion about the normal process we expect it to be valid only for slow to moderate speed reaction kinetics, which is the limit unable to be treated by the progress variable analog referred to earlier[32]. It should also be noted that use of Lundgren's closure leads to a reasonable value of the Kolmogoroff constant in the inertial subrange of the spectrum of a non-reacting passive scalar contaminant[28].

No definitive spectral results have yet been obtained for any reactive species by the method outlined here. If the computations succeed we expect, *inter alia*, to compare the results with predictions of various spectral ranges for linearly reacting species made by Corrsin two decades ago[5] without any computer assistance.

References

1. S. Corrsin, Phys. Fluids, 1, 42 (1958)
2. S. Corrsin, J. Aeronaut. Sci., 18, 417-423 (1951)

3. S. Corrsin, J. Appl. Phys., 22, 469-473 (1951)

4. A. N. Obukhoff, Izvest. Akad Nauk. S.S.R. Ser. Geograf. Geofiz, 13, 58 (1949).

5. S. Corrsin, J. Fluid Mech., 11, 407 (1961)

6. Y.H. Pao., AIAA. J., 2, 1550 (1964)

7. E. E. O'Brien, Phys. Fluids, 9, 215 (1966)

8. J. Lee., Phys. Fluids, 9, 363 (1966)

9. R. H. Kraichnan, Phys. Rev., 109, 1407 (1958)

10. C. H. Lin and E. E. O'Brien, Astronautica Acta, 17, 771 (1972)

11. C. Dopazo, Ph.D. Thesis, State University of New York at Stony Brook, Stony Brook, N.Y. (1973)

12. E. E. O'Brien, in Turbulent Reacting Flows, Eds. P. A. Libby and F. A. Williams, Vol. 44, Topics in Applied Physics, Springer-Verlag, Berlin, 1980

13. S. B. Pope, Phil. Trans., R. Soc., London, A291, 529 (1979)

14. R. E. Breidenthal, Ph.D. Dissertation, Cal. Inst. Tech., Pasadena, California (1978).

15. A. R. Karagozian and F. E. Marble, Western States Combustion Institute Meeting Paper 83-31 (1983)

16. R. J. Cantwell, Annual Review of Fluid Mechanics, 13, 457 (1981)

17. R. W. Bilger in Turbulent Reacting Flows, Eds. P. A. Libby and F. A. Williams, 65, Springer-Verlag (1980)

18. K. N. C. Bray in Turbulent Reacting Flows, Eds. P. A. Libby and F. A. Williams, 115, Springer-Verlag (1980)

19. T. S. Lundgren, Phys. Fluids, 24, 584 (1981)

20. S. B. Pope, Phys. Fluids, 24, 584 (1981)

21. S. A. Orszag, Statistical Theory of Turbulence in Proc., Les Houches Summer School, eds. R. Balean and J. L. Peake, 237, Gordon and Breach (1973)

22. R. H. Kraichnan, J. Fluid Mech., 47, 513 (1971)

23. R. H. Kraichnan, J. Fluid Mech., 5, 497 (1959)

24. R. H. Kraichnan, Phys. Fluids, 9, 1937 (1965)

25. J. R. Herring, D. Schertzer, M. Lesicur, G. R. Newman, J. P. Chollet and M. Larchevegue, J. Fluid Mech., 124, 411 (1982)

26. Y. Y. Kuo and E. E. O'Brien, Phys. Fluids, 24, 194 (1981)

27. V. M. Ievlev, Dokl. Akad. Nauk. S.S.S.R., 208, 104 (1973) [Sov. Phys.--Dokl., 18, 117 (1973)]

28. T. S. Lundgren in Statistical Models and Turbulence, Eds. H. Rosenblatt and C. van Atta, 70, Springer-Verlag, Berlin (1972)

29. E. O'Brien, CEAS Report #276, SUNY at Stony Brook, N.Y. (1977)

30. Y. Y. Kuo, Ph.D. Thesis, State University of New York, Stony Brook, N.Y. (1977)

31. E. O'Brien and C. F. Francis, J. Fluid Mech., 13, 309 (1962)

32. R. W. Bilger, Combustion Science and Technology, 13, 89 (1979)

33. N. N. Yanenko, The Method of Fractional Steps, Springer-Verlag, New York (1971)

34. J. R. Herring, et al., J. Fluid Mech., 124, 411 (1982)

35. C. E. Leith, J. Atmos. Sciences., 28, 145 (1971)

The Dynamics of Turbulent Spots

James J. Riley
University of Washington, Seattle, WA 98195, USA
Mohamed Gad-el-Hak
Flow Research Company, Kent, WA 98032, USA

In this article we present a critical review of the present state of knowledge of turbulent spots. We discuss the properties of both newborn as well as fully-developed spots. Both ensemble-averaged results and also the underlying structure of the spot are presented. It is shown that the ensemble-averaged results can be misleading, and that the spot has many features similar to those of a fully-turbulent boundary layer. The mechanisms by which a turbulent spot spreads into the surrounding fluid, which were first suggested by Corrsin and Kistler (1955), will be elaborated. Finally, similarities between spots and other flows will be discussed.

1. Introduction

In 1951 Emmons, while studying the flow of water over a nearly horizontal water table, observed that transition to turbulence occurred through the appearance of small, individual patches of turbulence in an otherwise laminar boundary layer. These patches, which he termed spots, started at random instants in space and time as very small regions of turbulence, and grew in an approximately linear manner as they were swept downstream. The number and size of the spots increased in the downstream direction until the spots amalgamated into a fully-turbulent boundary layer. Based upon these results, Emmons proposed that transition from laminar to turbulent flow occurs through the generation, growth, and amalgamation of turbulent spots.

The viewpoint that turbulent spots play a central role in boundary layer transition has become generally accepted. Furthermore it has been suggested that spots are dynamically similar to turbulent boundary layers but simpler, so that their study might be profitable in shedding light on the complex dynamics of turbulent boundary layers. It has been also suggested that turbulent boundary layers might be a composite of turbulent spots, so that the spots are the basic building block of the boundary layer. Thus the study of turbulent spots has become an essential part of the study of both boundary layer transition and turbulence.

Important dynamical questions with regard to turbulent spots are: (i) under what conditions in a laminar boundary layer can a spot exist; (ii) what are the internal dynamics of the spot; (iii) how does the spot grow; and (iv) how is the spot (re)generated. These questions have received considerable attention since Emmons' original study.

In 1955 Corrsin & Kistler performed an experimental and theoretical study of the freestream boundaries of turbulent flows. The principle objective was the study of the interface between the turbulent boundary layer fluid and the contiguous nonturbulent fluid in the freestream, and of how this interface propagates normal to the boundary. They found that the front consists of a very thin "superlayer" and propagates through the direct action of viscous forces transmitting random vorticity to the previously nonturbulent, irrotational fluid. This process, now known as entrainment, explains the propagation normal to the boundary of both the turbulent boundary layer and the turbulent spot. However, Corrsin and Kistler, in discussing the spreading of the turbulent spot tangential to the boundary, pointed out that "....... it is possible that the dominant turbulence propagation mechanism is different. In particular, it may be that a destabilization of the already rotational flow occurs in addition to a transmission of random vorticity by direct viscous action at the turbulent-laminar boundary". We will show in this article that the spread of the spot by entrainment normal to the boundary and destabilization tangent to the boundary are the principle mechanisms for spot growth.

Other research has shed considerable light on the questions raised above. A coherent picture of the spot is forming, although much still remains to be discovered. The objectives of this article are (i) to present a critical review of the present state of knowledge of turbulent spots; (ii) to explore possible relationships between spots and other flows; and (iii) to delineate areas needing further research.

In the next section we will discuss the initial stages of the spot (i.e., the incipient, or embryo spot); in the third section the average properties of the spot will be detailed; in the fourth section we will describe some of the underlying dynamics of the spot; in the fifth section the spot will be discussed in relation to several other flows; and finally in the last section we will present our opinions on the overall picture of the spot and possible directions for future research.

2. The Incipient Spot

The two dimensional linear stability theory (Tollmien, 1931) is now well understood and verified experimentally (Schubauer & Skramstad,

1948). However, this theory does not explain the mechanism of actual transition which is strongly three-dimensional in nature (Klebanoff, Tidstrom & Sargent, 1962; Kovasznay, Komoda & Vasudeva, 1962). In an attempt to include some of the three-dimensional effects, Benjamin (1961) and Criminale & Kovasznay (1962) calculated the fate of an isolated pulse-like disturbance in a laminar boundary layer with a Blasius velocity profile. They used linear stability theory combined with asymptotic methods to evaluate the solutions in the far field. However, because of the asymptotic approximations used, the overlap for making direct comparisons of laboratory experiments with theories was minimal.

Experimentally, Vasudeva (1967) applied a localized disturbance to a boundary layer to study its subsequent development both in its initial, linear phase and in the later, highly nonlinear range. This later stage compares with the breakdown pattern obtained in the experiments by Klebanoff et al. (1962) using doubly periodic excitation (periodic in time and in the spanwise direction). Vasudeva (1967) speculated that the laminar instability created a localized disturbance which further developed by nonlinear amplification into a local spot of incipient turbulence, whose form did not depend on the laminar instability theory.

Gaster (1975, 1978) and Gaster & Grant (1975) have performed carefully controlled experiments along with theoretical/numerical solutions which avoid asymptotic approximations. These studies have led to a fairly complete picture of the initial stages of growth of the spot, which we will now describe.

Gaster & Grant (1975) carried out experiments on a flat-plate boundary layer in a low turbulence-level wind tunnel. They created a disturbed patch or packet of waves by impulsively disturbing the laminar boundary layer at a point using an acoustic pulse. The initial impulsive disturbance excited a broad band of modes, and a wave packet formed through selective amplification and interference of the most unstable waves. This pulse excitation simulated natural transition better than a periodic wave maker, since excitation by freestream turbulence occurring during the early stages of natural transition also involves a wide spectrum of modes. Disturbances were detected with hot-wire anemometers placed slightly outside the boundary layer, near where the disturbance level was the highest. Gaster and Grant overcame fairly severe signal-to-noise ratio problems by both ensemble averaging and digitally filtering their signals. The temporal development of the wave system was measured for a range of locations in the streamwise (x) and lateral (z) directions, giving a fairly complete picture of the incipient spot development.

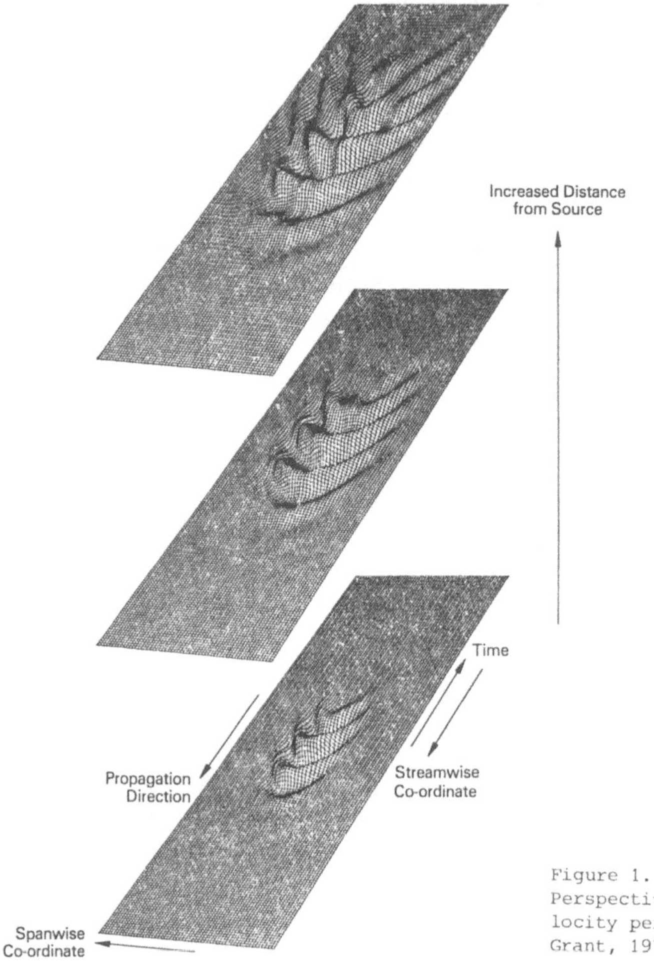

Figure 1. Structure of a wave packet: Perspective display of streamwise velocity perturbation. (From Gaster & Grant, 1975)

Near the source, the wave system appeared to be elliptically-shaped (Figure 1), and then gradually developed a distinctive bowed shape as it was swept downstream. In the early stages of development it appeared to be a wave packet which grew and changed due to amplitude modulation and dispersion of the unstable waves. However, further downstream the packet became somewhat distorted, with maxima appearing symmetrically off the centerline. This distortion appeared to be due to the rapid growth in a band of oblique unstable waves. Harmonics of these waves appeared, indicating that the dynamics of the wave packet were becoming nonlinear. Steep shear layers resulted (Gaster, 1978), with structures similar to hairpin vortices forming. The formation of these vortices is reminescent of the bursting process in turbulent boundary layers.

The steep shear layers led to a local violent breakdown of the flow, quite unlike that observed in experiments with periodic wave trains (Klebanoff, Tidstrom & Sargent, 1962). In fact, significantly, it was found that these modulated waves (wave packets) break down at much lower amplitude than periodic wave trains. The breakdown appeared to be part of a cascading process. The first nonlinearities introduced frequencies into the flow which were about an order of magnitude higher than the frequencies in the initial wave packet. Then these higher frequency instabilities were observed in turn to break down into disturbances of frequencies of an order of magnitude larger. It was suggested that this cascading process continued until arrested by viscosity.

Gaster (1975) also examined this flow theoretically. His method avoided the asymptotic analyses of former studies (Benjamin, 1961; Criminale & Kovasznay, 1962), and consisted of performing a numerical summation over a linear combination of unstable modes of all wavelengths and frequencies. His initial spectrum was assumed to be flat, in approximate agreement with the laboratory data, and he treated the boundary layer growth in an ad hoc manner. The overall shape and growth characteristics of the incipient spot were well predicted by this theory, as shown in Figure 2. There was also good agreement with the details of the flow in the early stages of development. However, the irregular distortions and subsequent breakdown were, of course, not predicted by this linear theory.

Amini (1978) conducted a wind tunnel experiment of the incipient spot. He introduced a weak point disturbance in an unstable laminar boundary layer developing on a flat plate. Amini used hot-wire anemometers to measure the streamwise velocity component of the disturbance prior to breakdown into a turbulent spot. The disturbance structure evolved rapidly in time, its length increasing by almost 100 percent during transit past a fixed probe. The contour of this structure in a plan

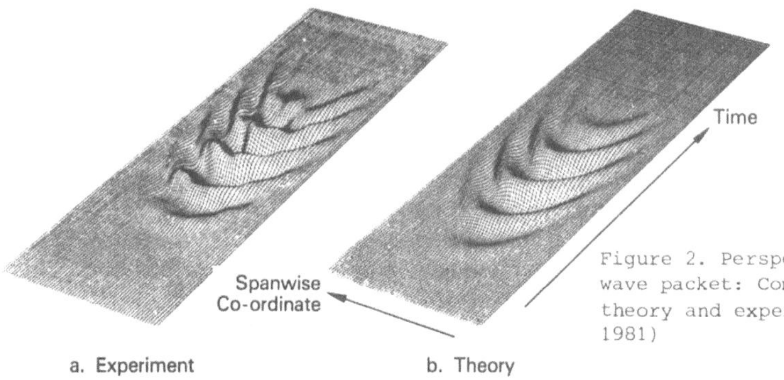

Figure 2. Perspective view of a wave packet: Comparison between theory and experiment. (From Gaster, 1981)

view was arrowhead in shape. Velocity measurements far from the wall and in the plane of symmetry of the structure indicated the presence of small "spikes", similar to the ones observed in the breakdown stage of Tollmien-Schlicting waves (Klebanoff et al., 1962). Close to the wall, a velocity excess occurred. In the spanwise direction alternate regions of streamwise momentum deficit and excess developed.

Leonard (1980, 1981) used a three-dimensional vortex filament description of the vorticity field to numerically simulate an incipient turbulent spot as it evolved from a localized disturbance in a laminar boundary layer. The filaments were marked with a series of node points which were tracked in a Lagrangian frame of reference. The numerical results were limited, since the model could satisfactorily represent only the very early time development of a spot away from the wall region. Nevertheless, good agreement with experiments were achieved for the gross properties of the embryo spot, including the velocities of the leading and trailing interfaces and the velocity perturbation away from the wall.

More recently, the present authors performed flow visualization experiments of the early stages of the spot. The study was conducted by towing a flat plate through a water channel. The spots were initiated by ejecting small amounts of water through a minute hole on the working surface. In order to visualize the embryo spot, closely-spaced fluorescent dye lines oriented parallel to the flow direction were released into the flow near the wall. The dye lines were excited with a sheet of laser light formed by projecting a laser beam onto a glass rod transverse to the axis of the beam. When observed from a plan view, these lines were first seen to be straight. However as the spot began to develop, these lines experienced a wave-type motion (Figure 3), with wavelength and phase speed approximately equal to the corresponding wavelength and phase speed of the most unstable wave (based upon linear stability theory). This wavy motion was followed by a very rapid breakdown into turbulence. As the incipient spot was developing, it was also moving downstream. However, the photographs in Figure 3 have been shifted accordingly, so that the transition events would be depicted in a frame of reference moving with the embryo spot.

Perry, Lim & Teh (1981) also used flow visualization techniques to study the early stages of development of the spot. Their experiments were carried out in a wind tunnel, using smoke to visualize the flow. The smoke was made visible using a laser sheet. Spots were initiated by a short pulse of air from a hole drilled in the tunnnel floor. The air pressure in the pulse was adjusted so that a spot would just form in the test section of the tunnel. If the air pressure was too large,

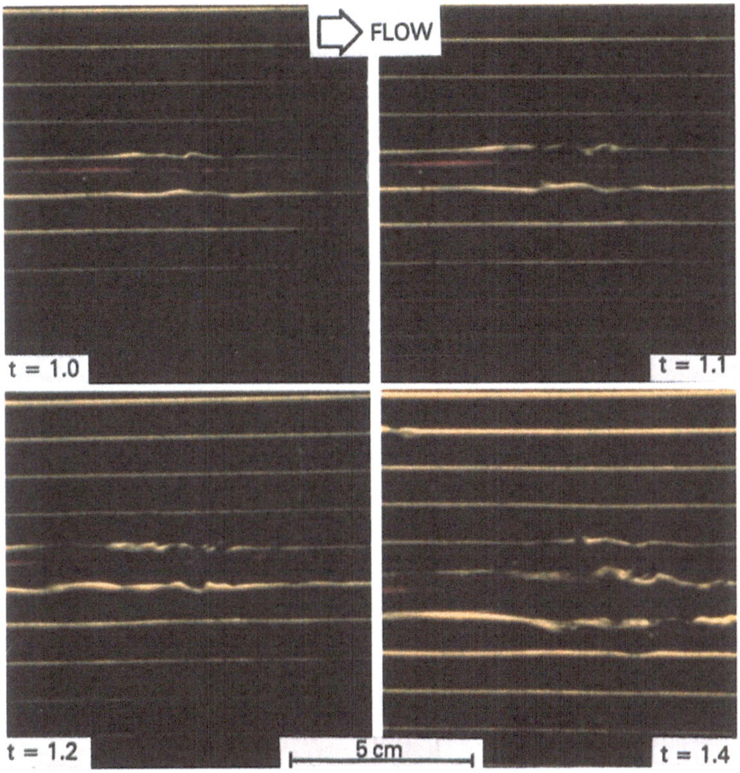

Fig. 3

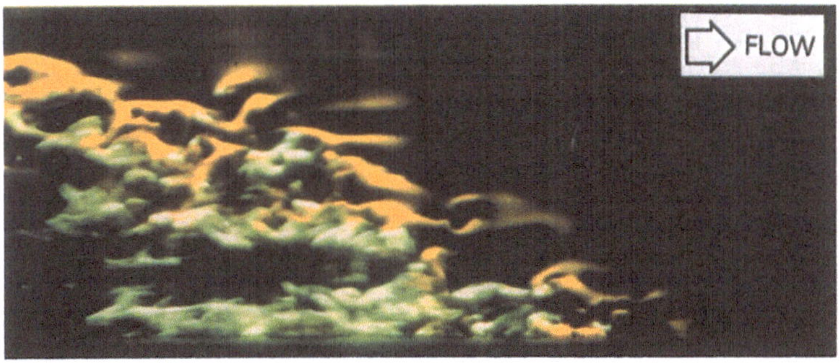

Fig. 4

Figure 3. Dye streaks visualization of an incipient spot; t is time in seconds after initiation, $R_x \approx 2 \times 10^5$

Figure 4. Magnified plan view of the leading edge of a turbulent spot in the early stages of development

the initial stages of the spot would be modified significantly, although the latter stages would be much the same as any spot. Perry et al. concluded from their visualization studies that the incipient spot appeared to be an array of Λ-shaped vortices (alternatively called hairpin vortices, horseshoe vortices or vortex loops) leaning forward in the flow direction. The vortices appeared to be very similar to the tightly-packed array of vortices suggested for the structure of fully-turbulent boundary layers by Theodorsen (1952, 1955) and observed recently by Head & Bandyopadhyay (1978). Figure 4 is a photograph taken by the present authors to depict the early stages of the spot. In this picture, fluorescent dye was painted across a section of the flat plate downstream of the spot generator. The dye was excited with a horizontal sheet of laser light at y = 0, and the camera was zoomed-in to reveal a magnified plan view of the leading edge of the turbulent spot in the early stages of development. Λ-shaped vortices can be observed, especially in the region directly upstream of the leading interface.

From their observations, Perry et al. hypothesized a sequence of events which explains the early development of the spot. The undisturbed boundary layer flow was considered as an array of vortex filaments oriented in the lateral (z) direction. The disturbance introduced by the wall air pulse caused a local bending of a (pack of) vortex filament(s), resulting in a V-shaped vortex, as shown in Figure 5a. Wavey

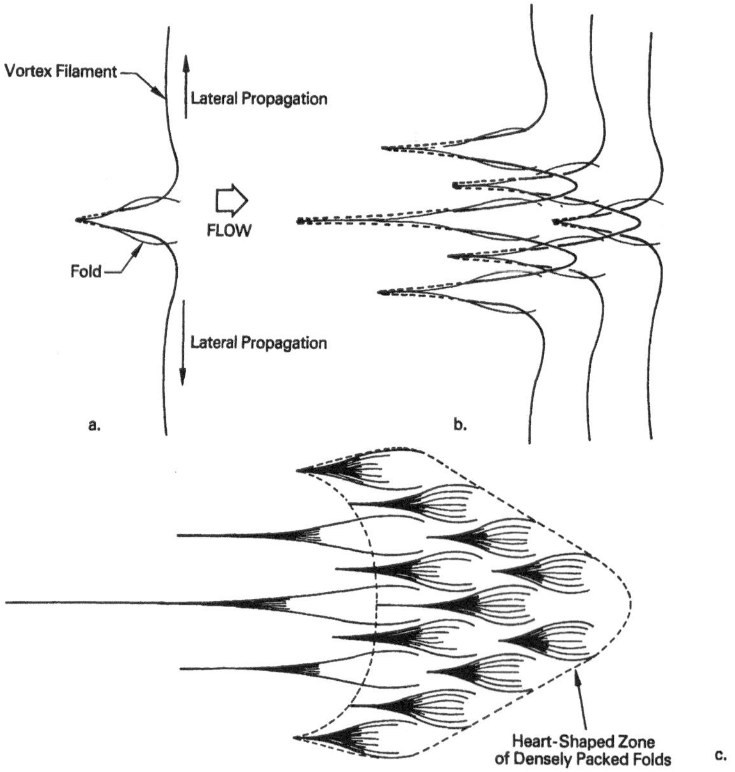

Figure 5. Incipient spot. (a) Vortex filament develops an Undulation. (b) The filament in front develops more lateral undulations and Induces Disturbances in the filaments further downstream. (c) Footprints corresponding to the folds in (b).
(From Perry et al., 1981)

disturbances propagated laterally, causing vortices to develop on each side of the primary vortex (Figure 5b). Λ-shaped vorticies formed, which were swept downstream over previously undisturbed flow. These vortices induced disturbances below them which caused the underlying boundary layer to break down, again in the form of Λ-shaped vortices. In this way the spot propagated itself forward and laterally at the same time. The "footprints" corresponding to this hypothesized model of the spot are the densely packed folds portrayed schematically in Figure 5c.

It is possible that the results of Perry et al. are closely related to the experiments of Gaster & Grant (1975). The lateral, wavy spread of the disturbance could be related to the Tollmein-Schlicting wave propagation in the wave-packet model of Gaster (1975). The breakdown into downstream leaning Λ-shaped vortices is at least qualitatively similar to the nonlinear breakdown described by Gaster (1978). Perhaps quantitative analysis of the visualizations of Perry et al. could determine this.

3. Average Properties of Spots

Many investigators have used fast-response probes to study artificially initiated turbulent spots. Schubauer & Klebanoff (1956) used hot-wire anemometers to measure the velocity signals and celerities associated with the turbulent spot. They determined the general shape of the spot, its spread angle and its propagation velocity. The lateral spread angle of a turbulent spot is typically 10° to each side of the plane of symmetry. This angle changes with the imposition of a pressure gradient, and has a weak dependence on the Reynolds number (Wygnanski, 1981; Wygnanski et al., 1982). In a plan view, the spot has an arrowhead shape, with a leading interface convecting downstream at a speed of about 90 percent of the freestream flow speed. Along the sides of the spot, this convection speed decreases monotonically to about 50 percent of the freestream speed at the extreme spanwise position. The trailing interface convects at a constant 50 percent of the ambient speed (Wygnanski et al., 1976).

The Schubauer & Klebanoff (1956) measurements indicated the presence of a "calmed region" immediately following the spot. The flow in this region is characterized by a relatively full velocity profile, so that it is more stable than the surrounding Blasius flow. Coles & Barker (1975) used conditional sampling techniques to study the spot's velocity field. By assuming a two-dimensional mean flow, they educed the streamline pattern in the center of the spot. They successfully produced a synthetic turbulent boundary layer by generating regular arrays of spots, and suggested a relation between turbulent spots and bursting events in a fully developed turbulent boundary layer.

Wygnanski, Sokolov & Friedman (1976) used V-type hot-wire anemometers to measure the longitudinal, normal and spanwise components of the mean velocity in the interior of spark-generated turbulent spots. As a turbulent spot passed by a fixed velocity probe near the surface, the probe registered an abrupt acceleration that continued up to the vicinity of the trailing edge. Accordingly, the skin friction increased towards the trailing interface. The mean velocity profiles in the interior of the spot were found to be the same as that in a turbulent boundary layer, and could be represented by the universal logarithmic distribution. The displacement thickness and momentum thickness changed rapidly within the spot; however, the shape factor remained at the constant value of 1.5.

Wygnanski et al. (1976) developed an elaborate scheme to obtain statistical data on the spot by ensemble averaging the more or less repeatable events. To avoid smearing out the sharpness of the spot's interface, they performed frequency filtering to yield the envelope of the turbulent fluctuation component of each individual spot. This was compared with a fixed threshold level to obtain the leading and trailing edge times. Wygnanski et al. (1976) have charted the shape of the spot in three-dimensions and have established that far downstream it becomes independent of the initial disturbance. An example of their results is shown in Figure 6, which portrays the development of the turbulent spot along its plane of symmetry. In Figure 6a, the spot generator was located 30 cm from the leading edge of a flat plate, while in Figure 6b it was placed 90 cm further downstream, the air freestream velocity being maintained at 940 cm/sec in both cases ($R_{\delta^*}|_{generator}$ = 508 and 1220, respectively). The numbers shown on the abscissa indicate the average distance of the leading interface from the generator position, integrated over the height of the spot and ensemble-averaged with respect to the number of events. The maximum height of the spot corresponds approximately to the thickness of a hypothetical turbulent boundary layer, originating at the spot generator's location, and with initial thickness equal to that of the laminar boundary layer at the same location. The height of the leading interface's overhang corresponds roughly to the thickness of the laminar boundary layer.

Wygnanski et al. (1976) used the velocity data to calculate the streamlines in the plane of symmetry of the spot in a frame of reference moving with respect to either the leading or the trailing interface of the spot. In the former frame of reference, they observed an extremly large eddy (vortex) extending in the vertical direction well beyond

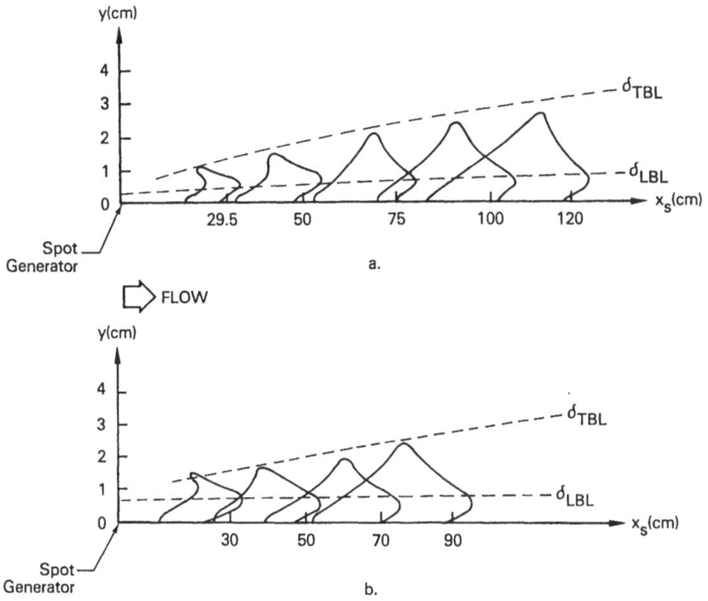

Figure 6. Elevation views of developing spots; numbers shown on the abscissa indicate the average distance of the leading interface from the spot generator. (a) Generator located 30 cm from leading edge. (b) 90 cm. (From Wygnanski et al., 1976)

the maximum height of the spot. This large eddy almost disappeared when the flow pattern was considered relative to the trailing edge of the spot.

Since the final transition to turbulence in a boundary layer is accomplished by the amalgamation of turbulent spots, the possibility exists that the spot and the large eddy structure (in the turbulent boundary layer) may have something in common. To explore this notion Zilberman, Wygnanski & Kaplan (1977) extended the Wygnanski et al. (1976) work to track the structure of the turbulent spot as it merged and interacted with a turbulent boundary layer. They showed that the spot structure tracked in the turbulent boundary layer retained its identity and suffered a negligible loss of intensity, in spite of the random buffeting by the surrounding turbulent fluid. The structure exhibited features in detailed agreement with those of the outer region of the turbulent boundary layer, such as a convection speed of 0.9 U_∞, and was consistent with existing two- and three-point space-time correlations taken in a turbulent boundary layer (Kovasznay, Kibens & Blackwelder, 1970).

Laser-Doppler velocity measurements for the flow in the plane of symmetry of a turbulent spot were conducted in a water tunnel by Cantwell, Coles & Dimotakis (1978). The primary objective of their work was to develop conditional sampling techniques which take into account the turbulent spot growth in appropriate similarity co-ordinates, thus removing a major objection to previous work (Coles & Barker, 1975; Wygnanski et al., 1976). Cantwell et al. (1978) graphically integrated the autonomous system of equations for the unsteady particle displacements, and obtained particle trajectories in an invariant form. Representative particle paths from their results are sketched in Figure 7. The particle paths are shown as dashed lines inside the body of the spot, which is indicated by the solid line. In this analysis, Cantwell et al. (1978) have neglected turbulent dispersion. From this figure, they concluded that an ensemble-averaged spot contains two vortex structures: the large transverse vortex identified by Coles & Barker (1975) and Wygnanski et al. (1976), and a secondary vortex close to the wall near the rear of the spot.

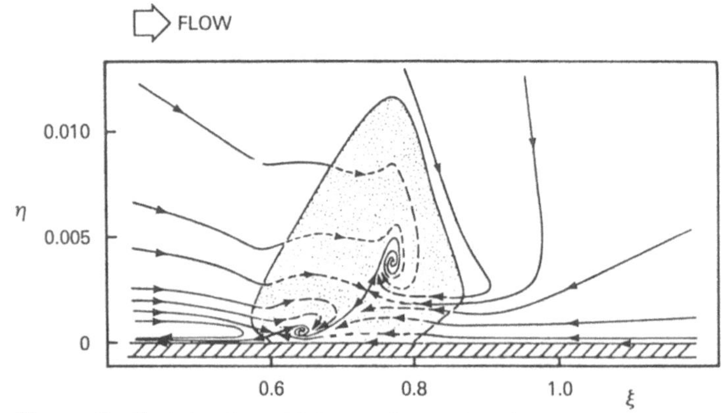

Figure 7. Sketch of particle trajectories in and around a turbulent spot; ξ and η are the concial variables $\frac{x-x_o}{U_\infty(t-t_o)}$ and $\frac{y}{U_\infty(t-t_o)}$, respectively. (From Cantwell et al., 1978)

Wygnanski, Haritondis & Kaplan (1979) conducted an experimental investigation in the region following the passage of an isolated turbulent spot. Their hot-wire measurements revealed the existence of a pair of oblique wave packets that trailed the tips of the spot. The packets were swept at an angle of approximately 40°, and exhibited frequency and wave speed characteristics in agreement with predictions made for oblique Tollmien-Schlichting waves. No waves existed near the centerline of the spot, not even in the calmed region where flow visualization methods indicated the existence of orderly streamwise vortices (Elder, 1960). Wygnanski et al. (1979) speculated that the shape of a turbulent spot, its rate of growth and its spreading angle are related to the accompanying pair of wave packets. These wave packets broke down by generating strong shear layers in both the spanwise and normal directions, giving rise to inflectional, highly unstable velocity profiles. The vertical shear layers had a characteristic dimension which was similar to the spacing between low-speed streaks in a fully developed turbulent boundary layer.

Although most of the information available on the structure of turbulent spots were obtained from flow visualization or velocity measurements, additional important data can be gleaned from measuring the fluctuations in the wall-pressure field or in the concentration of a scalar contaminant within a turbulent spot. Van Atta & Helland (1980) and Atonia, Chambers, Sokolov & Van Atta (1981) used temperature contamination to study the structure of a turbulent spot. They heated a flat plate to produce a surface temperature of about 100°C. Temperature measurements within a turbulent spot evolving in a laminar boundary layer revealed the existence of "cold" regions near the wall and "hot" regions far from the wall. The temperature disturbance relative to the laminar undisturbed values showed a structure strongly anticorrelated with contours of the longitudinal velocity component disturbance. Mautner & Van Atta (1982) measured the wall-pressure field associated with artificially generated turbulent spots in a flat-plate laminar boundary layer. The large-scale structure and predominant characteristics of the wall-pressure signatures were determined from ensemble-averaged wall-pressure distributions at several streamwise and transverse locations.

4. Underlying Structure of Turbulent Spots

From various experimental studies, especially those using probe measurements and ensemble averaging techniques, a fairly complete picture of the overall behavior of the turbulent spot has been obtained. This work does not, however, tell us much about the underlying physical mechanisms which determine the behavior of the spot. Such information can often be obtained more easily using flow visualization. In this section we will discuss some of the results which provide information about this underlying structure.

Emmons (1951) first observed spots through disturbances made by the spots on the water surface of his inclined, free-surface water table. From these visualizations he was able to determine the approximate shape of the spots as well as their lateral growth rates. Elder (1960) later studied turbulent spots in a boundary layer produced in a water tunnel, and used dye to visualize the spots. Dye visualization allows for the observation of many of the overall characteristics of the spot, but also reveals some of the detailed structures in the flow. Figure 8 is a photograph taken from Elder's work of the plan view of a turbulent spot using dye visualization methods. Food color dye was emitted from

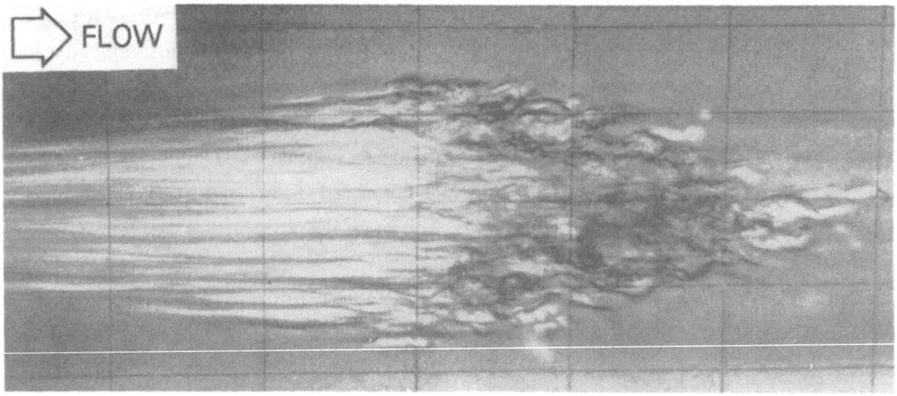

Figure 8. Conventional dye visualization of a turbulent spot; $R_x \approx 3 \times 10^5$. (From Elder, 1960)

a slot placed upstream of the initiation of the spot, and was illuminated with flood lights. The spot can easily be seen as it entrains dye from the wall region. The spot has a characteristic arrow-head shape, with the apex of the arrow oriented in the downstream direction (consistent with probe measurement results discussed in the previous section). Mitchner (1954) has noted, however, that a turbulent spot developing in a shallow, free-surface flow (e.g., a shallow water table) has a significantly different shape, with its apex oriented in the upstream direction.

In addition to the overall shape of the spot, Figure 8 also reveals streak-like structures near the wall. These structures are easiest to recognize near the rear of the spot. They are reminiscent of the streaks observed near the wall in turbulent boundary layers (see, e.g., Blackwelder & Eckelmann, 1979). These turbulent boundary layer studies have revealed that the streaks are accumulations of dye produced by longitudinally-oriented vortex pairs located very near the wall. Cantwell, Coles & Dimotakis (1978) used a different visualization method which allowed better observation of these structures in turbulent spots (Figure 9). They employed an extremely dense suspension of aluminium flakes in a water channel, and observed the flow near the boundary of a transparent wall. The aluminium flakes responded almost instantaneously to the local strain-rate field in the flow, giving an instantaneous picture, as opposed to the time-integrated view obtained from dye visualization. The depth-of-view in the figure is the lower part of the sublayer, so that the method also selects out for viewing only the flow very near the wall. The streaks, easily observed in this figure, are highly elongated in the flow direction. Analysis of the streak spacing gave a value of 86 wall units, close to the gener-

Figure 9. Aluminum flakes visualization of sublayer streaks for turbulent spots in water; $R_x \approx 2 \times 10^5$. (From Cantwell et al., 1978)

ally accepted value for turbulent boundary layers of 100. Perry et al. (1981) also estimated the streak spacing from their visualizations of the incipient spot and obtained values in the range of 80 to 100, again close to the value of 100 for turbulent boundary layers. Thus the behavior of the spots near the boundary appears to be very similar to that of a turbulent boundary layer.

Usual dye methods allow an overall view of the turbulent spot. However, the use of fluorescent dyes together with laser light allows one to observe a particular plane in the spot, and thus obtain better information on the detailed structure of the spot. Figure 10 shows a plan view of a turbulent spot moving from left to right. In this experiment (Gad-el-Hak et al., 1981) fluorescent dye was emitted from a spanwise slot upstream of the spot initiation point, and an x-z

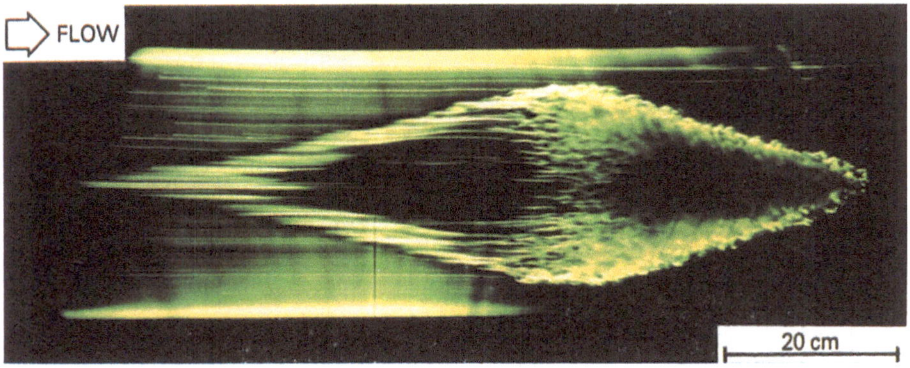

Figure 10. Visualization of a turbulent spot using fluorescent dye and a sheet of laser light at $y = 0$; $R_x \approx 5 \times 10^5$. (From Gad-el-Hak et al., 1981)

plane located at the wall ($y \approx 0$) was illuminated with a sheet of laser light. Dye in both the laminar boundary layer and also in the turbulent spot are easily distinguished in this figure. The light sheet illumination was provided from both sides (the top and bottom in

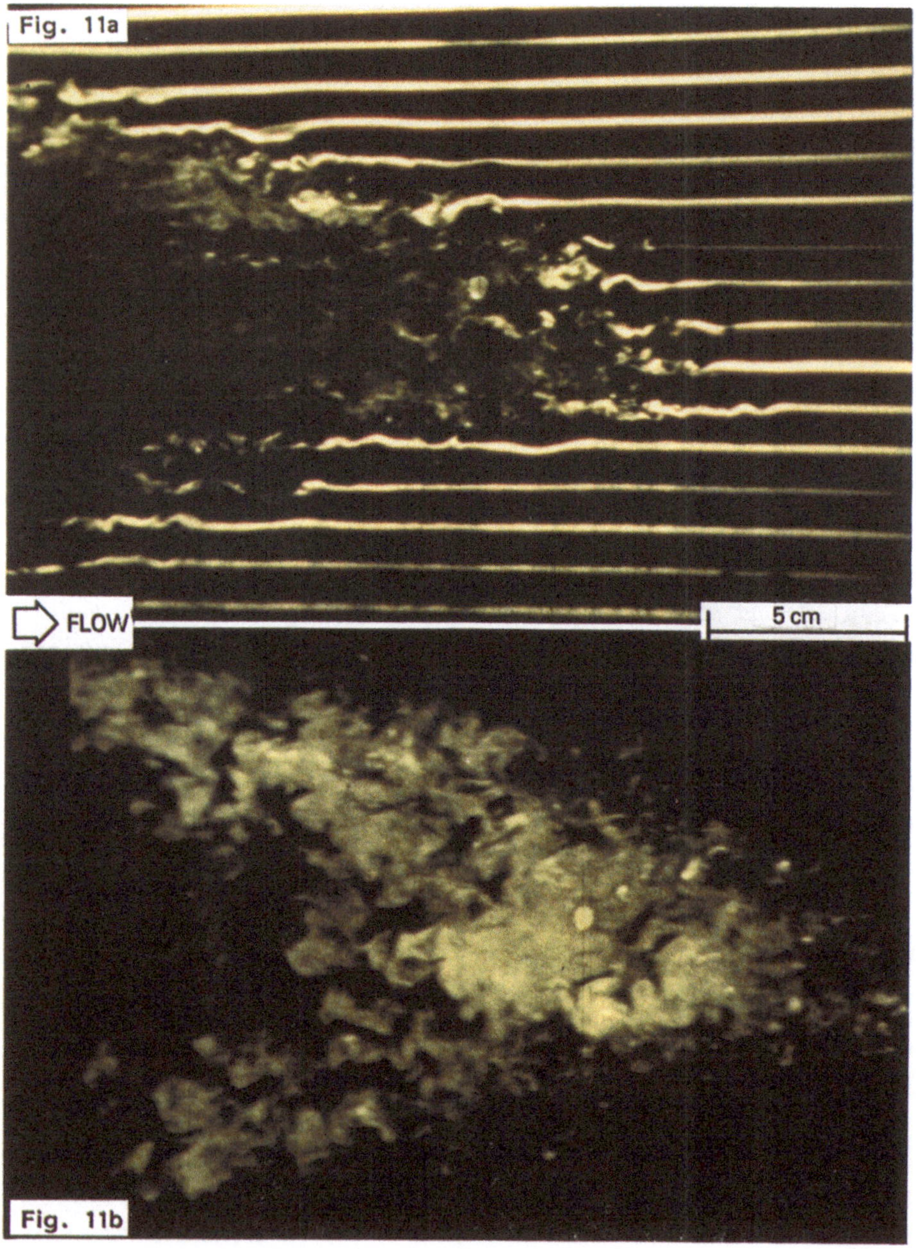

Figure 11. Simultaneous plan views of the spot; $R_x \approx 5 \times 10^5$. (a) Sheet of laser light at $y = 0$. (b) $y = \delta_L$

the figure). Since some light absorption occurs, the sides of the figure appear brighter than the center. As the spot moves over the sheet of dye, it scours dye from the wall region and diffuses it into the spot. The scouring leaves little dye at the wall in the calmed region behind the spot, resulting in the dark, triangular-shaped region to the left. To the right, the general arrowhead shape of the active part of the spot can be seen. It is interesting to note, however, that in the calmed region the dye is seeping in from the left and forming streak-like structures, indicating that counter-rotating vortices still persist at the wall in this region. Some evidence of streaks also appears at the rear of the active region of the spot; however, these are better observed with the visualization methods discussed above.

Gad-el-Hak et al. (1981) also conducted experiments using other elevations of the light sheet (see, e.g., Figure 11). For example, in one run an x-z plane located about two laminar boundary layer thicknesses above the plate was illuminated. Ciné films from this experiment revealed distinct structures in the spot as they moved away from the wall and penetrated the light sheet. First several eddies appeared in this plane. Then, as the spot developed, these eddies remained relatively coherent, and new ones appeared towards the rear of the spot. There was a slight tendency for the new eddies to first appear near the wing-tips, and later in the middle of the spot. Both the streamwise and spanwise length scales of these eddies were approximately equal to a turbulent boundary layer thickness. Thus in the longitudinal direction, the growth of the spot appears to be due to the addition of new eddies on the upstream side of the spot. This manner of growth is consistent with the ensemble-averaged results of Cantwell et al. (1978).

While Figure 10 shows a plan view of a spot in a particular fixed x-z plane, Figure 11 shows two x-z cross-sections of the spot at an instant in time. Figure 11a gives the spot in a plane very near to the wall ($y \simeq 0$), while Figure 11b is a view in a plane approximately one laminar boundary layer thickness above the boundary. The extremely large overhang region, which was apparent in the ensemble measurements of Wygnanski et al. (1976) and Cantwell et al. (1978), is also apparent from these figures. This overhang region appeared to be rather passive, not growing very rapidly; but it exerted considerable influence on the laminar flow below it.

Figure 12 (from Gad-el-Hak et al., 1981) is a photograph of a plan view of the spot at a fraction of the boundary layer thickness above the boundary. In this experiment, instead of allowing the dye to issue in a sheet, the dye was seeped through small holes, creating dye lines in the flow. Figure 12 shows these dye lines as they are being in-

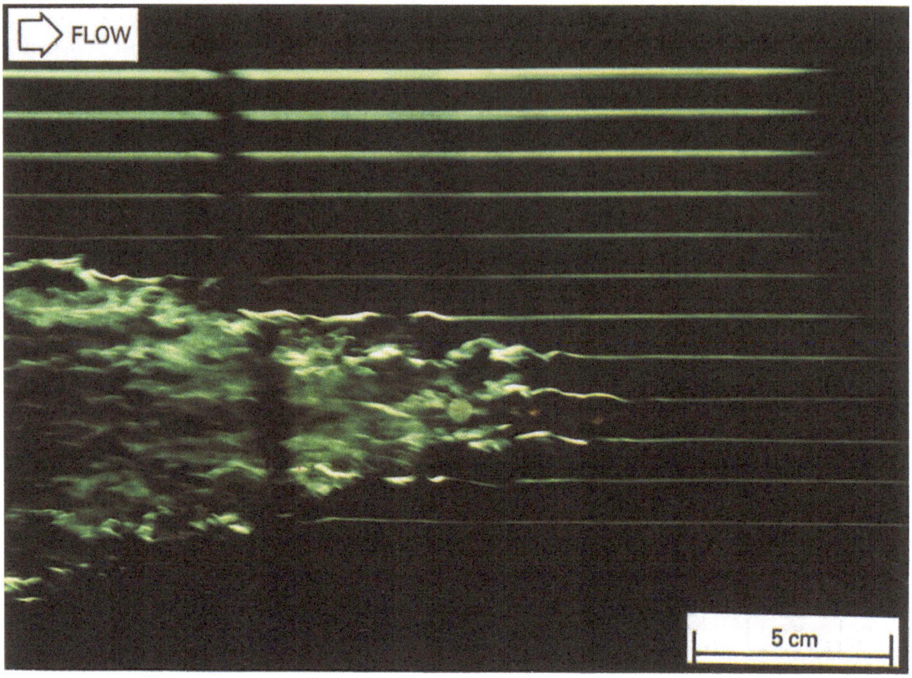

Figure 12. Fluorescent dye streaks visualization of a turbulent spot. The dye lines undergo large-amplitude oscillations before being overtaken by the spot; $R_x \approx 5 \times 10^5$. (From Gad-el-Hak et al., 1981)

gested into the spot. As the spot overtook the dye lines, large oscillations were observed along each dye line. These oscillations are perhaps similar to the wavy structures observed by Perry et al. (1981) near the perimeter of the incipient spot (see Section 2). In the ciné films from which Figure 12 was taken, these disturbances often appeared to be wave-like, with streamwise length scale slightly larger than a laminar boundary layer thickness. Points of constant phase moved downstream at approximately $0.4U_o$ to $0.5U_o$, and hence were rapidly overtaken and ingested by the spot. Similar fluctuations have been observed by Falco (1977) and Cantwell et al. (1978).

A possibly similar phenomena was observed with the light sheet oriented in an x-y plane. Using close-up photographs of the overhang region of the spot, the following interesting feature was found to consistently occur in this region. The overhang region would first be convected above previously laminar flow. At first the laminar flow near the wall would appear undisturbed. However, shortly after the appearance of the overhang, discrete lumps of dye form, indicative of vertical motion. (See the sketch in Figure 13). These lumps have a "wavelength" of approximately one or two laminar boundary layer thick-

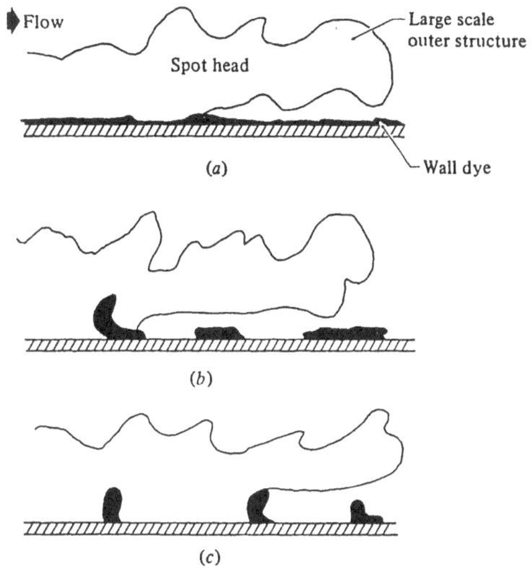

Figure 13. Events preceding the break-up of wall dye; (a) Overhang moving over wall dye layer; (b) Wall dye is separated into lumps; (c) The lumps lift up. (From Gad-el Hak et al., 1981)

nesses. Subsequently the dye lumps explode into turbulence, becoming a part of the spot, apparently without any entrainment occurring. This motion is consistent with that of Figure 12, and, as mentioned, with the observations and model of Perry et al. (1981). From these visualizations it appears that part of the spot is convected over laminar flow near the wall, inducing the laminar flow to break down into turbulence.

A better idea of the growth of the spot normal to the boundary can be obtained by further examining x-y cross-sectional views of the spot. Figure 14 presents an x-y plane at z=6 (approximately $10\delta_\ell$) obtained with the dye layer method (Gad-el-Hak et al., 1981). Preexisting dye sheets are illuminated by a laser light sheet allowing a close examination of the method of entrainment as well as the motion in the irrotational region above the spot. The general shape of the spot obtained was consistent with the ensemble averaged results of Wygnanski et al. (1976) and Cantwell et al. (1978). However no single dominant eddy was observed. Instead, the spot appeared to be composed of numerous eddies having length scales of typically a turbulent boundary layer thickness. The eddies grew and moved away from the plate, providing the growth of the spot normal to the boundary layer. The eddies seemed to move somewhat independently of each other, and, in fact, the interior of the spot closely resembled a turbulent boundary layer. Most of the growth and entrainment appeared to occur on the upstream (i.e., rear) of the spot, consistent with the observations discussed above, and also consistent with the ensemble-averaged results of Cantwell et al. (1978). The entrainment appeared to occur by "gulping", whereby large parcels of irrotational fluid are ingested

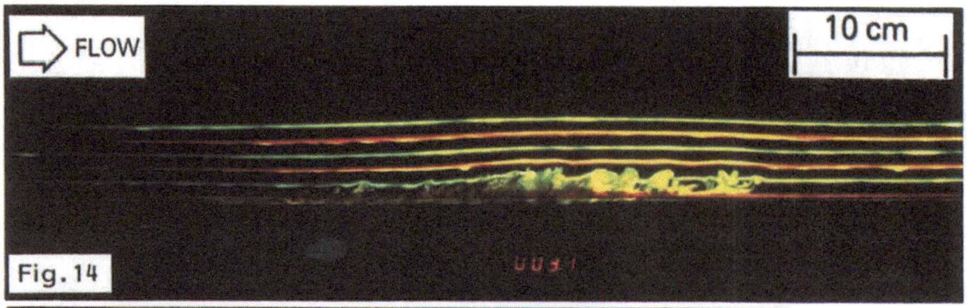

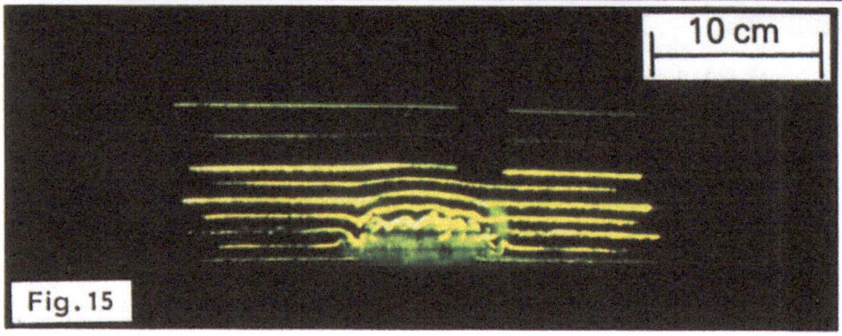

Figure 14. Cross-sectional view in the x-y plane of the turbulent spot; $z = 10\delta_L$; $R_x \approx 5 \times 10^5$. (From Gad-el-Hak et al., 1981)

Figure 15. Spanwise section of the spot in the y-z plane; Flow is out of the plane of the photograph; $R_x \approx 5 \times 10^5$. (From Gad-el-Hak et al., 1981)

into the boundary layer between two eddy structures. Similar conclusions have been drawn by Falco (1977) for entrainment in a fully-developed turbulent boundary layer at moderate Reynolds numbers. Some of the motion in the irrotational region in the vicinity of the spot can be inferred from the photograph in Figure 14. The outermost dye layer, located at approximately two turbulent boundary layer thicknesses from the boundary, is observed to have been displaced about 20 per cent as the spot passes below. This is consistent with the strong correlation in the normal velocity component extending into the potential region which was reported by Blackwelder and Kovasznay (1972) in a turbulent boundary layer. Downstream of the nose of the spot, dye lines at $y \simeq \delta_t$ are displaced downward, consistent with the ensemble-averaged results of Wygnanski et al. (1976).

A y-z cross-section of a spot is shown in Figure 15, taken from Gad-el-Hak et al. (1981). Again the spot appears to be more like a collection of random eddies. Also, from the displacement of dye lines in the irrotational region, it can be concluded that, above the spot, there is motion away from the boundary, while near the edges there is motion towards the boundary.

From conducting experiments using a variety of different visualization methods and from examining different aspects of the flow fields in turbulent spots, Gad-el-Hak et al. (1981) could find no strong evidence that the flows were dominated by one or two prominent vortical structures. To the contrary, the dynamics within the spot appeared to be controlled by many individual eddies, similar to those within a turbulent boundary layer.

The view that a turbulent spot consists of many vorticies is shared by Matsui (1980), who used hydrogen bubble techniques to visualize the spots in a water tunnel. Matsui suggested rather diverse possibilities for the configuration of vortex-like substructures that exist within the spot. He hypothesized that the generation of new vortices in the rear of a turbulent spot caused the trailing edge propagation speed to be lower than that of the leading edge. This velocity difference caused the longitudinal growth of the spot, while the generation and outward shift of new vortices on both sides of the spot caused its transverse growth.

Based on the recent flow visualization resuls, Wygnanski (1981) has concurred with the present authors' view that the treatment of the turbulent spot as a single entity (large horseshoe vortex) is an over-simplification. He states that the entrainment calculated by considering the average velocity in a spot gives only an overall integral quantity from which very little can be inferred about the kinematics of the process.

Wygnanski, Zilberman & Haritonidis (1982) provided additional data on the rate of growth of a turbulent spot. They also conclude that a similarity approach based on ensemble-averaged data is severely limited. It might be used to predict the overall scales and flow field, but much more sophisticated data-processing techniques are required to describe the structure of the spot. Itsweire (1983) and Itsweire & Van Atta (1983) attempted such a data-processing method. They developed a discriminative averaging technique to construct a "statistically most probable" spot with sufficient resolution to include some of the largest substructures detected in visual studies. They suggested the existence of considerable phase coherence between the structure of the most probable spot and vortical motions on the next smaller scale.

The picture of the spot that emerges from these various visualization and probe measurement studies is that (i) the interior of the spot is very similar to a turbulent boundary layer; (ii) that no large-scale predominate vortical structures are observed; (iii) that the growth of the spot normal to the boundary appears to be by turbulent entrainment, as in a turbulent boundary layer; and (iv) the growth of the spot laterally is due to some mechanism other than entrainment.

Gad-el-Hak et al. (1981) performed a series of visualization experiments to establish whether the lateral growth was by entrainment, or some other mechanism. The process of entrainment, as first studied in detail by Corrsin & Kistler (1955), has the following characteristics:

(1) The turbulent region is separated from the nonturbulent region by a distinct interface.
(2) The process of entrainment must occur by direct contact, i.e., by the local diffusion of vorticity, since the turbulent region is highly vortical and the nonturbulent region is irrotational.
(3) If a passive scalar is introduced into an entire turbulent region, then because of turbulent mixing it will also mark new fluid acquired by entrainment.

Gad-el-Hak et al. (1981) used this latter property of entrainment to determine whether the lateral growth of the spot was due to entrainment. In one experiment, a blue dye was added to the fluid in the solenoid valve used to generate the initial patch of turbulence, and hence mark the early spot. The same spot was also observed using a red dye seeped from the slot upstream of the spot initiation location. The latter dye marked the entire spot. As the spot grew, the region marked with the blue dye grew very little, while the spot was rapidly growing. This indicated clearly that fluid was being added to the spot by a process other that entrainment. In another experiment, the dye slot was partitioned into four different regions, and different color dyes were issued from each slot. The lateral diffusion of the dye was observed to be much slower than the lateral growth of the spot. Since entrainment proceeds at a rate similar to turbulent diffusion, this again indicated the presence of an alternative growth process. In order to investigate this mechanism further, Gad-el-Hak et al. (1981) also carried out studies of the turbulent wake of a roughness element placed in a laminar boundary layer. This work is discussed in the next section.
From the ensemble-averaged data of Cantwell et al. (1978), if h denotes a characteristic length scale of the spot in the direction normal to the wall, then the growth rate of h in the downstream direction is approximately 0.013. On the other hand, the growth rate of the spot in the lateral diection can be estimated from the maximum angle subtended by the spot as measured from its virtual origin. This growth rate is found to be approximately 0.18, an order of magnitude larger than the growth rate in the normal direction. Thus it is clear

that the spot does not grow laterally by entrainment, but by some other mechanism that provides much more rapid growth. From the visualization studies, this mechanism appears to be a breakdown of the laminar but unstable flow in the vicinity of the spot (as suggested by Corrsin & Kistler, 1955), and has thus been termed "Growth by Destabilization" by Gad-el-Hak et al. (1981).

5. Relationship to Other Flows

Turbulent spots in laminar boundary layers are dynamically similar to a number of other flows. Exploring these similarities can help in the understanding of the dynamics of turbulent spots; also, conclusions about turbulent spots have ramifications with regard to other flows.

Perhaps the flow most similar to a turbulent spot in a laminar boundary layer is a turbulent spot in an otherwise laminar plane channel flow. Carlson, Widnall & Peeters (1982) have carried out experiments in a channel flow at a Reynolds number of about 1000. The spots were generated artificially in an otherwise laminar flow, and were visualized using small, disc-shaped mica particles. This visualization approach, which is similar to that of Cantwell et al. (1978) discussed in the previous section, allows the viewing of the instantaneous flow field. The laminar channel flow has significantly different stability characteristics than a laminar boundary layer. Furthermore the Reynolds number is constant in the flow direction, while the Reynolds number in a laminar boundary layer grows continually in the flow direction.

The spots in the channel flow quickly filled the depth of the channel, so that growth normal to the channel boundaries became unimportant; only the lateral spread was of significance in determining the spot shape. This shape was more oval than the arrowhead shape observed in the boundary layer case. The spot subtended a half-angle of about 8°, slightly smaller than the corresponding angle in a laminar boundary layer. The front of the spot moved at about 2/3 of the centerline speed, while the rear interface moved at about 1/3 of this speed. These values are both somewhat less than corresponding values for a turbulent spot in a laminar boundary layer. Interestingly, after the spot had grown to about 35 times the channel depth, it split into two spots, a phenomenon not observed in a laminar boundary layer.

The lateral growth of the spot appeared to be due to the instability of the flow, similar to the boundary layer case. It was found that the spot died out if the flow speed was below the stability limit, a phenomenon which also was observed by the present authors for a spot in a laminar boundary layer. Strong oblique waves were found both ahead of and to the rear of the spot. The waves at the front of the spot were observed to break down into turbulence, thus increasing the size of the spot.

Other flows closely related to turbulent spots are turbulent slugs and puffs observed in otherwise laminar pipe flows. Wygnanski & Champagne (1973) studied pipe flow at Reynolds numbers in the range of onset to turbulence. They observed intermittant regions of turbulence in otherwise laminar flow. These turbulent regions, termed turbulent slugs, were found to naturally occur for Re > 5,000, but could be initiated at Re > 3,200 by introducing disturbances at the inlet. The slugs occupied the entire cross-section of the pipe, and grew rapidly as they proceeded downstream. The leading and trailing fronts of the slugs were very sharp and well-defined, while the interior of the slugs appeared identical to fully-developed, turbulent pipe flow. It is the present authors' opinion that these slugs grow in the flow direction in much the same manner as do turbulent spots, i.e., by instabilities in the flow and not solely by entrainment. Near the leading and trailing fronts of the slug, velocity profiles develop inflections, hinting that instabilities are important. Furthermore, the slugs cannot be maintained below a critical Reynolds number. And finally, Wygnanski and Champagne found that, in addition to turbulent diffusion, pressure diffusion was necessary to explain the propagation of the front. The pressure diffusion could be related to pressure fluctuations in the unstable laminar flow induced by turbulence in the nearby slug, causing the flow to break down into turbulence, and thus propagate the front.

For Reynolds numbers in the range of 2,000 to 3,200, Wygnanski & Champagne (1973) and also Wygnanski et al. (1975) found that large disturbances still produce a turbulent region, although one that is less vigorous than that for a slug. These regions, termed turbulent puffs, do not have well-defined leading and trailing interfaces, and are convected at a speed lower than the mean speed. Thus at the front of the puff it may not be growing, but possibly relaminarizing. Puffs, which have sometimes been observed to decay, have also been observed to split, reminiscent of spot-splitting in a channel flow.

Another phenomenon closely related to a turbulent spot is the turbulent wake of a roughness element placed in otherwise laminar boundary layer. When laminar boundary layers have speeds above the laminar stability limit, turbulent, V-shaped wedges are often observed, for example, issuing from small imperfections in the surface of airfoils, or from other fixed obstructions in the flow. Charters (1943) was the first to report the observation of these turbulent wedges. He found, while studying boundary layer transition on flat and curved surfaces, that turbulent, V-shaped wedges occurred behind local fixed disturbances in the flow. These wedges spread laterally at an approximately constant rate, and, once generated, appeared to be independent of the generating source. Because these turbulent regions contaminated his laminar boundary layer by spreading transversely to the boundary, Charters referred to this phenomenon as transition by transverse contamination. He found that the wedges grew at a half-angle of approximately 9.5°, and suggested that vortical fluctuations in the wedge contaminated adjacent regions, causing the wedges to spread transversely.

Schubauer & Klebanoff (1956), as part of a study of turbulent spots, investigated the turbulent wake of obstructions placed on the boundary surface. They found that at lower speeds the wakes possessed a turbulent core subtending a half-angle of about 6.4°, with a highly intermittant region extending out to 10.6°. For higher speeds or larger obstacles, the fully-developed core extended out to 10.6°. In the intermittant regions, the turbulence would arrive with an abrupt increase in velocity, and end with a gradual decrease, a behavior very similar to the signals they had measured as a spot passed. Within the turbulent core they found that the mean profiles were identical to those in a fully-turbulent boundary layer. Noting the similiarities between turbulent spots and turbulent wedges, they suggested that a turbulent wedge was just a succession of turbulent spots.

In addition to studying turbulent spots, Gad-el-Hak et al. (1981) also examined the wakes of roughness elements placed in laminar boundary layers. In one experiment a small roughness element, coated with a red dye, was placed in an otherwise laminar boundary layer. The wake of the rougness element was easily visualized from the dye, and found to subtend a half-angle of approximately 2°. The total turbulent region behind the roughness element was observed under indentical conditions by seeping blue dye from a spanwise slot upstream of the roughness element. The turbulent region extended significantly beyond the roughness wake, subtending a half-angle of approximately 6°. Since the growth of the wake region (red dye) is due to turbulent diffusion, which proceeds at a rate closely related to that of turbulent entrainment, this demonstrates that some other mechanism is causing the rapid lateral growth in the turbulent (blue dye) region. When experiments were carried out at towing speeds below the critical speed predicted by linear stability theory, the turbulent region could not be sustained. This indicated that the additional mechanism was an instability process, probably one in which the turbulence eddies near the inside boundaries of the wedge (or spot)induce strong disturbances in the adjacent unstable laminar boundary layer, causing the flow to break down into turbulence and become part of the wedge.

In the films taken frome these wake experiments, Gad-el-Hak et al. noted that structures which appeared similar to the wing tips of spots would occasionally pass by. These structures subtended a half-angle of approximately 10°, and were probably the same intermittant phenomenon observed by Schubauer & Klebanoff (1956) with their probe measurements.

In order to obtain more quantitative information about the wake of roughness elements, Gad-el-Hak et al. (1981) studied the wake of a heated roughness element placed in a wind tunnel boundary layer. Both the temperature wake (corresponding to the red-dye wake) and the velocity wake (corresponding to the blue-dye wake) were explored by traversing a thermometer and a hot-wire probe in the transverse, z, direction. Transverse profiles of the mean and fluctuating temperature and velocity fields were obtained for a range of downstream locations. The structures of the temperature and velocity wakes were significantly different, the temperature wake being much narrower than the velocity wake. Figure 16 gives the result for the widths of the fluctuating temperature and velocity wakes taken from these experiments. The wake width was defined as the location at which the profile had decreased to one-half its maximum value. The thermal wake spread at a rate having a half-angle of about 1.3°, while the velocity wake spread at a half-angle of approximately 8°, consistent with the visualization results.

The sequence of events associated with a turbulent spot initiation (Figure 3) is similar to the transition route of a decelerating

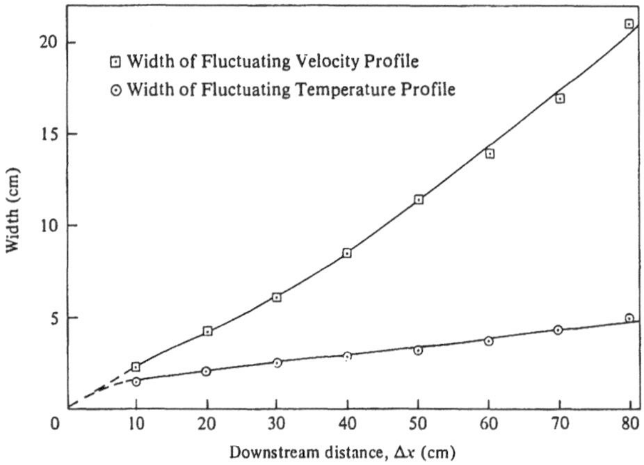

Figure 16. Growth of the thermal wake and the turbulent wedge. (From Gad-el-Hak et al., 1981)

laminar boundary layer as described by Gad-el-Hak, Davis, McMurray & Orszag (1984). Their Figure 4 depicts a laminar boundary layer that becomes unstable to two-dimensional waves when decelerated. These waves break down into three-dimensional patterns, hairpin vortices and finally turbulent bursts when the vortices lift off the wall. Liepmann, Brown & Nosenchuck (1982) observed a similar transition process initiated by a dynamic-heating technique.

The above described transition events have strong resemblance to the intermittent events that characterize fully-developed turbulent boundary layers; namely the bursting cycle (Blackwelder & Kaplan, 1976). As discussed in the previous section, a common structure that can be readily identified in both transitional and turbulent boundary layers is the low-speed streak (Blackwelder & Eckelmann, 1979). The streaks' presence in a turbulent spot is evidenced in the schematic in Figure 5c, and the dye photograph in Figure 8. Low-speed streaks are low-momentum regions existing near the wall and are believed to be caused by the pumping action of the counter-rotating streamwise vortices that are known to exist in the wall region (Bakewell & Lumley, 1967). The origin of these vortices developing in a boundary layer is presently unknown, with the exception of a boundary layer developing on a concave wall where the generation mechanism has been identified as a Görtler instability (Görtler, 1941). Previous attempts to establish a quantitative analogy between the low-speed streaks in transitional and turbulent boundary layers have failed. However, Blackwelder (1983) argues that the reason for the apparent lack of success is the use of a boundary layer thickness as a length scale. Blackwelder has shown that, when viscous scales are used to normalize the length and other parameters, striking similarities do indeed exist between transitional and turbulent eddy structures near the wall.

6. Conclusions

A significant amount of information concerning turbulent spots has been obtained since their discovery by Emmons in 1951. This information, almost totally experimental, gives both the (ensemble) average picture of the spot as well as details of the the underlying structure, and has led to at least a qualitative understanding of the basic dynamics of turbulent spots.

Most of the overall features of a turbulent spot are summarized in Figure 17, taken from Gad-el-Hak et al. (1981). This figure presents a qualitative sketch of an x-y cut through the spot. For dynamical considerations it is useful to divide the flow field in and near the spot into at least 5 different regions. It has been found that the

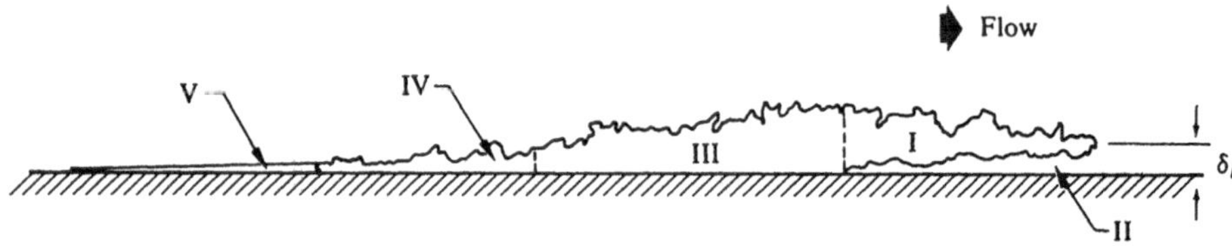

Figure 17. Schematic view of an x-y cut through the turbulent spot. (From Gad-el-Hak et al., 1981)

primary internal part of a spot (Region III in the Figure) is dynamically very similar to a fully-developed turbulent boundary layer, having the same mean velocity and turbulence intensity profiles, the same growth rates away from the boundary, and internal structure that visually appears the same. Even streaks of approximately the same length have been observed, although no study has as yet compared bursts in the spot with those in a turbulent boundary layer. Thus, our understanding of fully-developed turbulent boundary layers can probably be carried over to understand the internal dynamics of the spot.

The unique features of the spot occur at or near its boundaries. Region I, the overhang region, originally was part of Region III until it was convected over the laminar boundary layer. It appears to be turbulent, although separated from the boundary, and hence from its primary source of energy. Region I appears to induce diturbances in Region II, the laminar boundary layer below and ahead of the spot, causing the flow in this region to breakdown into turbulence and hence become part of the spot. The method by which this breakdown occurs is still not clear. Region IV appears to contain somewhat active turbulence, which was once part of Region III, but whose upper part has been convected away due to shear. Finally, Region V is the "calmed" region behind the spot, which also was once part of Region III, but now mainly consists of longitudinal streaks along the boundary. An important region which is not shown in this x-y cut is the region in the vicinity of the wingtips (see, e.g., Figure 10). The dynamics in this region govern the overall lateral spread (and hence maximum angle subtended by the spot). There is apparently little or no overhang in this region, and the spot appears to spread through instabilities in the adjacent flow, perhaps as suggested by Matsui (1980) or Perry et al. (1981). Again the dynamics of this process are not well understood.

From ensemble-averaged measurements of the velocity field in the spot (see, e.g., Figure 7), the spot appears to consist of two dominant vortices which control the dynamics of the spot. Further experiments, however, especially visualization experiments, show that this view of the spot may be misleading. These vortices are never observed in the visualizations, and, although the vortices exist in the ensemble-averaged sense, the fluctuating motions about this average appear to be the controlling dynamic features.

To better understand and predict the behavior of spots, there is a need for further experimental, numerical, and theoretical work. Experimentally, there is a need for ingeneous experiments to elucidate the nature of the instability process(es) by which the spot grows laterally. For example, is the breakdown due to the growth of Tollmien-Schlictung waves, or is it a more catastrophic breakdown into turbulence? The present authors think it is probably the latter. Also, more understanding could be obtained by subjecting the spot to a variety of different conditions, for example, favorable or adverse pressure gradients, and accelerating or decelerating boundaries. The present understanding implies certain specific responses; for example, a favorable pressure gradient could stabilize the boundary layer and thus inhibit the spot growth. Some work along these lines has been carried out by Gaster (1967) and Wygnanski (1981). Numerical experiments, e.g., Leonard (1981) and Orszag (private communication), also have the potential for shedding considerable light on the underlying mechanisms.

There has been very little theoretical work on the spot. Early stability theories, for example, Criminale and Kovasznay (1962) and Gaster (1975), have led to accurate prediction of the linear stage of the incipient spot. But at the present there is no predictive theory for the fully-developed spot. For example, there is no theory which correctly predicts even the proper angle of spread. Such a theory probably may have to inherently be nonlinear, and may have to include the pressure disturbances induced in the laminar flow by adjacent turbulent flow in the spot, and the subsequent nonlinear breakdown of this laminar flow.

Present turbulence closure models (e.g., k-ε models) probably cannot handle turbulent spots, because the physics of the growth by destabilization is not included in the models. However, recent modifications of these models to take into account the interface between turbulent and nonturbulent flows using the intermittancy function (e.g., Dopazo & O'Brien, 1979; Kollman, 1984) might be able to accurately treat the spot dynamics, although clearly a number of ad hoc assumptions would still be required.

This work is sponsored by the Air Force Office of Scientific Research Contract No. F49620-78-C-0062, and the National Science Foundation Grant No. MEA-8319804.

References

Amini, J. (1978) "Transition Contrôlée en Couche Limite: Etude Expérimentale du Développement d'une Perturbation Tridimensionnelle Instantanée," Docteur-Ingénieur Thése, L'Institut National Polytechnique de Grenoble.

Antonia, R. A., Chambers, A. J., Sokolov, M. & Van Atta, C. W. (1981) "Simultaneous Temperature and Velocity Measurements in the Plane of Symmetry of a Transitional Turbulent Spot," J. Fluid Mech. 108, p. 317.

Bakewell, H. P. & Lumley, J. L. (1967) "Viscous Sublayer and Adjacent Wall Region in Turbulent Pipe Flow," Phys. Fluids 10, p. 1880.

Benjamin, T. B. (1961) "The Development of Three-Dimensional Disturbances in an Unstable Film of Liquid Flowing Down an Inclined Plane," J. Fluid Mech. 10, p. 401.

Blackwelder, R. F. (1983) "Analogies between Transition and Turbulent Boundary Layers," Phys. Fluids 26, p. 2807.

Blackwelder, R. F. & Eckelmann, H. (1979) "Streamwise Vortices Associated with the Bursting Phenomenon," J. Fluid Mech. 94, p. 577.

Blackwelder, R. F. & Kaplan, R. E. (1976) "On the Wall Structure of the Turbulent Boundary Layer," J. Fluid Mech. 76, p. 89.

Blackwelder, R. F. & Kovasznay, L. S. G. (1972) "Time Scales and Correlations in a Turbulent Boundary Layer," Phys. Fluids 15, p. 1545.

Cantwell, B., Coles, D. & Dimotakis, P. (1978) "Structure and Entrainment in the Plane of Symmetry of a Turbulent Spot," J. Fluid Mech. 87, p. 641.

Carlson, D. R., Widnall, S. E. & Peeters, M. F. (1982) "A Flow-Visualization Study of Transition in Plane Poiseuille Flow," J. Fluid Mech. 121, p. 487.

Charters, A. C. (1943) "Transition between Laminar and Turbulent Flow by Transverse Contamination," N.A.C.A. Tech. Note No. 891.

Chen, C.-H. P. & Blackwelder, R. F. (1978) "Large-Scale Motion in a Turbulent Boundary Layer: a Study Using Temperature Contamination," J. Fluid Mech. 89, p. 1.

Coles, D. & Barker, S. J. (1975) "Some Remarks on a Synthetic Turbulent Boundary Layer," in Turbulent Mixing in Nonreactive and Reactive Flows, ed. S. N. B. Murthy, Plenum, p. 285.

Corrsin, S. & Kistler, A. L. (1955) "Free-Stream Boundaries of Turbulent Flows," N.A.C.A. Rep. No. 1244. (Supersedes N.A.C.A. TN 3133.)

Criminale, W. D. & Kovasznay, L. S. G. (1962) "The Growth of Localized Disturbances in a Laminar Boundary Layer," J. Fluid Mech. 14, p. 59.

Dopazo, C. & O'Brien, E. E. (1979) "Intermittancy in Free Turbulent Shear Flows," in Turbulent Shear Flows I, eds. F. Durst et al., Springer, p. 6.

Elder, J. W. (1960) "An Experimental Investigation of Turbulent Spots and Breakdown to Turbulence," J. Fluid Mech. 9, p. 235.

Emmons, H. W. (1951) "The Laminar-Turbulent Transition in a Boundary Layer," J. Aero. Sci. 18, p. 490.

Falco, R. E. (1977) "Coherent Motions in the Outer Region of Turbulent Boundary Layer," Phys. Fluids 20, p. S124.

Fiedler, H. E. & Head, M. R. (1966) "Intermittency Measurements in a Turbulent Boundary Layer," J. Fluid Mech. 25, p. 719.

Gad-el-Hak, M., Blackwelder, R. F. & Riley, J. J. (1980) "A Visual Study of the Growth and Entrainment of Turbulent Spots," in Laminar-Turbulent Transition, eds. R. Eppler and H. Fasel, Springer, p. 297.

Gad-el-Hak, M., Blackwelder, R. F. & Riley, J. J. (1981) "On the Growth of Turbulent Regions in Laminar Boundary Layers," J. Fluid Mech. 110, p. 73.

Gad-el-Hak, M., Davis, S. H., McMurray, J. T. & Orszag, S. A. (1984) "On the Stability of the Decelerating Laminar Boundary Layer," J. Fluid Mech. 138, p. 297.

Gaster, M. (1967) "On the Flow Along Swept Leading Edges," Aero. Quart. 18, p. 165.

Gaster, M. (1975) "A Theoretical Model of a Wave Packet in the Boundary Layer on a Flat Plate," Proc. Roy. Soc. A 347, p. 271.

Gaster, M. (1978) "The Physical Processes Causing Breakdown to Turbulence," 12th Naval Hydrodynamics Symposium, Washington, D.C., p. 22.

Gaster, M. (1981) "On Transition to Turbulence in Boundary Layers," in Transition and Turbulence, ed. R. E. Meyer, Academic Press, p. 95.

Gaster, M. & Grant, I. (1975) "An Experimental Investigation of the Formation and Development of a Wave Packet in a Laminar Boundary Layer," Proc. Roy. Soc. A 347, p. 253.

Görtler, H. (1941) "Instabilität Laminarer Grenzschichten an Konkaven Wanden Gegenü ber Gewissen Dreidimensionalen Störungen," Zeitschrift für Angewandte Math. & Mech. 21, p. 250.

Head, M. R. & Bandyopadhyay, P. (1978) "Combined Flow Visualization and Hot-Wire Measurements in Turbulent Boundary layers," in Coherent Structure of Turbulent Boundary Layers, ed. C. R. Smith & D. E. Abbott, Lehigh Univ., p. 98.

Itsweire, E. C. (1983) "An Investigation of the Coherent Structures Associated with a Turbulent Spot in a Laminar Boundary Layer," Ph.D. Thesis, Univ. of California - San Diego.

Itsweire, E. C. & Van Atta, C. W. (1983) "The Effect of Different Similarity Growth Transformations on Ensemble Mean Particle Paths in Turbulent Spots," J. de Physique-Letters 44, p. 917.

Klebanoff, P. S., Tidstrom, K. D. & Sargent, L. M. (1962) "The Three-Dimensional Nature of Boundary Layer Instability," J. Fluid Mech. 12, p. 1.

Kollman, W. (1984) "Intermittant Turbulent Flows," in these proceedings.

Kovasznay, L. S. G., Kibens, V. & Blackwelder, R. F. (1970) "Large-

Scale Motion in the Intermittent Region of a Turbulent Boundary Layer," J. Fluid Mech. 41, p. 283.

Kovasznay, L. S. G., Komoda, H. & Vasudeva, B. R. (1962) "Detailed Flow Field in Transition," Proc. Heat Transfer and Fluid Mech. Inst., Stanford Univ. Press, p. 1.

Leonard, A. (1980) "Vortex Simulation of Three-Dimensional Spotlike Disturbances in a Laminar Boundary Layer," in Turbulent Shear Flows II, eds. J. S. Bradbury et al., Springer, p. 67.

Leonard, A. (1981) "Turbulent Structures in Wall-Bounded Shear Flow Observed via Three-Dimensional Numerical Simulations," in Lect. Notes in Physics 136, ed. J. Jimenez, Springer, p. 119.

Liepmann, H. W., Brown, G. L. & Nosenchuck, D. M. (1982) "Control of Laminar-Instability Waves Using a New Technique," J. Fluid Mech. 118, p. 187.

Matsui, T. (1980) "Visualization of Turbulent Spots in the Boundary Layer Along a Flat Plate in a Water Flow," in Laminar-Turbulent Transition, eds. R. Eppler & H. Fasel, Springer, p. 288.

Mautner, T. S. & Van Atta, C. W. (1982) "An Experimental Study of the Wall-Pressure Field Associated with a Turbulent Spot in a Laminar Boundary Layer," J. Fluid Mech. 118, p. 59.

Mitchner, M. (1954) "Propagation of Turbulence from an Instantaneous Point Disturbance," J. Aero. Sci. 21, p. 350.

Perry, A. E., Lim, T. T. & Teh, E. W. (1981) "A Visual Study of Turbulent Spots," J. Fluid Mech. 104, p. 387.

Schubauer, G. B. & Klebanoff, P. S. (1956) "Contributions on the Mechanics of Boundary Layer Transition," N.A.C.A. Rep. No. 1289.

Schaubauer, G. B. & Skramstad, H. K. (1948) "Laminar Boundary Layer Oscillations on a Flat Plate," N.A.C.A. Rep. No. 909.

Theodorsen, T. (1952) "Mechanism of Turbulence," Proc. 2nd Midwestern Conf. Fluid Mech. Ohio State Univ. p. 1.

Theodorsen, T. (1955) "The Structure of Turbulence," in 50 Jahre Grenzschichtforschung, ed. H. Görtler & W. Tollmien, Braunschweig: Vieweg & Son, p. 55.

Tollmein, W. (1931) "The Production of Turbulence," N.A.C.A. TM No. 609.

Townsend, A. A. (1976) "The Structure of Turbulent Shear Flow," Cambridge Univ. Press.

Van Atta, C. W. & Helland, K. N. (1980) "Exploratory Temperature-Tagging Measurements of Turbulent Spots in a Heated Laminar Boundary Layer," J. Fluid Mech. 100, p. 243.

Vasudeva, B. R. (1967) "Boundary-Layer Instability Experiment with Localized Disturbance," J. Fluid Mech. 29, p. 745.

Wygnanski, I. (1981) "The Effects of Reynolds Number and Pressure Gradient on the Transitional Spot in a Laminar Boundary Layer," in Lecture Notes in Physics 136, ed. J. Jimenez, Springer, p. 304.

Wygnanski, I. & Champagne, F. H. (1973) "On Transition in a Pipe. Part 1: The Origin of Puffs and Slugs and the Flow in a Turbulent Slug," J. Flud Mech. 29, p. 281.

Wygnanski, I., Haritonidis, J. H. & Kaplan, R. E. (1979) "On Tollmien-Schlichting Wave Packet Produced by a Turbulent Spot," J. Fluid Mech. 92, p. 505.

Wygnanski, I., Sokolov, M. & Friedman, D. (1975) "On Transition in a Pipe. Part 2: The Equilibrium Puff," J. Fluid Mech. 69, p. 283.

Wygnanski, I., Sokolov, M. & Friedman, D. (1976) "On a Turbulent 'Spot' in a Laminar Boundary Layer," J. Fluid Mech. 78, p. 785.

Wygnanski, I., Zilberman, M. & Haritonidis, J. H. (1982) "On the Spreading of a Turbulent Spot in the Absence of a Pressure Gradient," J. Fluid Mech. 123, p. 69.

Zilberman, M., Wygnanski, I. & Kaplan, R. E. (1977) "Transitional Boundary Layer Spot in a Fully Turbulent Environment," Phys. Fluids 20, p. S258.

Spectral and Statistical Characteristics of Breaking Waves

O.M. Phillips

Department of Earth and Planetary Sciences, The Johns Hopkins University
Baltimore, MD 21218, USA

In the equilibrium range of wind-generated waves, it is postulated that the processes of energy (or action) input from the wind, loss by wave breaking and net transfer by non-linear resonant wave interactions are of comparable importance throughout the range. Consideration of the action spectral density balance then indicates that the wave-number spectrum in this range is proportional to $(\cos\theta)^{1/2} u_* g^{-1/2} k^{-7/2}$, where θ is the angle between the wind and the wave-number k, and the frequency spectrum is of the form found empirically by Toba (1973), namely $u_* g^{-4}$. These forms have also been derived by Kitaigorodskii (1983) though on a quite different physical basis. The spectral rate of energy loss by wave breaking is found to be proportional to $(\cos\theta)^{3/2} u_*^3 k^{-2}$ and the spectral rate of momentum loss from the waves to $(\cos\theta)^{5/2} g^{-1/2} u_*^3 k^{-3/2}$. As the wave field develops with increasing fetch or duration, the total rate of energy input to the water turbulence by wave breaking increases as $\rho_a u_*^3 \ln(k_1/k_0)$ where k_1 and k_0 are the upper and lower wave-number limits to the range; the total momentum flux increases also but asymptotes to a fixed fraction of $\rho_a u_*^2$. The various constants of proportionality are found in terms of Toba's constant and a coefficient expressing the rate of energy input from wind to waves.

The statistical distribution of breaking fronts is also considered. The average total length of breaking fronts per unit area with speeds of advance between c_0 and c_1 is proportional to $u_*^3 g (c_0^{-5} - c_1^{-5})$; because of the strong dependence on c, easily visible whitecaps for which c is above a threshold value c_T constitute only a small fraction of the total breaking events. The total length of whitecap fronts per unit area is proportional to $u_*^3 g c_T^{-5}$ when the speed of the fastest breaking fronts is significantly larger than c_T; the number of actively breaking whitecaps passing a given point per unit time to $u_*^3 g c_T^{-4}$ and the whitecap coverage to $u_*^3 g T c_T^{-4}$, where T is the average duration of a bubble patch.

1. Introduction

The breaking of waves is a process that is ubiquitous over two-thirds of the surface of the globe. It is clearly responsible for part of the transfer of mechanical energy and of momentum from the atmosphere to ocean currents and turbulence, for the enhancement of heat transfer and especially the exchange of gases between the atmosphere and the ocean as well as augmenting substantially but locally the drag of the air on the water itself (Banner and Melville, 1976).

In the past few years a great deal of attention has been paid to the dynamics of breaking and the search for criteria under which waves might be expected to break. The remarkable and pioneering theory and numerical experiments of Longuet-Higgins and Cokelet (1976) have traced the evolution of finite amplitude irrotational waves on deep water, either as a result of their intrinsic instabilities or of impulsive forcing to the point of wave breaking and just beyond. Less fundamental have been attempts to find a single threshold variable such as local vertical acceleration, or combination of such variables, which determine the probability of breaking of an individual wave crest. This concept lay behind the original idea that led to the simple $g^2\sigma^{-5}$ saturation spectrum proposed a number of years ago. The idea has been taken a great deal further in other directions with interesting success in a series of three papers in 1983 written by Snyder, Kennedy and Smith in various combinations. It does remain difficult, though, to associate any single local variable with the examples of breaking calculated by Longuet-Higgins and Cokelet; it seems that the recent time history of the surface configuration is more pertinent than a single local threshold variable.

In this paper, a rather different approach is taken, more in the spirit of Hasselmann (1974) in which the detailed configuration at the point of incipient breaking is ignored - it disappears anyway as soon as the wave breaks - while concentrating on the statistical consequences of the ensemble of breaking events at various points on the sea surface. The initial goal is to use simple dynamical reasoning to provide as reliable an estimate as possible for the average rate of spectral energy loss resulting from breaking; in turn, this leads to the form of the high frequency spectrum of gravity waves that was inferred empirically by Toba (1973) and also to a series of simple expressions for quantities such as the average length of breaking lines per unit area at any instant and the fraction of surface area turned over per unit time.

2. The Statistical Equilibrium of Short Waves

The spectrum of a random distribution of surface waves can be specified by

$$\Psi(\underset{\sim}{k}) = (2\pi)^{-2} \int \overline{\zeta(\underset{\sim}{x})\,\zeta(\underset{\sim}{x}+\underset{\sim}{r})}\; e^{-i\underset{\sim}{k}\cdot\underset{\sim}{r}}\, d\underset{\sim}{r},$$

where ζ represents the local surface displacement and the integral is over the entire separation ($\underset{\sim}{r}$) plane. The dynamics of the field is, however, more conveniently described (particularly when wave-current interactions are involved) by the balance of action spectral density

$$N(\underset{\sim}{k}) = (g/\sigma)\,\Psi(\underset{\sim}{k}) = (g/k)^{1/2}\,\Psi(\underset{\sim}{k}), \tag{2.1}$$

where σ is the intrinsic frequency and the water density is divided out throughout. Following energy paths (see, for example, Phillips, 1980),

$$\frac{dN}{dt} = \frac{\partial N}{\partial t} + (\underset{\sim}{C}+\underset{\sim}{U})\cdot\nabla N = -\nabla_k\cdot\underset{\sim}{T}(\underset{\sim}{k}) + S_w - D, \tag{2.2}$$

where $\underset{\sim}{C}$ is the local group velocity. The various processes that modify the action spectral density following a wave group are represented on the right. $T(\underset{\sim}{k})$ represents the spectral flux of action resulting from resonant wave-wave interactions. These exchanges are conservative for gravity waves and the integral of this term over all wave-numbers vanishes. The rate of spectral input of wave action from the wind is expressed schematically by the term S_W and D represents the rate of loss by wave breaking and possibly the formation of parasitic capillaries at large gravity wave-numbers.

For those components at wave-numbers large compared with that of the spectral peak, in a well-developed wave field under the continued action of the wind, the time scales of their growth are long compared with the internal time scales involved in wave-wave interactions, action input from the wind and loss by breaking, so that for these components the spectral balance reduces to

$$-\nabla_k\cdot\underset{\sim}{T}(\underset{\sim}{k}) + S_w - D = 0. \tag{2.3}$$

In this equilibrium range, the detailed functional forms of each of these terms would be expected to depend on the nature of the spectrum N(k) in this range and it is of interest to enquire what spectral characteristics are associated with the possible balances among the three terms of (2.3).

The spectral re-distribution of wave action has been the subject of pioneering investigations by Hasselmann (1962, 1968) and others; it can be represented as a "collision integral" over sets of four resonantly interacting gravity waves:

$$- \nabla_k \cdot \underset{\sim}{T}(\underset{\sim}{k}) = \iiint Q^2 \{[N(\underset{\sim}{k}) + N(\underset{\sim}{k}_1)] N(\underset{\sim}{k}_2) N(\underset{\sim}{k}_3) - $$
$$- [N(\underset{\sim}{k}_2) + N(\underset{\sim}{k}_3)] N(\underset{\sim}{k}) N(\underset{\sim}{k}_1)\} \times$$
$$\times \delta(\underset{\sim}{k} + \underset{\sim}{k}_1 - \underset{\sim}{k}_2 - \underset{\sim}{k}_3)\, \delta(\sigma + \sigma_1 - \sigma_2 - \sigma_3)\, d\underset{\sim}{k}_1 d\underset{\sim}{k}_2 d\underset{\sim}{k}_3 \tag{2.4}$$

where the coupling coefficient Q is a complicated homogeneous function of the wave numbers $k, \ldots, k_3$ and is of order k^3 and δ represents the Dirac delta function. Later work by Fox (1976) and Sell and Hasselmann (1972) suggests that the interactions are primarily local in the wave-number plane, so that the net action transfer to a given wave-number interval is determined primarily by the action spectral density in this vicinity. Near the spectral peak, of course, the flux to neighboring wave-numbers is dominated by the peak itself, but in the equilibrium range, the net flux to or from a wave-number band should scale with the local value of N, i.e. N(k). Consequently, since (2.4) is cubic in N and since $Q^2 \sim k^6$, the net spectral flux divergence scales as

$$- \nabla_k \cdot \underset{\sim}{T}(\underset{\sim}{k}) \sim Q^2 N^3 k^4/\sigma = N^3(\underset{\sim}{k})\, k^{19/2} g^{-1/2}, \tag{2.5}$$

as given by Kitaigorodskii (1983). This can be expressed equivalently in terms of the dimensionless function, the "degree of saturation"

$$B(\underset{\sim}{k}) = g^{-1/2} k^{9/2} N(\underset{\sim}{k}) = k^4 \Psi(\underset{\sim}{k}),$$

defined by the author (1984), in terms of which (2.5) becomes

$$- \nabla_k \cdot \underset{\sim}{T} \sim g k^{-4} B^3(\underset{\sim}{k}). \tag{2.6}$$

The rate of action (or energy) input from the wind has been the subject of many theoretical and experimental investigations over the past twenty years which have, if nothing else, demonstrated the complexity and variety of the detailed processes involved. In order to give a simple expression for S_w in (2.3) the best guide seems to be provided by the analysis of careful experiments interpreted in the light of only very general theoretical considerations. Plant (1982) suggests from a survey of such measurements that

$$S_w \doteq 0.04 \cos\theta\, \sigma\, (u_*/c)^2 N(\underset{\sim}{k}), \tag{2.7}$$

where θ is the angle between the wave-number $\underset{\sim}{k}$ and the wind, u_* is the friction velocity of the air flow over the water surface and $c = (g/k)^{1/2}$, the phase velocity of the component concerned. This form has been suggested by others as well; Mitsuyasu and Honda (1984) give a numerical coefficient of 0.05 and Gent and Taylor's (1976) calculation gives approximately 0.07. The form of (2.7) might also be justified on general dynamical grounds. The action and energy fluxes from wind to waves result from variations in surface stresses in phase with the orbital velocities at the surface; with stress variations of order $\rho_a u_*^2$ times the local wave slope, and orbital velocities also proportional to the slope, the net transfer rate must vary as $\rho_a u_*^2 N(k)$. For dimensional consistency, then $S_W \propto (\rho_a/\rho_W)\,\sigma\,(u_*/c)^2\, N(k)$, which, apart from the numerical constant and the directional factor (less certain anyway) reduces to (2.7). In terms of the degree of saturation, this becomes

$$S_w = m \cos\theta \, g k^{-4} (u_*/c)^2 B(\underset{\sim}{k}), \qquad (2.8)$$

where m = 0.04, but may be rather larger.

The development of an expression for the rate of spectral action dissipation is more tentative. The author has argued (1984) that this will depend on the spectral level, represented by B (rather than the wind stress directly) since the occurrence of local breaking and the consequent energy loss is the result of a local excess of energy or action, however this excess is produced. It may, for example, arise from a local convergence in an underlying current which increases the local degree of saturation and consequently the intensity of breaking. In an active wind-generated wave field where wave-current interactions are negligible, the degree of saturation may be enhanced by the wind stress, but the extent to which wave breaking occurs still has as its primary causative property, the degree of saturation B. In the equilibrium range, B may be expected to vary only slowly with wave-number magnitude k, so that in spite of the localness in physical space of the dissipation process, the spectral rate of dissipation of wave action at a given wave-number $\underset{\sim}{k}$ in this range may be considered to be a function of B at that wave-number:

$$D(\underset{\sim}{k}) = g k^{-4} f(B(\underset{\sim}{k})). \qquad (2.9)$$

In summary, then, we have three physical processes that are pertinent to the equilibrium range in an active wind-generated sea, which balance among themselves and which scale as follow:

$$\left.\begin{array}{ll} \text{Spectral flux divergence} & g k^{-4} B^3(\underset{\sim}{k}) \\ \text{Wind input} & m \cos\theta \, g k^{-4} (u_*/c)^2 B(\underset{\sim}{k}) \\ \text{Dissipation} & g k^{-4} f(B(\underset{\sim}{k})) \end{array}\right\} \qquad (2.10)$$

The form of the spectrum in this range depends upon the balances that may exist among these processes, and several alternatives may be visualized.

Kitaigorodskii (1983) has proposed the existence of a Kolmogoroff type of equilibrium range in wind-generated waves in which the energy input from the wind is assumed to occur primarily at the energy-containing scales with dissipation at much larger scales. This then postulates the existence of a range of wave-numbers over which the spectral flux divergence, wind input and dissipation are all negligible; the spectral energy flux ε_0 is constant over this range and the spectral form must be such as to accommodate this constant flux. On similarity grounds he gives for the (directionally averaged) energy spectrum

$$F(k) = \int_{-\pi}^{\pi} \Psi(k,\theta)\, d\theta \sim \varepsilon_0^{1/3} g^{-1/2} k^{-7/2}, \tag{2.11}$$

and for the frequency spectrum

$$S(\sigma) \sim \varepsilon_0^{1/3} g \sigma^{-4}. \tag{2.12}$$

Arguing further that $\varepsilon_0 \propto (\rho_a/\rho_w)U^3$, where U is the mean wind speed, or, approximately that $\varepsilon_0 \propto u_*^3$, he obtains wave-number and frequency spectra of the forms $u_* g^{-1/2} k^{-7/2}$ and $u_* g \sigma^{-4}$ respectively for wave-numbers and frequencies above those at which energy input from the wind occurs and below those for which dissipation is regarded as important.

The principal conceptual difficulty with Kitaigorodskii's argument is the need to postulate that the energy input from the wind is concentrated at wave-numbers close to those of the spectral peak. To be sure, the air flow over the dominant waves may modify the rate of energy input to smaller waves superimposed on the longer ones, but it is difficult to see why it should be suppressed entirely. Indeed, according to (2.7), the time scale for wind energy input is (for $\theta \sim 0$)

$$N(\underset{\sim}{k})/S_w \simeq 25\,(c/u_*)^2 \sigma^{-1} = 25\, g^2/u_*^2 \sigma^3,$$

which decreases rapidly as the frequency increases. Yet very careful measurements of the frequency spectra of wind-generated waves by Toba (1973) and more recently by Forristall (1981), Kahma (1981) and Donelan et al.(1982) indicate strongly that over a considerable range of frequencies higher than that of the spectral peak, the spectrum is much better represented as $g u_* \sigma^{-4}$ than as the $g^2 \sigma^{-5}$ saturation form proposed in 1958 by the author on much simpler dimensional grounds. It seems that the matter demands re-consideration -- twenty-five years is a pretty fair lifetime for a simple idea.

The basic point of this paper is to indicate how an equilibrium spectrum of the Toba type can be derived from a very different assumption about the dynamical balances in the equilibrium range and, as a bi-product, to infer a number of simple properties concerning the statistics of the breaking events themselves. In contrast to the hypothesis made by Kitaigorodskii, let us suppose that in the equilibrium range of an active wind-generated sea, all of the three processes represented in (2.10), namely the spectral flux divergence resulting from wave-wave interactions, wind input and dissipation by wave breaking, are comparable throughout the range. Under this assumption

$$B^3(\underset{\sim}{k}) \propto m\cos\theta\,(u_*/c)^2\,B(\underset{\sim}{k}) \propto f(B(\underset{\sim}{k})), \tag{2.12}$$

whence it follows immediately that

$$B(\underset{\sim}{k}) = \beta\,(\cos\theta)^{1/2}(u_*/c), \tag{2.13}$$

and

$$f(B) = a\,B^3(\underset{\sim}{k}), \tag{2.14}$$

where β and a are numerical constants. (Note that, from the definition, $B(\underset{\sim}{k}) = B(-\underset{\sim}{k})$; in (2.13) $-\pi/2 < \theta < \pi/2$).
This then leads to a wave-number spectrum in the equilibrium range

$$\begin{aligned}\Psi(\underset{\sim}{k}) &= k^{4}B(\underset{\sim}{k}) = \beta\,(\cos\theta)^{1/2}(u_*/c)\,k^{-4},\\ &= \beta\,(\cos\theta)^{1/2}\,u_*\,g^{-1/2}\,k^{-7/2},\end{aligned} \tag{2.15}$$

similar to that given by Kitaigorodskii on a quite different basis. The freqency spectrum can be found from (2.15), although care must be taken to restrict the range of frequencies to those below which the advection by the dominant waves (and the consequent Doppler shifting) becomes significant. The orbital speed of the dominant waves is approximately $2(\overline{\zeta^2})^{1/2}\sigma_0$, where σ_0 is the frequency at the spectral peak, so that Doppler shifting becomes significant for components whose intrinsic phase velocity g/σ is not large compared to this. Accordingly, the frequency spectrum

$$\begin{aligned}\Phi(\sigma) &= 2\int_{-\pi/2}^{\pi/2} k\,\Psi(\underset{\sim}{k})\,(\partial\sigma/\partial k)^{-1}\,d\theta\,\Big|_{k=\sigma^2/g}\\ &= \alpha\,u_*\,g\,\sigma^{-4}, \quad \sigma_0 \ll \sigma \ll g/2(\overline{\zeta^2})^{1/2}\sigma_0,\end{aligned} \tag{2.16}$$

where $\alpha = 4\beta\int_{-\pi/2}^{\pi/2}(\cos\theta)^{1/2}d\theta = 9.36\beta$. This is the form found empirically by Toba (1973) from wind tunnel data and confirmed in field observations by Kawai, Okada and Toba (1977), Donelan et al. (1982) and others. The constant of proportionality measured by Toba in a wind-tunnel was approximately 0.02 and Donelan et al.'s field measurements are consistent with this, although Kawai, Okada and Toba's later field work gives a value of 0.06 ± 0.01. Kawai et al. give some explanation for the difference between this result and Toba's earlier estimate, though the reasons for the discrepancies may still not be well understood.

The expressions (2.13) and (2.14) allow us to estimate the spectral rates of dissipation of wave action, wave energy and wave momentum in the wind direction which are, respectively,

$$\begin{aligned} D(\underset{\sim}{k}) &= g k^{-4} f(B) \\ &= a\beta^3 (\cos\theta)^{3/2} g k^{-4} (u_*/c)^3 \\ &= a\beta^3 (\cos\theta)^{3/2} g^{-1/2} u_*^3 k^{-5/2}, \end{aligned} \quad (2.17)$$

$$\varepsilon(\underset{\sim}{k}) = \sigma D(\underset{\sim}{k}) = a\beta^3 (\cos\theta)^{3/2} u_*^3 k^{-2}, \quad (2.18)$$

and

$$\tau(\underset{\sim}{k}) = (\varepsilon(\underset{\sim}{k})/c)\cos\theta = a\beta^3 (\cos\theta)^{5/2} g^{-1/2} u_*^3 k^{-3/2}. \quad (2.19)$$

In the absence of more complete observational verification, not too much significance should be ascribed to the directional factors given in these expressions, but it is interesting to note that the directional distribution of the equilibrium range energy density that they indicate is quite broad.

The total fluxes of energy and momentum from the wind to the sea occur in three separate pathways: (a) directly by the mean shear stress on the water surface, (b) from wind to waves, resulting in wave growth and radiation from the generating area and (c) from wind to the waves of the equilibrium range, from which it is lost locally from the waves by breaking. The last of these can now be estimated from (2.18) and (2.19). If k_0 represents the lowest wave-number associated with active wave breaking (which may be coincident with that of the spectral peak, but may be somewhat higher) and k_1 is the upper limit to this range, then the total rate of energy loss from the waves by breaking, or, equivalently, the rate of energy input to the surface layer turbulence in this way is

$$\begin{aligned} \varepsilon_0 &= 2\int_{-\pi/2}^{\pi/2}\int_{k_0}^{k_1} \varepsilon(\underset{\sim}{k})\, k\, dk\, d\theta , \\ &= 3{\cdot}42\, a\beta^3 u_*^3 \ln(k_1/k_0), \end{aligned} \quad (2.20)$$

in which the directionality factor in (2.18) is taken at face value. This quantity is more usually expressed in terms of the air density; restoring the density factors we have

$$\varepsilon_0 = (3{\cdot}42\, a\beta^3 \rho_w/\rho_a)\cdot \rho_a u_*^3 \ln(k_1/k_0). \quad (2.21)$$

The total momentum flux to the surface layer by wave breaking is likewise

$$\tau_w = (5.64\, a\beta^3 \rho_w/\rho_a) \cdot \rho_a g^{-1/2} u_*^3 (k_1^{1/2} - k_o^{1/2}). \quad (2.22)$$

According to Banner and Phillips (1974), freely travelling gravity waves for which $c < u_*$ (or $k > g/u_*^2$) are strongly suppressed by the wind drift induced by the direct shear stress at the water surface; if $k = g/u_*^2 \gg k_o$, then from (2.22),

$$\tau_w = (5.64\, a\beta^3 \rho_w/\rho_a) \cdot \rho_a u_*^2 , \quad (2.23)$$

which must, of course, be less than $\rho_a u_*^2$.

Accordingly, as a wave field develops from, say, an initial state of rest, the momentum flux to the surface layer by wave breaking, initially zero, increases as the equilibrium range covers a wider and wider interval of wave-numbers, approaching asymptotically a fixed fraction of the total wind stress. The energy flux to the turbulence of the surface layer by wave breaking continues to increase, albeit logarithmically. With $k = g/u_*^2$ and $k_o = g/c_o^2$

$$\mathcal{E}_o = (3.42\, a\beta^3 \rho_w/\rho_a) \ln (c_o/u_*)^2 \rho_a u_*^3 , \quad (2.24)$$

and if the fetch and duration of the field are sufficient to generate dominant waves moving at the wind speed,

$$\mathcal{E}_o \simeq (3.42\, a\beta^3 \rho_w/\rho_a) \ln (C_D^{-1}) \rho_a u_*^3 , \quad (2.25)$$

where C_D is the drag coefficient.

3. Constraints on the Constants of Proportionality

Among the interesting consequences of the analysis of the previous section are the relations it provides among various numerical coefficients that have been inferred from independent sets of measurements, though none to high precision. The quantity m of (2.8) expressing the rate of energy input from the wind is about 0.04 but may be rather larger; Toba's constant α of (2.16) specifying the spectral level of the frequency spectrum in the equilibrium range may be bracketed by the values 0.02 and 0.06 found in different experiments. The constant involved in the wave-number spectrum in the saturation range has not yet been measured directly, but the $(\cos\theta)^{1/2}$ directionality factor gives $\alpha = 9.36\beta$; different but reasonable directional distributions may give of up to fifty per cent or so in the coefficient. One firm constraint that we have from (2.23) is that

$$5.46\, a\beta^3 \rho_w/\rho_a < 1$$

or

$$a\beta^3 < 1.8 \times 10^{-4}. \quad (3.1)$$

Now, in the action spectral density balance (2.12), f(B) certainly represents a loss and the wind input a gain; the calculations of Sell and Hasselmann (1972), although not too reliable at these large wave-numbers, indicate that the net spectral flux also represents a gain. Consequently, the rate of dissipation aB^3 must be greater than or equal to the rate of wind input, so that

$$a\beta^2 \geq m \simeq 0.04 \qquad (3.2)$$

If $\alpha = 0.02$ then $\beta = 2 \times 10^{-3}$ and $a\beta^3 > 8 \times 10^{-5}$, comfortably satisfying (3.1) and suggesting that fifty per cent or more of the total wind stress is communicated to the surface layer by wave breaking. On the other hand, if $\alpha = 0.06$, then $\beta = 7 \times 10^{-3}$ and $a\beta^3 > 2.8 \times 10^{-4}$, which is inconsistent with (3.1). We conclude therefore that either a value of 0.06 for Toba's constant, or a value of 0.04 for the wind-wave coupling coefficient (or both) are too high. Nevertheless, even with somewhat smaller values one can also conclude that in a well-developed wind-wave field, (1) a substantial fraction of the total wind stress is communicated to the surface layer by wave breaking and (2) the energy flux as turbulence to the surface layer by wave breaking is a modest multiple ($\ln C_D^{-1} = 6.5$) of $\rho_a u_*^3$, and is certainly greater than the energy flux by the mean surface shear stress acting on the wind-induced mean drift.

4. Some Statistical Characteristics of Breaking Events

As the wind blows over the water surface, at any instant the fronts of the breaking waves define a distribution of isolated line or arc segments. The scales of the breaking waves may cover a very wide range, from very short gravity waves in which a converging, moving stagnation point is marked by a group of capillary ripples, through intermediate scales (15-30 cm or so) where the breaking is unsteady but only a few bubbles are produced, to actual whitecaps in which the breaking and the generation of turbulence is so vigorous that extensive patches of foam are generated. There is clearly some association of the breaking events with waves of different scales, but it is difficult to make the association in an unambiguous way if we consider only the surface configuration at one given instant - the breaking crest may indeed be a local maximum in the instantaneous surface configuration, but there is no guarantee that a local wave length can be defined clearly. It seems more satisfactory to use the velocity $\underset{\sim}{c}$ of the breaking front as a measure of the scale of the breaking, since this is a well defined quantity that might (conceptually at any rate) be measured from movie images of the sea surface. In practice, this may be obtained relatively easily for those breaking events that generate whitecaps, though it may be difficult to distinguish the many smaller scale, fugitive occurrences of breaking that do not generate discernable bubble trains but which still turn over the water surface as they advance.

In any event, let us define a distribution $\Lambda(\underset{\sim}{c})$ such that $\Lambda(\underset{\sim}{c})d\underset{\sim}{c}$ represents the average total length per unit surface area of breaking fronts that have velocities in the range $\underset{\sim}{c}$ to $\underset{\sim}{c} + d\underset{\sim}{c}$. The total length of breaking front per unit area is then $\int \Lambda(\underset{\sim}{c})d\underset{\sim}{c}$.

What is the rate of energy loss from the waves to turbulence per unit length of front in these breaking events? This question has been examined by Duncan (1981) in a series of laboratory experiments; he showed that in an active breaker in deep water, the breaking zone extends down the forward face of the wave over a fixed fraction of its amplitude and that its shape is geometrically similar for waves of different scales. Furthermore, he found that the breaking waves themselves are geometrically similar, so that the cross-sectional area of the breaking zone is proportional to the square of the local wavelength, or to $(c^2/g)^2$. The weight of the breaking zone per unit length of the breaking front provides a tangential force proportional to c^4/g that acts on the oncoming stream whose speed is approximately c. Consequently, the rate of energy loss per unit length of front is $\gamma\, c^5/g$, where γ is a numerical constant estimated by Duncan from his experiments as approximately 0.06.

It is interesting to observe that the rate of turbulent energy production by breaking increases very rapidly with the characteristic speed of advance c of the breaking wave. A few large scale breaking events can produce as much energy loss from the wave field and input to the turbulence as many small ones. Nevertheless, the characteristic time scale for the duration of a breaking event, the ratio of the wave energy in one wavelength to the rate of loss by breaking, is proportional to the wave period, so that both large and small scale breaking events are equally transient.

The average rate of energy loss per unit area by breakers with speeds between $\underset{\sim}{c}$ and $\underset{\sim}{c} + d\underset{\sim}{c}$ is then

$$\gamma \Lambda(\underset{\sim}{c})\, c^5/g \; d\underset{\sim}{c} \,. \tag{4.1}$$

If we now identify the scales of waves that are breaking by the speeds with which the fronts advance, then, for wave-numbers below those seriously affected by Doppler shifting, $k = g/c^2$. An element of area $d\underset{\sim}{k}$ on the wave-number plane is related to the element $d\underset{\sim}{c}$ on the velocity plane by

$$d\underset{\sim}{k} = k\,dk\,d\theta = -(2g^2/c^6)\,c\,dc\,d\theta ,$$
$$= -(2g^2/c^6)\,d\underset{\sim}{c} , \qquad (4.2)$$

the negative sign being associated with the fact that integration to large k corresponds to integration to smaller c. The distribution (2.18) of energy loss by wave breaking per unit area $d\underset{\sim}{k}$ can be re-written in terms of the distribution with respect to velocity as

$$\mathcal{E}(\underset{\sim}{c})\,d\underset{\sim}{c} = 4\alpha\beta^3 (\cos\theta)^{3/2} u_*^3\, c^{-2}\, d\underset{\sim}{c} , \qquad (4.3)$$

one factor of two arising from (4.2) and another from the equal contributions to (2.18) for $\underset{\sim}{k}$ and $-\underset{\sim}{k}$. Consequently, from (4.1) and (4.3), dropping the negative sign,

$$\Lambda(\underset{\sim}{c}) = \frac{4a\beta^3}{\gamma} (\cos\theta)^{3/2} u_*^3\, g\, c^{-7}. \qquad (4.4)$$

This distribution of total length of breaking fronts on the wave-number plane is very strongly weighted towards those with small speed c or small scales - those fronts that produce whitecaps are evidently a very small fraction of the whole.

The total density of breaking fronts (length per unit area) with speeds between c_0 and c_1 is therefore

$$L(c_0, c_1) = \frac{4a\beta^3}{\gamma} u_*^3 g \int_{-\pi/2}^{\pi/2} (\cos\theta)^{3/2} d\theta \int_{c_0}^{c_1} c^{-7} \cdot c\,dc$$

$$= 1{\cdot}3 \frac{a\beta^3}{\gamma} u_*^3 g\,(c_0^{-5} - c_1^{-5}). \qquad (4.5)$$

Now, if we consider only those breaking fronts that generate a trail of bubbles, the breaking event then being identified as a whitecap, then only those breaking zones with a rate of energy release $\gamma c^5/g$ exceeding some threshold ε_t, say, will contribute. Consequently, the lower limit of the integral is such that $c_0 = c_T$ where

$$\varepsilon_T = \gamma c_T^5 / g .$$

If the longest waves that are breaking are shorter than those with speed c_T, then virtually no whitecaps will be formed; if they are only somewhat larger, $c_1^{-5} \ll c_T^{-5}$ and (4.5) reduces to

$$L = 1{\cdot}3 \frac{a\beta^3}{\gamma} u_*^3 g\, c_T^{-5} , \qquad (4.6)$$
$$= 1{\cdot}3\, a\beta^3 \varepsilon_T^{-1} u_*^3 \qquad (4.7)$$

where $a\beta^3 \sim 10^{-4}$ and $\gamma \simeq 0.06$.

On the other hand, if we consider all breaking events, even the very small scale ones that involve no air entrainment or bubbling at all, then the total length of front per unit area is very much larger, strongly dependent on the minimum scales that are breaking but hardly at all on the density of the much rarer, but far more vigorous whitecapping events. If the slowest moving waves that are breaking have speeds $c_0 \simeq u_*$, then

$$L = 1\cdot 3 \frac{\alpha\beta^3}{\gamma} g/u_*^2 \tag{4.8}$$

which actually decreases with u_*, since the increase in the speed of the smallest breaking waves more than compensates for the increased density of them (on the $\underset{\sim}{c}$-plane).

These results are clearly related to "whitecap coverage", the fraction of surface area covered by bubbles. If bubbles, once generated, persist for an average time T, on the surface, then the average length of a foam streak is cT and the whitecap coverage is

$$W = \int cT\Lambda(\underset{\sim}{c})\, d\underset{\sim}{c}$$

Now T is likely to depend on the temperature, humidity and various surface properties not well understood; if we simply suppose it to be constant then by a similar calculation

$$W \simeq \frac{\alpha\beta^3}{\gamma} T g c_T^{-4} u_*^3 , \tag{4.9}$$

where c_T is the slowest speed of fronts capable of producing whitecaps, provided, of course, that the fastest breaking waves are moving significantly more rapidly than this. Although the accuracy of this result is likely to be low (matching, indeed, the considerable scatter in reported measurements of the whitecap coverage), the wind speed dependence that it exhibits, u_*^3, is close to, but a little less than those found empirically. Fitting observations to the form $W \propto U^n$, Monahan (1971) found $n \sim 3.4$, Tang (1974) gives $n \sim 3.2$, Wu, (1979) 3.75 and Monahan and Muircheartaigh (1980), 3.52 or 3.41.

Finally, and somewhat more reliably, the fraction of sea surface area turned over per unit time can be established. The area per unit area swept up per unit time by breaking fronts, or the number of active breaking fronts passing a given point per unit time is

$$\begin{aligned} \tau &= \int c\,\Lambda(\underset{\sim}{c})\, d\underset{\sim}{c} \\ &= \frac{\alpha\beta^3}{\gamma} u_*^3 g c_0^{-4} , \end{aligned} \tag{4.10}$$

where c_0 represents the speed of the slowest breaking fronts, or, if we are concerned only with the number of whitecaps passing a given point per unit time, c_0 is the threshold front speed.

It is unfortunately difficult to compare these results in detail with the observations that have been made to date, largely because of the sensitivity to c_0 (which has not been measured) and the observational difficulty of identifying precisely in a given experiment the speed of the smallest breaking fronts that are detected or counted. The lowest wind speed at which whitecaps occur is about 2.5 m/s, so that presumably the threshold phase speed for breaking is somewhat smaller than this. Nonetheless, it is hoped that these results will stimulate further careful observations in which these questions can be resolved.

It is a pleasure to acknowledge the support of the Fluid Dynamics Branch of the Office of Naval Research under contract N00014-76-C-0184.

References

Banner, M.L. and O.M. Phillips, 1974: On the incipient breaking of small scale waves. J. Fluid Mech., 65, 647-56.

Banner, M.L. and W.K. Melville, 1976: On the separation of air flow over water waves. J. Fluid Mech., 77, 825-42.

Donelan, M.A., J. Hamilton and W.H. Hui, 1984: Directional spectra of wind-generated waves. Phil. Trans. Roy. Soc., A, xxx.

Duncan, J.H., 1981: An experimental investigation of breaking waves produced by a towed hydrofoil.

Forristall, Z., 1981: Measurements of a saturation range in ocean wave spectra. J. Geophys. Res., 86, 8075-84.

Fox, M.J.H., 1976: On the nonlinear transfer of energy in the peak of a gravity wave spectrum - II. Proc. Roy. Soc., A. 348, 467-83.

Gent, P.R. and P.A. Taylor, 1976: A numerical model of the air flow above water waves. J. Fluid Mech., 77, 205-28.

Hasselman, K., 1962: On the non-linear energy transfer in a gravity-wave spectrum. Part 1. General Theory. J. Fluid Mech., 12, 481-500.

Hasselmann, K., 1968: Weak interaction theory of ocean waves. Basic Developments in Fluid Dynamics, Vol. 2. M. Holt, Ed. Academic Press. 117-82.

Hasselmann, K., 1974: On the spectral dissipation of ocean waves due to whitecapping. Boundary-Layer Meteorol., 6, 107-27.

Kahma, K.K., 1981: A study of the growth of the wave spectrum with fetch. J. Phys. Oceanogr., 11, 1503-15.

Kawai, S., K. Okada and Y. Toba, 1977: Field data support for three-seconds power law and $gu_* \sigma^{-4}$ spectral form for growing wind waves. J. Oceanogr. Soc. Japan, 33, 137-50.

Kennedy, R.M. and R.L. Snyder, 1983: On the formation of whitecaps by a threshold mechanism. Part 11: Monte Carlo Experiments. J. Phys. Oceanogr., 13, 1493-1504.

Longuet-Higgins, M.S. and E.D. Cokelet, 1976: The deformation of steep surface waves. Proc. Roy. Soc., A 350, 1-26.

Mitsuyasu, H. and T. Honda, 1984: The effects of surfactant on certain air-sea interaction phenomena. Wave Dynamics and Radio Probing of the Ocean Surface, Plenum Press, N.Y.

Monahan, E.C., 1971: Oceanic whitecaps. J. Phys. Oceanogr. 1, 139-44.

Monahan, E.C. and I. Muircheartaigh, 1980: Optimal power-law description of oceanic whitecap coverage dependence on wind speed. J. Phys. Oceanogr., 10, 2094-99.

Phillips, O.M., 1980: The Dynamics of the Upper Ocean. Cambridge University Press, pp.336

Phillips, O.M., 1984: On the response of short ocean wave components at a fixed wave-number to ocean current variations. J. Phys. Oceanogr., 14, xxx.

Plant, W.J., 1982: A relationship between wind stress and wave slope. J. Geophys. Res., 87, 1961-67.

Sell, W. and K. Hasselmann, 1972: Computation of nonlinear energy transfer for JONSWAP and empirical wave spectra. Rep. Inst. Geophys., Univ. Hamburg.

Snyder, R.L. and R.M. Kennedy, 1983: On the formation of whitecaps by a threshold mechanism. Part 1: Basic Formalism. J. Phys. Oceanogr., 13, 1482-92.

Snyder, R.L., L. Smith and R.M. Kennedy, 1983: On the formation of whitecaps by a threshold mechanism. Part III: Field experiment and comparison with theory. J. Phys. Oceanogr., 13, 1505-18.

Tang, C.C.H., 1974: The effect of droplets in the air-sea transition zone as the mean brightness temperature. J. Phys. Oceanogr. 4, 579-93.

Toba, Y., 1973: Local balance in the air-sea boundary processes, III. On the spectrum of wind waves. J. Oceanogr. Soc. Japan, 29, 209-20.

Wu, J., 1979: Oceanic whitecaps and sea state. J. Phys. Oceanogr., 9, 1064-8.

How Do Liquid Drops Spread on Solids?

S. Rosenblat

Department of Mathematics, Illinois Institute of Technology, Chicago, IL 60616, USA

S.H. Davis

Department of Engineering Sciences and Applied Mathematics, Northwestern University
Evanston, IL 60201, USA

1. Introduction

The coating of a solid with a liquid and the displacement of oil in a porous medium by water are processes that involve the motion of a contact line, the three-phase line common to three materials. In the above illustrations the contact line involves liquid, gas and solid and liquid, liquid and solid, respectively.

Because moving contact lines arise so frequently in practice, an understanding of the associated mechanics is required. Underlying such an understanding is the observation that the contact line does not consist of the same points at all times. As shown in Figure 1, the contact line is the site of a splitting of the fluid trajectories which gives rise to a singularity in the flow field. Dussan V. and Davis (1974) show that the motion of the contact line together with the imposition of the no-slip boundary condition necessarily makes this singularity one of infinite force precluding a local study of moving contact-line mechanics. Thus, local slip at the fluid-solid boundaries is posed and leads to tractable boundary-value problems that have predictive capability.

The realization that effective slip occurs near the contact line allows one to solve problems involving mutual displacement. One such prototype problem involves the spreading of a liquid drop on a solid, the spreading occurring spontaneously upon the placement of the drop on the solid. Such a model was posed by Greenspan (1978)

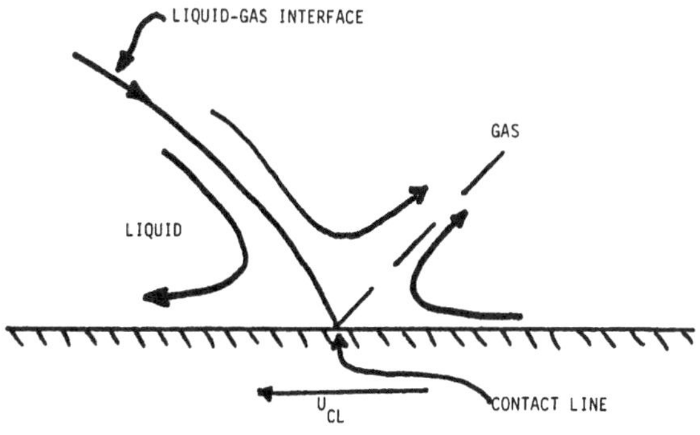

Figure 1: Kinematics of flow near a moving contact line (after Dussan V. and Davis, 1974). Coordinate system moves with the contact line at speed u_{CL}.

for the creeping flow of a thin, Newtonian liquid on a smooth solid. In the present paper, we shall reexamine this model in order to probe more deeply into the roles of viscous and surface tension forces and the wetting characteristics of the solid. We shall discuss two distinct contributions to the spreading characteristics, which we call "capillary push" and "contact-line pull". Since many applications of spreading studies involve the coating by polymer liquids, we shall extend our study to viscoelastic liquids and identify which non-Newtonian properties dominate in the spreading flows of interest.

2. Formulation

We consider a drop of viscous liquid on a smooth rigid plane. We use a cylindrical polar coordinate system (r^*,ϕ^*,z^*) to describe axisymmetric drops in which the rigid plane is located at $z^* = 0$, and the z^* axis points into the liquid. Since the drop is axisymmetric, all quantities are independent of the azimuthal coordinate ϕ^*; body forces are ignored.

Initial State

The initial shape of the drop is taken to have the form

$$z^* = h_0^*(r^*) \quad , \quad 0 \leq r^* < a_0 \tag{1}$$

with the edge condition

$$h_0^*(a_0) = 0 \quad . \tag{2}$$

The initial contact angle θ_0 is given by

$$\tan \theta_0 = -\frac{dh_0^*}{dr^*}(a_0) \quad , \tag{3}$$

and the volume is

$$V^* = 2\pi\int_0^{a_0} r^* h_0^*(r^*)dr^* \quad . \tag{4}$$

Drop Dynamics

At time $t^* > 0$ we denote by $a^*(t^*)$ the radius of the drop. Thus we have the initial condition

$$a^*(0) = a_0 \quad . \tag{5}$$

We take the shape of the drop at time t^* to have the form

$$z^* = h^*(r^*,t^*) \quad , \quad 0 \leq r^* < a^*(t^*) \quad \text{with} \tag{6}$$

$$h^*(a^*(t^*),t^*) = 0 \quad . \tag{7}$$

The contact angle at time t^* is denoted by $\theta^* = \theta^*(t^*)$ and is given by

$$\tan \theta^* = -\frac{\partial h^*}{\partial r^*}(a^*(t^*),t^*) \quad . \tag{8}$$

Conservation of the volume of the drop over time gives the additional constraint

$$V^* = 2\pi \int_0^{a^*(t^*)} r^* h^*(r^*,t^*) dr^* \quad . \tag{9}$$

The rate of change of the quantity a^* at any instant t^* is taken, on empirical grounds, to depend on θ^*, $\dot{a} = G(\theta^*)$, which incorporates all the wetting properties of the solid. Dussan V. (1981) discusses these issues in detail. For convenience we use a linear law

$$\frac{da^*}{dt^*} = \kappa(\theta^* - \theta_A) \tag{10}$$

where $\kappa > 0$ is an empirically determined constant, and where $\theta_A > 0$; θ_A is the advancing contact angle which corresponds to static equilibrium when the drop is on the point of spreading.

Equations and Boundary Conditions

The motion is governed by the Navier-Stokes and continuity equations,

$$\rho\left(\frac{\partial \underline{v}^*}{\partial t^*} + \underline{v}^* \cdot \nabla^* \underline{v}^*\right) = -\nabla^* p^* + \nabla^* \cdot \underline{\underline{S}} \tag{11}$$

$$\nabla^* \cdot \underline{v}^* = 0 \tag{12}$$

where ρ is the density, $\underline{v}^* = (u^*,0,w^*)$ is the velocity vector, p^* is the pressure and $\underline{\underline{S}}^*$ is the extra-stress tensor. A constitutive relation between the stress and the deformation-rate will be given below.

The boundary conditions are as follows:

(i) The normal velocity component is zero on the rigid plane,

$$w^* = 0 \quad \text{on} \quad z^* = 0 \quad ; \tag{13}$$

(ii) As follows from Dussan V. and Davis (1974), the usual no-slip condition at the rigid boundary is modified to avoid the appearance of a singularity at the contact line. Following Greenspan (1978) we take the condition to be

$$(\gamma^{*2}/\mu_0)S_{13}^* = h^* u^* \quad \text{on} \quad z^* = 0 \tag{14}$$

where μ_0 is the zero-shear-rate viscosity, γ^* is a slip length (distance from the contact line over which slip takes place) and S_{13}^* represents the r^*z^*-component of the extra stress.

(iii) The kinematic condition at the free surface is

$$w^* = \frac{\partial h^*}{\partial t^*} + u^* \frac{\partial h^*}{\partial r^*} \quad \text{on} \quad z^* = h^*(r^*,t^*) \quad . \tag{15}$$

(iv) The dynamic boundary condition at the free surface is

$$-[p^*]\underline{n} + [\underline{\underline{S}}^* \cdot \underline{n}] = 2H^* \sigma \underline{n} \quad \text{on} \quad z^* = h^* \tag{16}$$

where $[p^*]$ denotes the pressure difference across the interface, σ is the surface tension, $\underline{n}$ is the outward unit normal to the surface, and H^* is the mean curvature of the interface.

Rheology

A constitutive equation that is suitable for polymer solutions is the generalized Maxwell model,

$$\underline{\underline{S}}^* + \tau\{\frac{\partial \underline{\underline{S}}^*}{\partial t^*} + (\underline{v}^* \cdot \nabla^*)\underline{\underline{S}}^* + \frac{1}{2}(\underline{\underline{\omega}}^* \cdot \underline{\underline{S}} - \underline{\underline{S}}^* \cdot \underline{\underline{\omega}}^*) - \frac{1}{2}\beta(\underline{\underline{S}}^* \cdot \dot{\underline{\underline{\gamma}}}^* + \dot{\underline{\underline{\gamma}}}^* \cdot \underline{\underline{S}}^*)\} = \mu\dot{\underline{\underline{\gamma}}}^* \tag{17}$$

where

$$\dot{\underline{\underline{\gamma}}}^* = \nabla^*\underline{v}^* + (\nabla^*\underline{v}^*)^T \tag{18}$$

$$\underline{\underline{\omega}}^* = \nabla^*\underline{v}^* - (\nabla^*\underline{v}^*)^T \quad . \tag{19}$$

τ is the relaxation time of the liquid and β is a number, which in practical situations can range between -1 and +1. (See, for example, Petrie (1979).) When $\beta = 0$ the model, equation (17) reduces to the well-known corotational Maxwell model. In steady unidirectional shear flow this model yields shear thinning and both first and second normal stress differences, and yields stress relaxation in unsteady simple shear. When $\beta = 1$, equation (17) reduces to the upper convected Maxwell model. Here there is no shear thinning in simple shear, the viscosity remaining constant at its zero-shear-rate value, but first and second normal stress differences, as well as stress relaxation, are present.

3. Lubrication Approximation

We proceed on the basis of the assumption that the initial angle θ_0 is very small. This enables us to use the lubrication approximation, in which all quantities are appropriately scaled, and then the equations and boundary conditions are expanded in powers of θ_0. The first-order problem is retained in the limit $\theta_0 \to 0$.

A dimensionless time t is defined by

$$t = t^*\sigma\theta_0^3/a_0\mu_0 \tag{20}$$

and dimensionless coordinates are defined by

$$z = z^*/(a_0\theta_0) \quad , \quad r = r^*/a_0 \quad . \tag{21}$$

The dimensionless shape of the drop becomes $h(r,z)$ where

$$h = h^*/(a_0\theta_0) \quad . \tag{22}$$

The radius of the drop is a(t), given by

$$a(t) = a^*(t^*)/a_0 \tag{23}$$

and the contact angle is

$$\theta(t) = \theta^*(t^*)/\theta_0 \tag{24}$$

with the final equilibrium (advancing) contact angle given by

$$\theta_F = \theta_A/\theta_0 \quad . \tag{25}$$

We also define a dimensionless volume by

$$V = V^*/a_0^3\theta_0 \quad . \tag{26}$$

Dimensionless velocity components (u,w) and pressure p are given by

$$u = u^*/(\kappa\theta_0) \quad , \quad w = w^*/(\kappa\theta_0^2) \quad , \quad p = p^*(a_0\theta_0/\mu_0\kappa) \quad . \tag{27}$$

Finally we have a dimensionless stress tensor $\underline{\underline{S}}$ defined by

$$\underline{\underline{S}} = \underline{\underline{S}}^*(a_0/\mu_0\kappa) \quad . \tag{28}$$

Initial State

In our non-dimensionalization the drop has unit radius initially, and the initial contact angle $\theta(0)$ is also unity.

The initial shape has the form

$$z = h_0(r) \quad , \quad 0 \leq r < 1 \tag{29}$$

with end condition

$$h_0(1) = 0 \quad . \tag{30}$$

Using the approximation $\theta_0 \to 0$, we find the condition for the initial angle to be

$$\frac{dh_0}{dr}(1) = -1 \quad . \tag{31}$$

We also have the volume condition

$$V = 2\pi\int_0^1 rh_0(r)dr \quad . \tag{32}$$

Drop Dynamics

The shape of the drop at time t is

$$z = h(r,t) \quad , \quad 0 \leq r < a(t) \quad \text{with} \tag{33}$$

$$h(a(t),t) = 0 \quad . \tag{34}$$

In the lubrication approximation the contact angle θ (t) is given by

$$\theta(t) = -h_r(a(t),t) \quad . \tag{35}$$

The volume conservation condition (9) becomes

$$V = 2\pi\int_0^{a(t)} rh(r,t)dr \quad . \tag{36}$$

Equations (10) and (35) combine to give the following differential equation for the edge dynamics of the drop:

$$\frac{da(t)}{dt} = C[-h_r(a(t),t) - \theta_F] \quad \text{where} \tag{37}$$

$$C = \frac{\mu_0\kappa}{\sigma\theta_0^2} \quad . \tag{38}$$

Equation (38) is subject to the initial conditions

$$a(0) = 1 \quad . \tag{39}$$

Here C is the effective capillary number for the wetting characteristic in that it measures the slope κ in equation (10).

Equations and Boundary Conditions

In the lubrication limit $\theta_0 \to 0$ the Navier-Stokes and continuity equations reduce to

$$-p_r + S_{13,z} = 0 \tag{40}$$

$$-p_z = 0 \tag{41}$$

$$(ru)_r + (rw)_z = 0 \quad . \tag{42}$$

Note that S_{13}, which is associated with shear thinning in viscoelastic materials, is the only stress component that remains in the reduced system (40)-(42). Components such as S_{11}, S_{22}, S_{33}, associated with normal stress differences, are not explicitly present in the equation of motion in the lubrication limit.

The boundary conditions are as follows:

(i) zero normal velocity at the rigid plane

$$w = 0 \quad \text{on} \quad z = 0 \tag{43}$$

(ii) modified slip condition at the rigid plane

$$\gamma^2 S_{13} = hu \quad \text{on} \quad z = 0 \quad \text{where} \tag{44}$$

$$\gamma^2 = (\gamma^*/a_0\theta_0)^2 \quad . \tag{45}$$

(iii) kinematic interfacial condition

$$h_t + C(uh_r-w) \quad \text{on} \quad z = h(r,t) \quad . \tag{46}$$

(iv) dynamic boundary conditions on the interface are

$$h_{rr} + \frac{1}{r}h_r + Cp = 0 \quad \text{on} \quad z = h(r,t) \quad \text{and} \tag{47}$$

$$S_{13} = 0 \quad \text{on} \quad z = h \quad . \tag{48}$$

Rheology

Using the scalings indicated above, we find that the constitutive relation (17) becomes, in dimensionless form,

$$\underline{\underline{S}} + \lambda \frac{\partial \underline{\underline{S}}}{\partial t} + \varepsilon C\{(\underline{v}\cdot\nabla)\underline{\underline{S}} + \tfrac{1}{2}(\underline{\underline{\omega}}\cdot\underline{\underline{S}}-\underline{\underline{S}}\cdot\underline{\underline{\omega}}) - \tfrac{1}{2}\beta(\underline{\underline{S}}\cdot\dot{\underline{\underline{\gamma}}} + \dot{\underline{\underline{\gamma}}}\cdot\underline{\underline{S}})\} = \dot{\underline{\underline{\gamma}}} \qquad (49)$$

where in the lubrication approximation

$$\dot{\underline{\underline{\gamma}}} = \begin{pmatrix} 0 & u_z \\ u_z & 0 \end{pmatrix} , \quad \underline{\underline{\omega}} = \begin{pmatrix} 0 & -u_z \\ u_z & 0 \end{pmatrix} \qquad (50)$$

and where λ,ε are relaxation parameters defined by

$$\lambda = \frac{\tau\sigma\theta_0^3}{a_0\mu_0} , \qquad (51)$$

$$\varepsilon = \frac{\tau\sigma\theta_0^2}{a_0\mu_0} . \qquad (52)$$

Note that

$$\lambda/\varepsilon = \theta_0 . \qquad (53)$$

The form of equation (49) shows that λ is a measure of stress relaxation while ε is a measure of shear thinning. Hence, relation (53) implies that stress relaxation only becomes important when shear thinning is absent. Henceforth, we retain only the effects of shear thinning by taking $\beta = 0$ and considering the corotational Maxwell model.

The components of equation (30) are

$$S_{11} - \varepsilon C u_z S_{13} = 0 \qquad (54)$$

$$S_{13} + \tfrac{1}{2}\varepsilon C u_z(S_{11} - S_{33}) = u_z \qquad (55)$$

$$S_{33} + \varepsilon C u_z S_{13} = 0 . \qquad (56)$$

From these we obtain

$$S_{13} = \frac{u_z}{1 + \varepsilon^2 C^2 u_z^2} \qquad (57)$$

which implies that the effective viscosity for simple shear is

$$\frac{1}{1 + \varepsilon^2 C^2 u_z^2} . \qquad (58)$$

4. Evolution Equation

We integrate equation (41) and use the boundary conditions (24) and (25) to obtain the equation

$$\frac{\partial h}{\partial t} + \frac{C}{r}\frac{\partial}{\partial r}(hQ) = 0 \qquad (59)$$

where

$$hQ = \int_0^{h(r,t)} r\, u(r,z,t)dz \quad . \tag{60}$$

From integration of equation (41) we obtain

$$p = \overline{p}(r,t) \tag{61}$$

so that the boundary condition (28) gives

$$\frac{1}{r}(rh_r)_r + C\overline{p}(r,t) = 0 \quad . \tag{62}$$

Substituting into (40), we obtain

$$\frac{1}{r}(rh_r)_r + CS_{13,z} = 0 \quad . \tag{63}$$

Integrating this with respect to z and using condition (48), we obtain

$$CS_{13} = (h-z)\ \{\frac{1}{r}[rh_r]_r\}_r \quad . \tag{64}$$

We now combine equations (59) and (57) to obtain

$$\frac{Cu_z}{1 + \varepsilon^2 C^2 u_z^2} = (Dh)(h-z) = F \quad , \quad \text{say} \tag{65}$$

from which we deduce that

$$Cu_z = \frac{1 - \sqrt{1-4\varepsilon^2 F^2}}{2\varepsilon^2 F} \quad . \tag{66}$$

Here

$$Dh \equiv [\frac{1}{r}(rh_r)_r]_r \quad . \tag{67}$$

Equation (58) shows that the effective viscosity tends to zero as the shear-thinning parameter ε tends to infinity. Such models are known to be reliable only for small values of ε. We therefore assume ε to be sufficiently small that an expansion of equation (66) is possible, whereupon we obtain

$$Cu_z = (Dh)(h-z) + \varepsilon^2[(Dh)(h-z)]^3 + 0(\varepsilon^4) \quad . \tag{68}$$

Integrating and using the boundary condition (44) we obtain

$$Cu = (Dh)\{\gamma^2 + \frac{1}{2}h^2 - \frac{1}{2}(h-z)^2\} + \frac{1}{4}\varepsilon^2(Dh)^3\{h^4 - (h-z)^4\} \quad . \tag{69}$$

Integrating again over 0 to h, we obtain

$$ChQ = (Dh)(\gamma^2 h + \frac{1}{3}h^3) + \frac{1}{5}\varepsilon^2(Dh)^3 h^5 \quad . \tag{70}$$

Substituting into relation (59) we now obtain

$$h_t + \frac{1}{r}\{r(Dh)(\gamma^2 h + \frac{1}{3}h^3) + \frac{1}{5}\varepsilon^2 r(Dh)^3 h^5\}_r = 0 \quad . \tag{71}$$

Equation (71) is the sought-after evolution equation to be solved subject to the appropriate side conditions.

5. Newtonian Liquids

We wish to consider the evolution equation (71) for Newtonian liquids (ε=0),

$$h_t + \frac{1}{r}\{r(Dh)[\gamma^2 h + \frac{1}{3}h^3]\}_r = 0 \qquad (72)$$

subject to the edge conditions

$$\left.\begin{aligned} \frac{da}{dt} &= C(-h_r - \theta_F) \\ h &= 0 \end{aligned}\right\} \quad \text{at } r = a(t) \quad , \qquad (73)$$

the symmetry condition

$$h_r = 0 \quad , \quad \text{at } r = 0 \quad , \qquad (74)$$

and the initial condition

$$h(r,t) = h_0(r) \quad \text{at} \quad t = 0 \quad . \qquad (75)$$

Furthermore, the volume of the drop is V,

$$V = 2\pi\int_0^{a(t)} r\, h(r,t)dr \quad . \qquad (76)$$

System (72)-(76), apart from minor redefinitions and rescaling, is identical to that of Greenspan (1978). It contains four parameters: the capillary number C, the slip parameter γ, the equilibrium contact angle θ_F, and the volume V. Before we discuss the behavior of the drop, recall that the time t in system (72)-(76) is scaled in units of the viscous-capillary scale $a_0\mu/\sigma\theta_0^3$.

System (72)-(76) governs the creeping-flow of a spreading drop in which the spreading occurs through two distinct mechanisms.

(i) If the initial contact angle were equal to its equilibrium value θ_A, but the meniscus were not an equilibrium shape, then there would still be relative motion in the drop. Surface tension on the interface would produce capillary pressure gradients that drive a viscous flow, change θ from θ_A, and cause spreading. Call this mechanism "capillary push".

(ii) If the drop shape were a spherical cap, and hence a static meniscus, but had a contact angle $\theta > \theta_A$, then the edge of the drop would necessarily move outward until $\theta = \theta_A$. Call this mechanism "contact-line pull". The total influence of the wetting characteristics of the solid surface are incorporated in the relation (10), and they are responsible for the "contact-line pull".

In general, the drop spreads through the joint mechanisms of "capillary push" and "contact-line pull", the proportion of which is governed by the number C. Here C is a non-dimensional version of κ in equation (10). If C is small, then the contact angle is very sensitive to contact-line speed. If C is large, then it is not.

We can now turn to system (72) and examine the spreading characteristics of drops having either small or large C. In order to do this we introduce a new time scale

$$T = Ct \tag{77}$$

in which the unit of time is independent of μ_0 and σ. We can then rewrite equations (72) and (73) in terms of this as follows:

$$Ch_T + \frac{1}{r}\{r(Dh)[\gamma^2 h + \frac{1}{3}h^3]\}_r = 0 \tag{78}$$

$$\frac{da}{dT} = -h_r - \theta_F \quad \text{at} \quad r = a(t) \quad . \tag{79}$$

We consider first the case $C \ll 1$. From equation (73), when $t = O(1)$, $\frac{da}{dt} \sim 0$, so that the contact line remains fixed but the shape of the drop, governed by equation (72), readjusts to its initial shape through "capillary push". This readjustment occurs until $t = O(C^{-1})$; equation (78) shows that $\frac{1}{r}\{[\gamma^2 h + \frac{1}{3}h^3]\}_r \sim 0$ whose solution $Dh \sim 0$ gives a constant curvature meniscus. This static meniscus is subjected to appreciable "contact-line pull" through equation (79). This lasts until the final equilibrium shape evolves.

We now consider the case $C \gg 1$. From equation (78) we see that $h_T \sim 0$ leads to a meniscus that is quasi-static. However, the limit $C \to \infty$ is singular and there is a boundary layer near $r = a(t)$ of width $O(C^{-1/4})$ in which there are rapid spatial changes. This is consistent with equation (79) which shows that there is appreciable "contact-line pull" to the edge. This situation holds for $t = O(C^{-1})$. After this small initial time equation (79) shows that this readjustment occurs with fixed contact angle.

We define $t^{(2)}$ as the time for a drop to double its initial radius; the corresponding doubling time on the scale defined by equation (77) is denoted by $T^{(2)}$. Figure 2 shows the shapes of the drop at $t = 0$ and $t = t^{(2)}$ for $C = 1$. Figure 3 shows $T^{(2)}$ as a function of volume V for various values of C. One sees

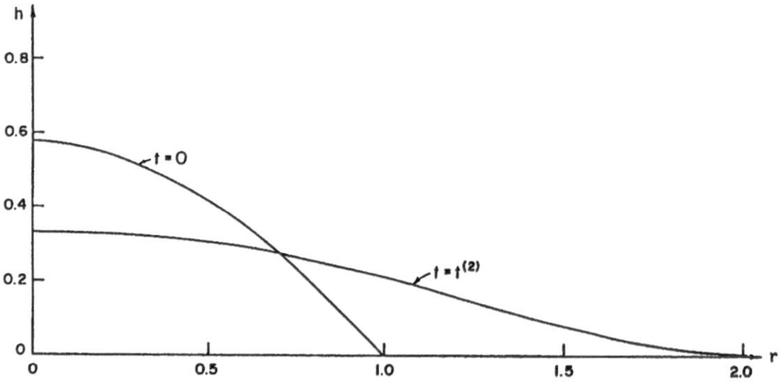

Figure 2: Axisymmetric drop, shapes at $t = 0$ and $t = t^{(2)}$ for $C = 1$ $\gamma = 0.01$ and $\varepsilon = 0$.

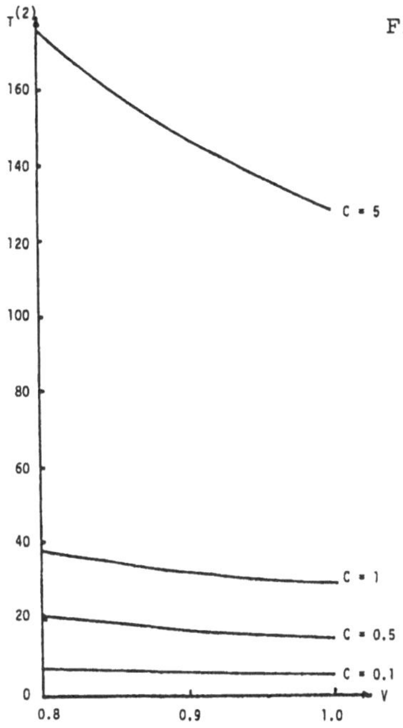

Figure 3: Axisymmetric drop, doubling time as a function of V for various C, with $\varepsilon = \theta_F = 0$ and $\gamma = 0.01$.

from Figure 3 that for a fixed V, $T^{(2)}$ decreases with decreasing C. A combination of forms (20) and (77) reveals that the time scale T is independent of the material properties of the liquid, and therefore one infers from Figure 3 that the doubling time increases with increasing viscosity and decreases with increasing surface tension. The curves in Figures 2 and 3 are obtained from numerical solutions to (72)-(76) computed by Rosenblat and Davis (1983).

6. Non-Newtonian Liquids

In the case of non-Newtonian liquids the evolution equation (71) must be used in place of the Newtonian version (72), but the boundary and initial condition (73)-(76) continue to apply without modification. An additional parameter ε is present; this parameter is a measure of the degree of shear thinning caused by the presence of polymer additives, for example.

The two basic physical mechanisms of "capillary push" and "contact-angle pull" remains in force for non-Newtonian liquids, and the effects of viscoelasticity are essentially limited to producing quantitative changes in derived quantities such as the spreading rate. Although the parameter ε enters the evolution equation (71) in a somewhat complicated way, so that its role is not obvious in this equation, it is relatively easy to see the part played by viscoelasticity by looking at equation (58); it causes a decrease in viscosity compared with the latter's zero-shear-rate value. This in turn results in decreased viscous resistance to capillary pressure gradients, thereby accelerating spreading rates.

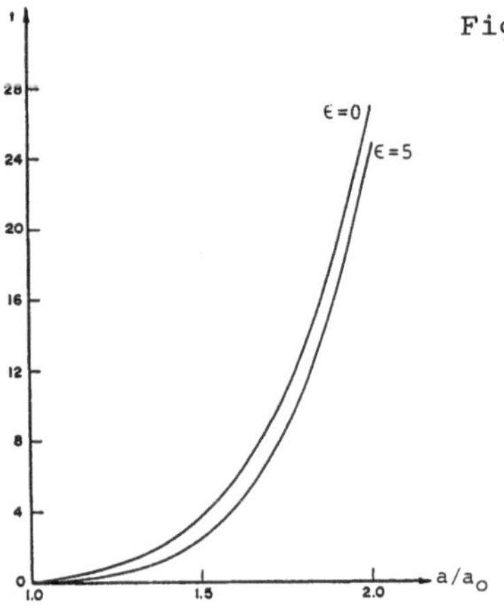

Figure 4: Axisymmetric drop, variation of radius with t for shear-thinning and non-shear-thinning cases, with C = V = 1, $\theta_F = 0$ and $\gamma = 0.01$.

Numerical solutions of the system (71) and (73)-(76) by Rosenblat and Davis (1983) confirm this result, and show that spreading rates increase with increasing ε when all other parameters are held fixed. This is shown in Figure 4. The effects, however, are very small in magnitude since the viscoelasticity only enters at $O(\varepsilon^2)$. As pointed out earlier, shear thinning is more significant than stress relaxation or normal stresses (at least in the lubrication approximation), so that these latter are very small indeed.

In practice the most important consequence of the addition of polymeric additives is the change in the zero-shear-rate viscosity μ_0. This quantity enters the capillary number C linearly and therefore, as shown in the previous section, dominates the various characteristic times of the drop dynamics. In general the addition of polymers tends to increase the zero-shear-rate viscosity, and therefore to slow down the spreading of the droplet which, of course, is opposite to the effect of shear thinning. This particular, probably predominant effect of viscoelasticity appears only in the value assigned to the capillary number C.

7. Conclusions

We have seen that the spreading of a Newtonian liquid drop depends on both "capillary push" and "contact-line spread" as measured by the effective capillary number C.

Non-Newtonian effects are dominated by shear for the thin drops governed by lubrication theory. Thus, shear thinning is the principal effect of viscoelasticity leading to non-Newtonian drops spreading faster than their Newtonian counterparts which have equal zero shear-rate viscosities. This viscoelastic effect is most pronounced near the contact lines where the strict condition for "sensibly" Newtonian behavior is $\varepsilon \ll \gamma$ rather than the usual condition $\varepsilon \ll 1$ appropriate to flows without contact lines. Thus, the fluid remembers the contact line.

References

Dussan V., E. B., Ann. Rev. Fluid Mech. 11, 371, 1979.

Dussan V., E. B. and Davis, S. H., J. Fluid Mech., 65, 71, 1974.

Greenspan, H. P., J. Fluid Mech., 84, 125, 1978.

Petrie, C. J. S., "Elongational Flows", Pitman, 1979.

Rosenblat, S. and Davis, S. H., Contractor Report ARCSL-CR-83062, Chemical Systems Laboratory, US Army Armament Research and Development Center, Aberdeen Proving Ground, Maryland, 1983.

Effects of Streamline Curvature on Turbulence

M.M. Gibson
Mechanical Engineering Department, Imperial College, London, Great Britain

1. Introduction

The effects of longitudinal curvature on the turbulence in thin shear layers has received a good deal of attention from experimenters and modellers and in many respects they are now well documented. The turbulence structure is highly sensitive to the additional mean strain rate introduced when the mean streamlines are curved in the plane of the mean shear. Turbulent energy and shear stress are reduced relative to rectilinear flow when the angular momentum of the mean flow increases in the direction of the radius of curvature, as in a two-dimensional boundary layer on a convex wall, and increased when the angular momentum decreases with increasing radius, as in concave wall flow. In contrast to laminar flow where the fractional changes in shear stress are of the same order as the shear layer thickness to the radius of curvature, measurements in turbulent boundary layes on curved walls show fractional changes an order of magnitude greater.

The extent of these changes is illustrated in Figure 1, where profiles of the shear stress in a number of convex wall boundary layers are plotted for comparison with Klebanoffs [1] flat plate data. In each of these experiments a boundary layer

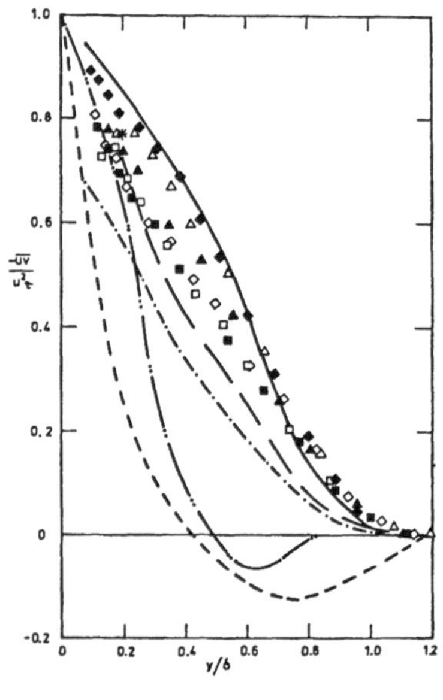

Fig. 1. Shear stress profiles in flat and convex wall boundary layers (from [4]). —— Klebanoff [1], flat: * [2], flat: — — [2], convex: — - — [3], convex; — - - — [5], convex; - - - [6], convex. Data points from [4] show changes from flat to mildly curved convex wall flow.

developed first on a flat wall upstream of a curved section. The effects of mild curvature, when the ratio of boundary layer thickness δ, to the radius of curvature of the wall R, is approximately 0.01, appears in the results from [2,3] and our own measurements [4] shown by the data points; it is enough to reduce the relative shear stress levels away from the wall so that the profiles become concave upward. These trends are accentuated when the stabilizing curvature is strong and prolonged as in the experiments with $\delta/R = 0.1$ reported by So and Mellor [5] and Gillis and Johnston [6]. In these flows the effects are so strong in the outer part of the layer that the shear stress is extinguished or reversed. At the same time the wall shear with which these data are normalized is itself reduced below levels appropriate to plane flow in the same conditions, and the growth rate of the layer is also reduced.

The recent literature contains numerous papers on the response of the fluctuating velocity field and a number of these were used to provide test cases at the 1981-2 AFOSR-HTTM Stanford Conferences on calculation methods for complex flows. The effects may be predicted either by modifying the mixing length distribution in ways suggested by the buoyancy analogy or, not altogether satisfactorily, by the numerical solution of modelled equations for the Reynolds stresses. The corresponding changes in the fluctuating scalar field have received much less attention at the level of detail needed to validate turbulence model hypotheses. This neglect is surprising in view of the evident importance of heat transfer in engineering practice. Early heat transfer measurements by Kreith [7], Thomann [8] and, more recently, by Mayle et al [9] show that changes in the Stanton numbers in curved wall flow are of the same order of magnitude as estimated changes in the skin friction coefficients which were not measured in these experiments. As a first approximation it might be reasonable to assume that Reynolds analogy is unaffected by changes in the mean strain or, alternatively, that the eddy diffusivity ratio (or turbulent Prandtl number) remains the same as in plane flow. Assumptions of this nature may well suffice for engineering calculations but they are insecurely based. Examination of the boundary layer conservation equations for the shear stress and cross-stream scalar flux shows that the generation terms that contain the additional mean strain rates are weighted differently in the two equations with respect to the main generation terms. Appeal to the buoyancy analogy also undermines simple assumptions of the Reynolds analogy description: the eddy diffusivity ratio in a horizontal density stratified shear layer is known to be sensitive to the degree of stratification.

The present short paper is largely based on recent experimental work by the author and his colleagues which originated in an unsuccessful attempt to predict scalar transfer in curved shear layers with a simple second moment closure scheme. The buoyancy analogy is first extended to cover heat transfer as well as momentum transfer, some of the results of measurements in curved heated wall flow are briefly discussed, and the paper concludes with some reflexions on second moment modelling related to Professor Corrsin's studies of homogenous shear flow.

2. The Analogy Between Streamline Curvature and Buoyancy

This analogy seems to have been recognised first by Prandtl [10] and revived by Bradshaw [11] who discusses the effects on the Reynolds stresses of gravitational sources and extra strain in longitudinally curved flow. It is worth while considering the simple extension of the analogy to the fluctuating scalar field.

An element of fluid in a buoyant flow which has a density of $(\rho + \rho')$ will experience a force $- \rho' g/\rho$ per unit mass in the vertical direction. A element of fluid in a curved flow which has velocity $(U + u)$ will experience an apparent centrifugal force (greater than the mean centrifugal force which is balanced by the mean radial pressure gradient) of $((U + u)^2 - U^2)/R \simeq 2Uu/R$ in the outward direction. The turbulent energy and shear stress equations for two-dimensional, (a) horizontally stratified and (b) longitudinally curved shear flow can be written with some approximations and the replacement of density by temperature as:

$$\frac{D}{Dt}\left(\frac{1}{2}\,\overline{q^2}\right) - \frac{\partial}{\partial y}\left(\frac{1}{2}\,\overline{vq^2} + \frac{\overline{pv}}{\rho}\right) = -\,\overline{uv}\,\frac{\partial U}{\partial y} - \epsilon + \left\{\begin{array}{l} g\overline{v\theta}/T \\ \\ \overline{uv}\;U/r \end{array}\right. \qquad (1)$$

$$\frac{D}{Dt}(\overline{uv}) - \frac{\partial}{\partial y}\left(\overline{uv^2} + \frac{\overline{pu}}{\rho}\right) = -\,\overline{v^2}\,\frac{\partial U}{\partial y} + \overline{\frac{p}{\rho}\left(\frac{\partial u}{\partial y} + \frac{\partial v}{\partial x}\right)} + \left\{\begin{array}{l} g\overline{u\theta}/T \\ \\ \overline{u^2}U/r \\ +(\overline{u^2}-\overline{v^2})U/r \end{array}\right. \qquad (2)$$

The additional analogous generation terms are bracketed together and, in the curvilinear coordinate system, y is the distance measured from a reference streamline whose local radius of curvature is R, and $r = R + y$. The extra sources of stress and energy in the buoyant flow involve the fluctuating density (temperature) field; in the curved flow only the extra mean strain rate U/r (for temperature differences of order ten degrees "centrifugal buoyancy" terms are an order of order of magnitude smaller). The curved flow sources are the sum of the "true" rate of generation from the mean strain and "apparent" generation terms which arise through the rotation of coordinate axes. It is important to make this distinction when modelling the pressure-strain terms of the Reynolds stress equations.

A measure of the influence of buoyancy is the flux Richardson number which is defined as (minus) the ratio of turbulent energy production by buoyancy forces to production by the mean shear:

$$R_f = \frac{g\overline{v\theta}/T}{\overline{uv}\;\partial U/\partial y} \qquad (3)$$

Bradshaw [11] defines an analogous flux Richardson number for curved flow as the ratio of (minus) the v-component energy production to the u-component shear production. A more convenient definition which arises naturally in the manipulation of the conservation equations is the ratio of (minus) the v-component energy production to the total mean shear production $-\overline{uv}(\partial U/\partial y - U/r)$. Thus:

$$R_f = \frac{2\,\overline{uv}\,U/r}{\overline{uv}(\partial U/\partial y - U/r)} = \frac{2S}{1-S} \tag{4}$$

where S is the strain-rate ratio $(U/r)/(\partial U/\partial y)$.

The corresponding equations for the streamwise and cross-stream fluxes are

$$\frac{D}{Dt}(\overline{u\theta}) - \frac{\partial}{\partial y}(\overline{uv\theta}) = -\overline{uv}\,\frac{\partial T}{\partial y} - \overline{v\theta}\,\frac{\partial U}{\partial y} + \overline{\frac{p}{\rho}\frac{\partial\theta}{\partial x}} + \begin{cases} 0 \\ -\overline{v\theta}\,U/r \end{cases} \tag{5}$$

$$\frac{D}{Dt}(\overline{v\theta}) - \frac{\partial}{\partial y}(\overline{v^2\theta} + \frac{\overline{p\theta}}{\rho}) = -\overline{v^2}\,\frac{\partial T}{\partial y} + \overline{\frac{p}{\rho}\frac{\partial\theta}{\partial y}} + \begin{cases} g\overline{\theta^2}/T \\ \overline{u\theta}\,U/r + \overline{u\theta}\,U/r \end{cases} \tag{6}$$

Here again a distinction is drawn between the (first) extra "true" generation terms and those which arise solely through the transformation from Cartesian to curvilinear coordinates.

It is instructive to examine the relative weighting of the extra source terms in the uv and vθ equations. The combined rates of production of these quantities may be rearranged in the following way:

SOURCE OF	$\overline{uv}$	$\overline{v\theta}$
BUOYANT	$-\overline{v^2}\,\frac{\partial U}{\partial y}(1 - \frac{\overline{u\theta}}{\overline{v\theta}}\,\frac{\overline{uv}}{\overline{v^2}}\,R_f)$	$-\overline{v^2}\,\frac{\partial T}{\partial y}\left\{1 - \frac{\overline{v^2}\,\overline{\theta^2}}{(\overline{v\theta})^2}\left[\frac{\overline{uv}}{\overline{v^2}}\right]^2\frac{K_H}{K_M}\,R_f\right\}$
CURVED	$-\overline{v^2}(\frac{\partial U}{\partial y} - \frac{U}{r})\left\{1 - (\frac{\overline{u^2}}{\overline{v^2}} - 1)R_f\right\}$	$-\overline{v^2}\,\frac{\partial T}{\partial y}(1 - \frac{\overline{u\theta}}{\overline{v\theta}}\,\frac{\overline{uv}}{\overline{v^2}}\,\frac{K_H}{K_M}\,R_f)$

where K_H/K_M is the ratio of the turbulence exchange coefficients, the reciprocal of the "turbulent Prandtl number":

$$\frac{K_H}{K_M} = \frac{\overline{v\theta}(\partial U/\partial y - U/r)}{\overline{uv}\partial T/\partial y} \tag{7}$$

It is now possible to form some idea of the relative strengths of these effects at small values of R_f by substituting plausible values of the stress and heat flux ratios. When this is done the factors multiplying the principal production terms appear as follows (it is appropriate to consider the factor multiplying the velocity gradient $\partial U/\partial y$ rather than the strain rate $(\partial U/\partial y - U/r)$ in curved flow):

	$\overline{uv}$	$\overline{v\theta}$
BUOYANT	$(1 - 2R_f)$	$(1 - 4R_f)$
CURVED	$(1 - 3.5R_f)$	$(1 - 2R_f)$

While these are all of the same order, and there is room for disagreement on the exact values inserted because the experimental data are so scattered, the conclusion that the generation rate of $\overline{v\theta}$ is more strongly affected by buoyancy interactions than that of $\overline{uv}$ appears to be supported by measurements. The homogeneous, buoyant, shear flow data of Young [12] reproduced in [13], though scattered, show decreasing values of K_H/K_M with increasing stability and the same trend is evident in the unstable regime ($R_f < 0$) of atmospheric surface layer measurements [14] reproduced from [13] in Figure 2 together with the behaviour predicted by the use of modelled second moment equations. In making these predictions B.E. Launder and the author argued that in the stable regime ($R_f > 0$) the effect of the ground on the fluctuating pressure interactions responsible for the destruction of shear stress and heat flux is modified by the attenuation of the length scales. The underlying trend for K_H/K_M to decrease with increasing R_f is then countered as the ground effect weakens and the structure more closely approaches that of free flow with presumed higher values of K_H/K_M in local equilibrium conditions. Two curves are shown in Figure 2: the original published predictions [13] shown by the broken curve, and the results obtained when the constants in the turbulence model are changed, as will be described, to be consistent with the treatment of curved shear layers.

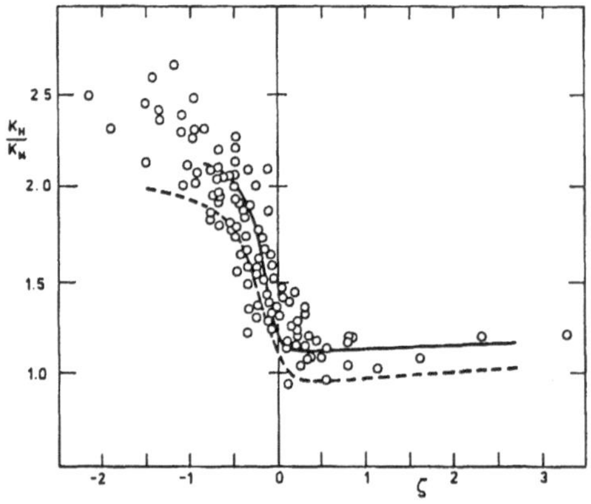

Fig. 2. Measured and predicted dependence of the ratio of exchange coefficients on stability in the atmospheric surface layer. Data from [14]; predictions [13] (broken line), revised 1983 (full line).

The factors multiplying the source terms in the curved flow equations show the opposite response to changes in R_f: the generation rate of $\overline{uv}$ is apparently more strongly affected than the generation rate of $\overline{v\theta}$. If this interpretation is correct the result might be expected to be an increase in K_H/K_M with increasing R_f in both stable and unstable regimes which could not be affected by the proximity of the wall in the same way as the buoyant flow. In fact, two sets of measurements [15,16] in heated boundary layers with stabilizing curvature, and in the internal temperature boundary layer in a convex wall jet [17], indicate, with some necessary qualifications, just the opposite response to a change in curvature. This unexpected result is related to the times taken for structural changes in the outer part of the layer, where the effects of extra strain on the large scale motion are strongest, to be communicated as changes in wall stress and heat flux.

Interactions with the fluctuating pressure field presumably also play a crucial role.

3. Measurements in Curved Wall Layers

Few experiments have been reported in which velocity and temperature field measurements have been made in suficient detail to guide the development of heat transfer models for curved flow. Simon and Moffat [15] have recorded surface flux and mean temperature distributions in the boundary layer on a ninety degree curved plate with a thickness to radius ratio of about 0.1 and for which some details of the fluctuating velocity field were available. Our own experiments [4, 16] on a more mildly curved convex wall flow (δ/R = 0.01) were conducted in the rig illustrated in Figure 3. To these we have since added measurements in the internal temperature layer in a convex wall jet under a uniform velocity free stream [17], and rather less detailed measurements in a mildly curved concave boundary layer. The main result of all these measurements is that the heat transfer through the curved layers is apparently more sensitive to the curvature than the turbulent velocity field.

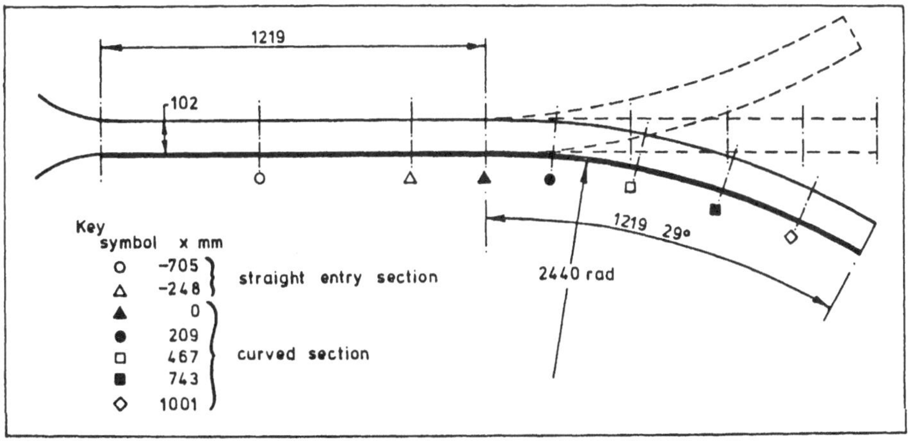

Fig. 3. Wind tunnel sections for heated wall boundary layer experiments [4, 16].

Figure 3 shows alternative working sections that were fitted to the contraction of a conventional open circuit blower wind tunnel. The boundary layer was tripped at exit from the contraction and then developed naturally on the floor of the constant section (305 mm x 102 mm) straight duct to a thickness δ of 19.8 mm 1.22m downstream of the trip. At this point alternative convex or concave curved sections could be fitted, each 1.22 m long with nominal test wall radius of curvature R = 2.44 m. The floors of the curved sections were of aluminium, instrumented with thermocoupoles and heat flux meters and heated by electric blankets to a uniform temperature 15 deg. C above that of the wind tunnel air. Comparative velocity field measurements were also made in an unheated straight section of the same length. Standard hot-wire and resistance thermometer measurement techniques were used with on-line data processing by micro-computer. The effects of this mild ($\delta/R = 0.01$) wall curvature on the Reynolds stresses are indicated by the data plotted in Figure 1; the changes in the thermal turbulence are of the same order. The mean field data plotted in Figure 4 shows, as expected, that the skin friction coefficient c_f and the Stanton number are depressed on the convex wall relative to the values shown for plane wall flow and increased on the concave wall, in accordance with the general rule [18] that the changes are an order of magnitude greater than δ/R.

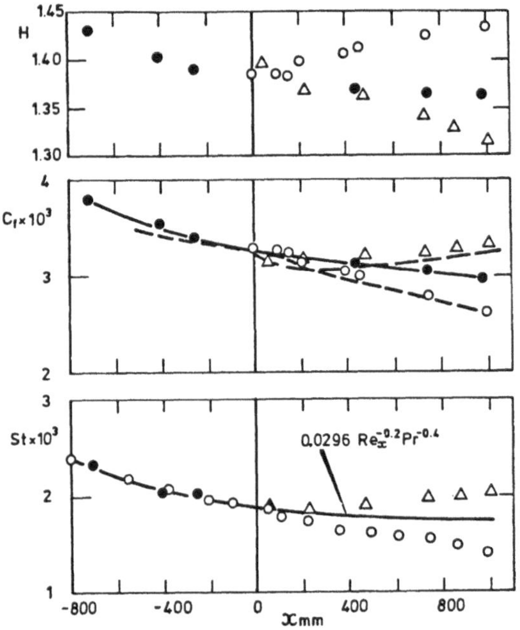

Fig. 4. Streamwise variation of the shape factor, skin friction coefficient and Stanton number in flat, o, mildly curved convex, ●, and concave, Δ, boundary layers. Broken line is Ludweig-Tillmann formula for c_f.

It was not expected, however, that the changes in St would be greater than those in c_f. The distortion of the mean flow is shown in the changes in the shape factor H which increases in the convex wall layer and decreases in the concave wall layer. A minor feature of interest is that the skin friction may be calculated quite accurately in these and other curved flows by the Ludweig-Tillmann formula, in which low c_f values are associated with high H and vice-versa, as for boundary layers in pressure gradients.

One reservation must be expressed: while the surface heat transfer has been measured independently, and in different ways in the two sets of experiments, the skin friction coefficient in these and in all previous curved wall experiments were obtained by Preston tube and Clauser methods which rely on the persistence of the "law of the wall" in the usual flat plate form. It is, however, easy to find a semi-logarithmic distribution of the measured mean velocities; very much more difficult to fit the temperature profiles. The Simon and Moffat data from the highly curved flow in particular show large departures from the logarithmic form.

Because the behaviour of the mean field near the wall is important in calculation methods which employ "wall functions" to resolve the sublayers we have experimented with various hypotheses for the flow in this region. The buoyancy analogy and subsequent experience suggests that the effects of mild curvature may be accounted for in calculation methods by multiplying the rectilinear flow mixing length distribution $\ell_o(y)$ by a factor linear in the strain rate ratio:

$$\ell = \ell_o(1 - \beta S) \tag{8}$$

where the coefficient β is generally of order ten; a maximum value of 14 is recommended in [18] for mild curvature effects while Adams and Johnston [19] have used a value of 6 to calculate the development of a highly curved flow equivalent to that of [6]. The response of ℓ to a change from flat to curved wall conditions is described by a simple rate equation for β

$$X \frac{d\beta}{dx} = \beta_o - \beta \tag{9}$$

in which the adjustment length X may be estimated [18] as the product of the mean velocity and the "memory time" of the stress containing eddies

$$X = Ut = U \frac{1}{2} \overline{q^2}/\epsilon \tag{10}$$

The mildly curved convex layer results, reproduced from [4] and [16] in Figures 5 and 6, broadly support this hypothesis (8) and its equivalent for the temperature length scale, at least in the inner region. In the outer zone the attenuation of length scales is considerably overestimated. In an adjustment length of 50δ it is appropriate to use the full values of β and β_θ to modify the flat plate mixing length profiles.

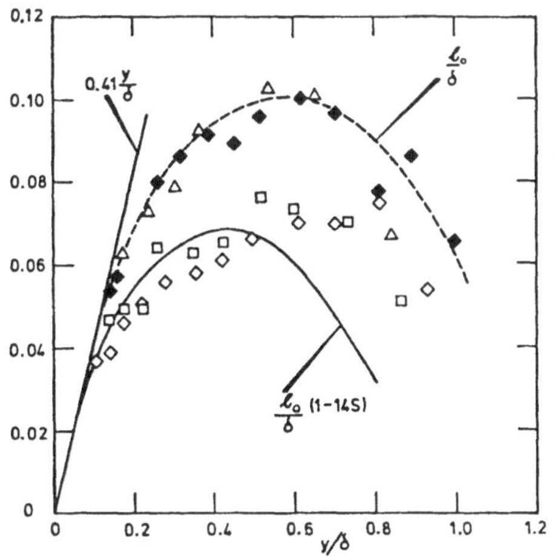

Fig. 5. Cross stream variation of the turbulence length scale in flat and convex wall boundary layers [4].

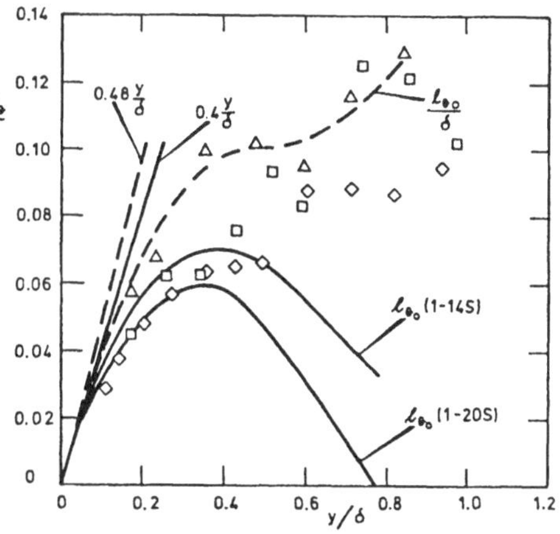

Fig. 6. Cross stream variation of the thermal length scale in flat and convex wall boundary layers [16].

The use of (8) with the usual mixing length formula for a constant stress inner layer produces, to first order, the modified law of the wall:

$$u^+ = \frac{1}{\kappa}\ \ell n y^+ + A + \beta\,\frac{y^+}{R^+}\left(\frac{1}{\kappa}\ \ell n y^+ + A - \frac{1}{\kappa}\right) \tag{11}$$

and its equivalent for the temperature distribution:

$$T^+ = \frac{1}{\kappa_\theta}\ \ell n y^+ + A_\theta + \beta_\theta\,\frac{\kappa}{\kappa_\theta}\,\frac{y^+}{R^+}\left(\frac{1}{\kappa}\ \ell n y^+ + A - \frac{1}{\kappa}\right) \tag{12}$$

The additional terms in (12) should be greater than those in (11). It is argued that time constant in a rate equation for β_θ is related to the response time of the thermal turbulence:

$$t_\theta = \frac{1}{2}\ \overline{\theta^2}/\epsilon_\theta \tag{13}$$

which is less than half of the mechanical time scale (10). The initial changes in Stanton number plotted in Figure 4 support this idea of a more rapid response; direct measurements of the thermal/mechanical time scale ratio in the convex boundary layer are reproduced from [16] in Figure 7. A value of about 0.34 was also obtained for this quantity in a heated homogeneous shear flow by Tavoularis and Corrsin [20]. Although the accuracy of this sort of measurement is not very high the results suggest a slight increase in the time scale ratio in the curved flow resulting from an expected decease in t while t_θ, like the temperature variance, remains substantially unaffected by the curvature.

The changes to the wall laws (11), (12), reflect only qualitatively but not quantitatively the observed trends, with slightly better agreement obtainable for the high curvature data of [15] then for own measurements. Better results have been obtained by C.A. Verriopoulos [21] using the mixing length correction with an assumed 1/7 power law for the velocity distribution to integrate from the wall through the sublayers. Figure 8 illustrates how the temperature profiles in the two convex boundary layers may be fitted in this way when β is taken as 14 and K_H/K_M

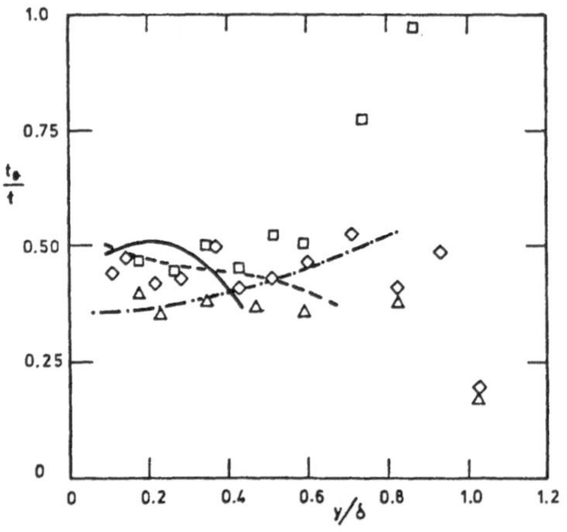

Fig. 7. Cross stream variation of the thermal/mechanical time scale ratio. Data points as in Fig. 3. Comparative data from plane flows cited in [16].

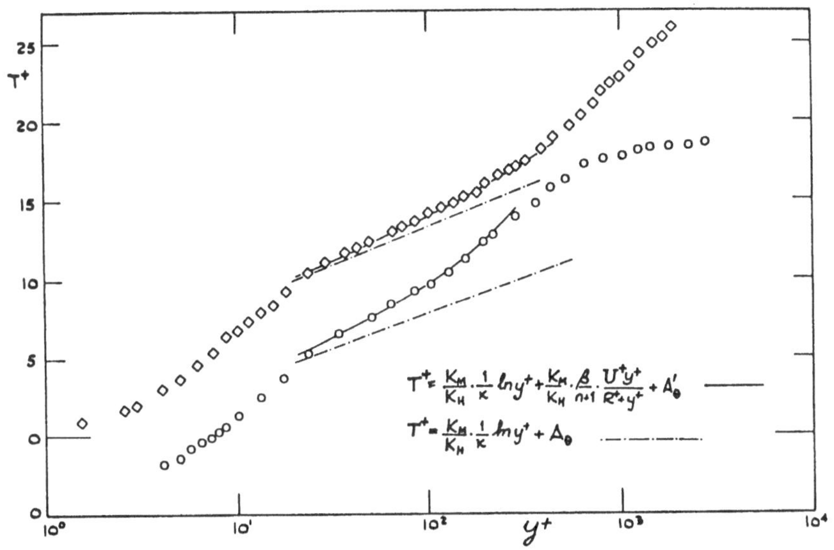

Fig. 8. Departure of the mean temperature profiles from the inner law in mildly curved [16] and strongly curved [15] convex wall flow. Empirical formulae by Verriopoulos [21].

is 0.89 and 0.84 of the global values obtained from the slopes of the temperature and velocity profiles in flat wall flow for weak and strong curvature respectively. The values chosen are, however, not unique and the data can be fitted over a range of β and K_H/K_M. Finally, Figure 9 shows the cross stream variation of K_H/K_M in the mildly curved boundary layer. After allowing for the usual considerable uncertainty which is inevitable in this sort of measurement, the trend seems to be that this quantity is decreased by stabilizing curvature. These results are at variance with the expectations based on the order of magnitude arguments of the last section. It would be useful to have additional confirmation from other sources, especially from curved wall experiments with the wall shear obtained independently of assumptions about the nature of the mean velocity profile. The

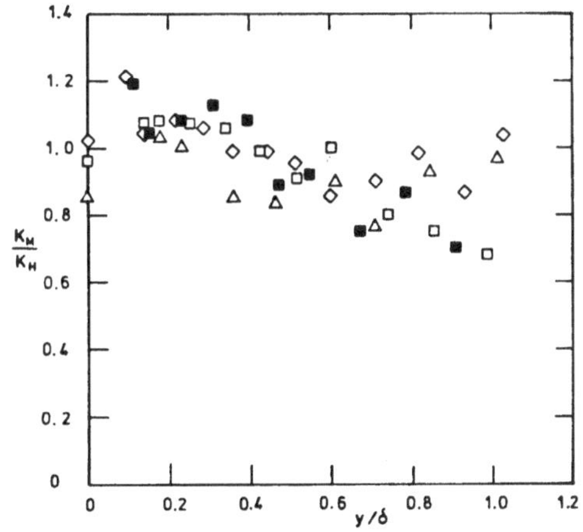

Fig. 9. Cross stream variation of the ratio of the turbulent exchange coefficients in flat and mildly curved convex wall flow. Data points as in Fig. 3. (from [16]).

observed response to the extra strain affecting the turbulent stresses and scalar fluxes also undermines assumptions commonly made in the construction of modelled equations for the second moments. It is to this topic that we now digress in the following section.

4. Modelling the Second Moment Equations

For the purpose of practical flow calculation in the forseeable future, the lowest level of turbulence closure which accounts properly for complex strain fields generally, and curvature effects in particular, is at second moment level. The application has not, however, been uniformly successful in this context. For example, Launder and Morse [22] found that their modelled equations for the Reynolds stresses actually gave a shear stress component with the wrong sign in a swirling jet, so that the calculated spreading rate was reduced rather than increased by the swirl. Launder and Morse identified two major weaknesses in their method: in the empirical auxiliary equation for the turbulent energy dissipation rate, and in modelling the pressure-strain correlations of the Reynolds stress equations. Similar weaknesses are exposed in the application of similar methods to the calculation of longitudinally curved flows.

With a few exceptions (see, for example, the discussion of intercomponent turbulent energy transfer in homogeneous shear flow by Harris, Graham and Corrsin [23]) most modellers have expressed the difficult pressure-strain terms in the stress equations as the sum of two components: one involving only the averages of fluctuating quantities, and the other containing explicitly the mean strain rate. A basic requirement is that the two components ought separately to satisfy the conditons set by the observed rate of return of anisotropic turbulence to isotropy in the absence of mean strain, and a theoretical result for the response of suddenly distorted isotropic turbulence. A third condition is that the model should realistically predict the shear stress and the partition of energy between components in simple shear flow.

Few, if any, pressure-strain models in general use actually satisfy all three conditions. Most authors have followed Launder's [24] example in modelling the two components separately and requiring that the mean strain or "rapid" part of the model satisfy the exact result of rapid distortion theory. The coefficient of the turbulence term is then adjusted, not necessarily to accord with the measured rates of return to isotropy, but to recover roughly the correct stress levels measured by Champagne, Harris and Corrsin [25] in a weakly sheared homogeneous shear flow. It is the incorrect balance between these two components in an over-simplified model of the pressure strain that contributes to the anomalous results obtained for curved and swirling flow.

To fix ideas consider the the widely used pressure-strain model [13,24]

$$\overline{\frac{p}{\rho}\left(\frac{\partial u_i}{\partial x_j} + \frac{\partial u_j}{\partial x_i}\right)} = - C_1 \frac{2\epsilon}{q^2}\left(\overline{u_i u_j} - \frac{1}{3}\,\delta_{ij}\,\overline{q^2}\right) - C_2\left(P_{ij} - \frac{2}{3}\,\delta_{ij}P\right) \quad (14)$$

The first term on the right hand side is Rotta's linear intercomponent energy transfer model; in the second, mean strain, term, P_{ij} and P are defined as the production rates of $\overline{u_i u_j}$ and $\frac{1}{2}\overline{q^2}$ respectively. This expression [14] necessarily represents a drastic over-simplification of the complex physical processes involved but it contains the two essential ingredients and gives results which in practice are not noticeably inferior to those obtained from more complicated formulations. The constant C_1 must be approximately 3.0 to fit the return to isotropy data and $C_2 = 0.6$ satisfies the result of rapid distortion theory.

For high Reynolds number shear flow in local equilibrium it is easily shown [26] that the stress ratios are functions only of $(1.-C_2)/C_1$ which, to give approximately the right stress levels for the homogeneous flow with weak shear [25], has to lie between 0.18 and 0.24. Figure 10 shows how the model constants have been adjusted by various authors so as to satisfy this condition. In our earlier calculations of buoyant shear flow in the atmospheric surface layer (Figure 2) we took care to satisfy the rapid distortion requirement and in consequence had to reduce C_1 from the return to isotropy value of about 3.0 to 1.8 to get $(1.-C_2)/C_1$ into the weakly sheared homogeneous flow range. It is now argued that the emphasis should be changed so as to reduce the contribution of the rapid component; indeed this change is necessary to obtain reasonable predictions of longitudinally curved flows which are independent of the choice of coordinate system. The predicted effects of swirl on the development of a jet are also substantially improved when the relative weighting of the rapid contribution is reduced; in this particular model [14] by setting C_2 equal to 0.3 and restoring C_1 to 3.0. The effects of gravity on the density stratified turbulence in the atmospheric surface layer have also been recalculated, with this and other minor changes, to give the results compared with our original predictions [13] in Figure 2.

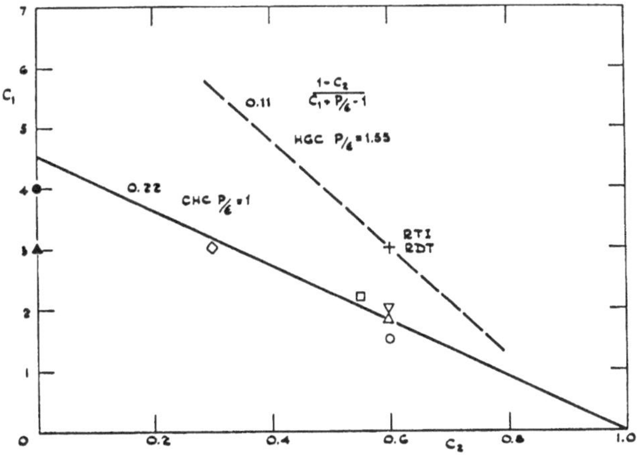

Fig. 10. Values of the constants in the pressure strain model (equation 14) needed to satisfy three basic conditions. Full line for weakly sheared homogeneous flow [25]; broken line for strongly sheared flow [23].

In requiring that a pressure-strain model given roughly the right stress levels in the weak shear flow [25] we, and most other authors, have ignored the later measurements by Harris, Graham and Corrsin [23] in homogeneous flow with strong shear which have implications for turbulence modelling discussed by Leslie [26]. On this occasion it is worth turning aside to look at the model results for the high shear case with a turbulent energy production to dissipation rate ratio, P/ϵ, of approximately 1.55 (from Leslie's analysis of the data). It turns out that when the first two conditions are satisfied by setting C_1 and C_2 equal to 3.0 and 0.6 so that $(1. - C_2)/C_1 = 0.13$, the stress ratios obtained by Leslie's method are not too far from the measured values with which they are compared in the followed table.

TABLE 1
Measured and predicted stress levels in highly sheared homogeneous turbulence

		C_1	C_2	$(1.-C_2)/C_1$	$\overline{u^2}/\overline{q^2}$	$\overline{v^2}/\overline{q^2}$	$\overline{w^2}/\overline{q^2}$	$\overline{uv}/\overline{q^2}$
	(a)	1.8	0.6	0.22	0.51	0.245	0.245	−0.18
Model	(b)	3.0	0.3	0.23	0.54	0.23	0.23	−0.19
	(c)	3.0	0.6	0.13	0.45	0.275	0.275	−0.155
Data	[23]	–	–	–	0.50	0.20	0.30	−0.15

Although the model constants (c) in the table give very nearly the correct shear stress to energy ratio of 0.15, the partition of energy between components, which may strongly influence the development of flows with complex strain fields, is not well predicted. We have not pursued this matter much further, only to ascertain that the spreading rate of a swirling jet is considerably underpredicted when these constants are used. When the pressure strain model [14] is substituted in the Reynolds stress equations for curved flow, and the transport terms are discarded for local equilibrium turbulence, the resulting set of algebraic equations is easily solved for the stress to energy ratios as functions of the curvature Richardson number. Calculated values of the structure function $-\overline{uv}/\overline{q^2}$ for the three sets of model constants are plotted in Figure 11. It is seen that although

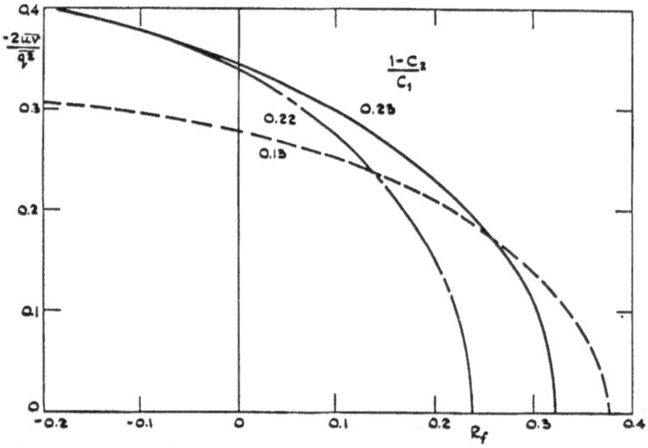

Fig. 11. Predicted extinction of the shear stress in curved shear flow in local equilibrium.

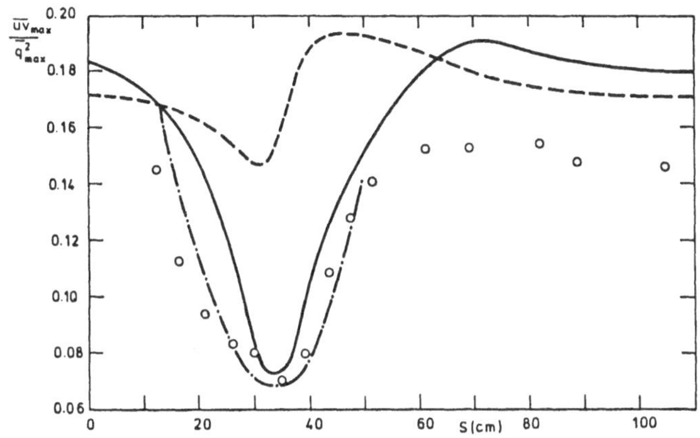

Fig. 12. Measured and predicted variation of the structure function along the centreline of a highly curved mixing layer [27].

the ratio $(1 - C_2)/C_1$ is nearly the same (0.22, 0.23) for sets (a) and (b) the critical Richardson numbers at which shear stress extinction is predicted are considerably different. The critical value for set (b), $R_f = 0.32$ is very close to the lowest value measured by Castro and Bradshaw [27] on the centreline of a highly curved mixing layer where, in nearly local equilibrium conditions, the shear stress fell to very low levels. Figure 12 shows the measured and calculated variation of the structure function on the centreline of this flow. The pressure strain model [14] has been used with the constants $C_1 = 3.0$, $C_2 = 0.3$ as in (b) above to give the results shown by the full line; also shown, by the chain dotted curve, are the results of calculations based on modified rapid distortion theory [28], and, by the broken line, those obtained using an uncorrected eddy viscosity model. This single example will suffice to show what can be done in the calculation of curved flow using a relatively simple model of the pressure strain. Fairly good results can also be obtained for highly curved boundary layers and wall jets with strong destablizing curvature in the outer mixing layer but not, unfortunately, for mildly curved wall layers. The modelled equations linked to wall functions based on the logarithmic law are apparently too "stiff" to respond correctly to relatively small changes in the mean strain rate.

The problem of closing the equations for the heat fluxes by modelling the pressure containing correlation has yet to be solved satisfactorily. For homogeneous flows the solution of a Poisson equation for the pressure can again be used to relate this correlation to a volume integral involving two-point quantities: one type containing only turbulence correlations and another containing the mean rate of strain. Received opinion is that a single-point closure model ought to contain similar separate terms, for example in the counterpart to the simple pressure-strain model [14]:

$$\overline{\frac{p}{\rho}\frac{\partial\theta}{\partial x_i}} = -C_{1\theta}\frac{2\epsilon}{\overline{q^2}}\overline{u_i\theta} + C_{2\theta}\overline{u_k\theta}\frac{\partial U_i}{\partial x_k} \qquad (15)$$

This conventional hypothesis that the pressure correlation is unaffected by the presence of a mean temperature gradient is questionable. Certainly the observed effects of longitudinal curvature on heat transfer are not accounted for by closing the conservation equations in this way and a less restricted approach seems to be needed. The theoretical arguments for the inclusion of the mean temperature gradient in the pressure correlation model which have been presented by Jones and Musonge [29] and Dakos [30] are a promising step in this direction, though they have yet to be tested in model studies of inhomogeneous flows.

Support from the Science and Engineering Research Council is gratefully acknowledged.

References

[1] Klebanoff, P.S.: N.A.C.A. Rep. 1247 (1955).
[2] Muck, K.C.: Ph.D. Thesis, Univ. London (1982).
[3] Ramaprian, B.R. and Shivaprasad, B.G.: J. Fluid Mech. 85, 273 (1978).
[4] Gibson, M.M., Verriopoulos, C.A. and Vlachos, N.S.: Expts. Fluids 2, 17 (1984).
[5] So, R.M.C. and Mellor, G.L.: J. Fluid Mech. 60, 43 (1973).
[6] Gillis, J.C. and Johnston, J.P.: Turbulent Shear Flows 2, 16 (Springer, Berlin 1980).
[7] Kreith, F.: Mechanical Engineering 77, 1247 (1955).
[8] Thomann, H.: J. Fluid Mech. 33, 283 (1968).
[9] Mayle, R.E., Blair, M.F. and Kopper, F.C.: Trans. A.S.M.E., J. Heat Transfer 101, 521 (1979).
[10] Prandtl, L.: N.A.C.A. TM 625 (1931).
[11] Bradshaw, P.: J. Fluid Mech. 36, 177 (1969).
[12] Young, S.T.B.: Univ. London, Queen Mary Coll. Rep. QMC-EP 6018 (1975).
[13] Gibson, M.M. and Launder, B.E.: J. Fluid Mech. 86, 491 (1978).
[14] Businger, J.A., Wyngaard, J.C., Izumi, Y. and Bradley, E.F.: J. Atmos. Sci. 28, 181 (1971).
[15] Simon, T.W. and Moffat, R.J.: A.S.M.E. paper 79-WA/GT-10 (1979).
[16] Gibson, M.M. and Verriopoulos, C.A.: Expts. Fluids 2, 73 (1984).
[17] Dakos, T., Verriopoulos, C.A. and Gibson, M.M.: J. Fluid Mech. 145, 339 (1984).
[18] Bradshaw, P.: AGARDograph 169 (1973).
[19] Adams, E.W. and Johnston, J.P.: A.S.M.E. paper 83-GT-80 (1983).
[20] Tavoularis, S. and Corrsin, S.: J. Fluid Mech. 104, 311 (1981).
[21] Verriopoulos, C.A.: Ph.D. Thesis, Univ. London (1983).
[22] Launder, B.E. and Morse, A.: Turbulent Shear Flows 1, 279 (Springer, Berlin 1979).
[23] Harris, V.G., Graham, J.A.H. and Corrsin, S.: J. Fluid Mech. 81, 657 (1977).
[24] Launder, B.E., Reece, G.J. and Rodi, W.: J. Fluid Mech. 68, 537 (1975).
[25] Champagne, F.H., Harris, V.G. and Corrsin, S.: J. Fluid Mech. 41, 81 (1970).
[26] Leslie, D.C.: J. Fluid Mech. 98, 435 (1980).
[27] Castro, I.P. and Bradshaw, P.: J. Fluid Mech. 73, 265 (1976).
[28] Townsend, A.A.: J. Fluid Mech. 98, 1 (1980) .
[29] Jones, W.P. and Musonge, P.: Proc. 4th Turbulent Shear Flows Symposium, Karlsruhe (1983).
[30] Dakos, T. and Gibson, M.M.: Unpublished note, Imperial College, London (1984).

Limitations of Second Order Modeling of Passive Scalar Diffusion

John L. Lumley and Ike Van Cruyningen

Sibley School of Mechanical and Aerospace Engineering, Cornell University
Ithaca, NY 14853, USA

The problem of modeling scalar variance from an elevated source is discussed at length. From a simple model, it is shown that observed behavior is entirely a result of the growth of the instantaneous plume width relative to the mean plume width. From a model valid for large sources, we suggest a simple explanation for the apparent dependence of asymptotic levels of variance on source size. Pope (1983) has pointed out that second order models of passive scalar transport do not retain the superposability of the primitive equations. Explanations for this, and possible consequences are explored. It is shown how second order models for the rapid terms, the time scales of which are determined by the mean velocity (rather than inertial spectral transfer), can be constructed in a superposable manner (Shih, 1984). Difficulties are described in simultaneously satisfying requirements imposed on the cross-dissipation by realizability and by superposability. We present computations of the data of Warhaft (1984) using a model in which the transport is based on first principles, and much of the rest of the model satisfies realizability. The model is able to reproduce the data of Warhaft satisfactorily if the initial plume width is larger than the Kolmogorov microscale. Calculations are carried out of a plume superposed on a background; satisfactory results are achieved when the background level is low, despite the lack of formal superposability, leading to the conclusion that this may not be a serious problem.

1. Introduction

One of the most important areas of application for turbulence modeling is the prediction of the dispersal of passive and active contaminants in the atmosphere and ocean. Whatever the turbulence

*Supported in part by the U.S. Office of Naval Research under the following programs: Physical Oceanography (Code 422PO), Power (Code 473); in part by the U.S. National Science Foundation under grant No. ATM 79-22006; and in part by the U.S. Air Force Geophysics Laboratory. Prepared for presentation at "Lectures on the Fundamentals of Fluid Mechanics", a symposium honoring Stanley Corrsin, June 11, 12 1984 at Northwestern University. Parts of this paper will appear in the M. S. Thesis of IVC.

modeling technique used, it is important to verify that it is capable of reproducing the important qualitative and quantitative features of the observations. In this section, we will use crude models to display the qualitative features of dispersion that are evident primarily in the measurements of Warhaft (1984) (and to a lesser extent in those of Fackrell & Robbins (1982)), and that we feel are important.

We are particularly interested in the second-order modeling technique, the so-called one-point closures. In this technique, equations are carried for the means, variances and fluxes, and these equations are closed by representing the third order quantities and certain other terms such as the pressure correlations, in terms of the second order quantities.

One of the aspects of contaminant dispersal that we will find is important concerns the time scale ratio of a plume from an elevated source. As a paradigm for plume dispersal from an elevated source, we may consider a passive plume from a line source in a homogeneous, decaying ($\langle q^2\rangle \propto x^{-1.35}$, $\ell \propto x^{0.325}$), isotropic turbulence, a model for Warhaft's (1984) experimental situation. We may suppose it self-preserving, with an average half-width proportional to ℓ_θ, and a maximum (centerline) value of θ_0. If the plume is self-preserving, we may expect that

$$\begin{aligned} \langle\varepsilon_\theta\rangle &= h(\eta)\theta_0{}^2 v'/\ell_\theta, \\ \langle\theta'^2\rangle &= \theta_0{}^2 g(\eta) \end{aligned} \tag{1}$$

where $g(\eta)$ and $h(\eta)$ are universal functions, and $\eta = y/\ell_\theta$. The first relation may be obtained from equating the thermal production to the thermal dissipation at the production peak. v' is the r.m.s. turbulence intensity. We may also write $\langle\varepsilon\rangle = v'^3/\ell$, and $\langle q^2\rangle = 3v'^2$, from which we may obtain the time scale ratio

$$r = \langle\varepsilon_\theta\rangle\langle q^2\rangle/\langle\theta^2\rangle\langle\varepsilon\rangle = (\ell/\ell_\theta)3h(\eta)/g(\eta) \tag{2}$$

As we move upstream closer and closer to the source, ℓ/ℓ_θ, and hence r, becomes larger and larger. If assumptions like those in (1) are substituted in the equation for the mean temperature, and the coefficients in the equation are required to be constant, we find that if v' = const, then $\theta_0 \propto x^{-1}$, while if $v'^2 \propto x^{-1.35}$, then $\theta_0{}^2 \propto x^{-0.65}$. This can be interpreted as saying that, if we are close to the elevated source, we expect linear spread, while if we are far from the source, so that both plume and field appear to have the same origin, the spread is $\propto x^{0.325}$, which is the same as ℓ. Hence, if the plume really begins from an elevated line source, r will descend from infinity proportional to inverse distance from the source; farther from the source, when ℓ_θ and ℓ appear to have nearly the same origin, they will be growing together, and r will tend to an asymptotic value. It is a coincidence that the initial linear behavior of ℓ_θ predicted here on the basis of self-preservation matches reasonably well the initial linear behavior predicted for a line source during the short time that the velocity field remains perfectly correlated, since the shape of the plume during this period is not self-preserving.

Although our analysis is strictly valid only for a line source, it is clear that an expression similar to (2) will hold for a more general situation. Hence, it is clear that for a source which is of finite size, the initial value of r will be bounded, and will be smaller the larger the source is.

We may also examine the behavior of the concentration variance. Suppose, for example, that at a given distance from the source the instantaneous waving plume can be represented as

$$\theta(t) = \theta_s f((y-y_s)/\ell_s) \tag{3}$$

(c.f. Fackrell & Robins, 1982) where $f(\eta)$ is the profile of the instantaneous plume, with ℓ_s proportional to its half width, and θ_s its centerline value. We will suppose for convenience that $f(\eta)$ has a Gaussian shape, with ℓ_s its standard deviation. We will suppose that y_s has a Gaussian distribution $\beta(y_s)$ with standard deviation σ_s, in accordance with observation. This simple model leads to a gradient transport form, so that $\langle\theta' v\rangle = -\partial_y\langle\theta\rangle d_t(\sigma_s^2/2)$ and $\langle\theta'^2 v\rangle = -\partial_y\langle\theta'^2\rangle d_t(\sigma_s^2/2)$, so that both second and third moments have the same transport coefficient, given traditionally by the rate of spread of half the plume dispersion variance. In particular, we can write for the mean value

$$\langle\theta(y)\rangle = \theta_s \int \beta(y_s) f((y-y_s)/\ell_s) dy_s \approx \theta_s \beta(y) \ell_s (2\pi)^{1/2} \tag{4}$$

if we take $\ell_s \ll \sigma_s$. Proceeding in exactly the same way, we obtain for the variance

$$\langle\theta'^2\rangle = \theta_s^2 \ell_s \beta(y)\sqrt{\pi}(1 - 2\ell_s\beta(y)\sqrt{\pi}) \tag{5}$$

We may easily obtain a more exact expression without assuming that the instantaneous plume is thin relative to the mean width, but it is not qualitatively different, and is not worth the trouble. By differentiating equation (5) we can find that $\langle\theta'^2\rangle$ can have off-axis peaks, but only so long as $\sigma_s/\ell_s 2\sqrt{2} < 1$, i.e.- when the instantaneous plume is relatively thick compared to the mean plume. The condition corresponds to $\ell_s/\sigma_s > 0.35$, which is a bit marginal, but does not seriously violate our condition of thinness of the instantaneous plume.

Now, we expect that ℓ_s will initially grow as $x^{1/2}$, while it is spreading by molecular action. When ℓ_s enters the inertial subrange (supposing that the Reynolds number is high enough), the instantaneous plume width will grow according to Richardson's nearest neighbor statistics; close to the source, when the background can be regarded as non-decaying, $\ell_s \propto x^{3/2}$, while farther from the source, when the background and the plume can be regarded as having the same source, it will grow as $x^{0.325}$. When it is no longer governed by nearest neighbor statistics, it will also grow as $x^{0.325}$.

Hence, if the plume is initially of width smaller than the Kolmogorov microscale, σ_s/ℓ_s will initially grow as $x^{1/2}$; when the plume width enters the inertial subrange, it will begin to shrink as $x^{-1/2}$ and, as distance from the source increases, will approach a constant. Hence, we expect the ratio

$$\sqrt{\langle\theta'^2\rangle_m}/\langle\theta\rangle_m = [(1 - \sqrt{2}\ell_s/\sigma_s)/(\sqrt{2}\ell_s/\sigma_s)]^{1/2} \quad (6)$$

(where the subscript m designates the centerline value) to grow initially (for a very small source); when the plume width enters the inertial subrange, the ratio will begin to shrink, and will finally approach a constant. This is in agreement with the data of Warhaft (1984). Note that, if σ_s/ℓ_s shrinks below 2.8 (roughly) as it approaches its asymptotic value, then off-axis peaks will again appear. This is observed in the data of Warhaft (1984).

Although our reasoning is really only valid for a line source, we can see qualitatively what will happen for a source of finite size. If the source size is initially within the inertial subrange, both the length scale ratio, and the ratio of standard deviation to mean will initially shrink, ultimately approaching a constant. The initial value will depend on the size of the source. All of this is in good agreement with the observations of Fackrell & Robins (1982).

Csanady (1967a, b) also considered this problem, and concluded that (6) would approach a constant in shear flows and in the atmosphere; we will question this below.

One final aspect of plume behavior remains. Fackrell & Robins (1982), Chatwin & Sullivan (1979) and Durbin (1980) have suggested that the ultimate value of the centerline ratio of standard deviation to mean concentration is a function of the source size. Fackrell & Robins (1982) data appear to support this, but the experimental values are not really far enough downstream to be absolutely sure. At first glance this is a rather startling conclusion, and it deserves some discussion.

The model we are using is not valid in the very far downstream regime, nor for large sources. We may construct a somewhat better model for a large source, if we imagine a collection of equally spaced point sources within the large source region. If we are far enough downstream, it is legitimate to consider that the wandering plumes from these individual point sources are far enough apart to wander independently. If the source is very large, several integral scales in diameter at the point of origin, the individual plumes will wander essentially independently from the beginning. Under these circumstances, we find that the wandering of the centroid of the combined plume is a smaller fraction of the mean spread of the plume, the larger the source, and that the ratio is not a function of distance from the source. This is the reason why our simplified model is no longer valid, since it derives its major contribution from the wandering of the centroid. In a plume at great distances from the source, the major contribution to the variance comes from convergence and divergence of regions of the plume, creating local highs and lows of concentration. In our composite model, this is modeled by the approaching or drawing apart of the individual plumes. We can easily compute the mean concentration and concentration variance from this composite source. We find that the mean concentration is the sum of expressions like (4), one from each individual source, and the variance is the sum of expressions like (5), one from each source. The ratio of standard deviation to mean will thus be inversely proportional to the square root of the number of sources (this is easiest to see very far downstream, when the differences in the source location can be neglected). Since the number of sources is proportional to the cube of the ratio of the source size to the integral scale, the ratio of standard deviation to mean is inversely proportional to the 3/2 power of this ratio. Although this crude demonstration is valid only for rather large sources, it establishes the principle that a persistent dependency of the ratio on source size is possible. The trend from this crude model is in accord with Fackrell & Robins (1982), and with the conclusions of Durbin (1980).

Our models above are not really valid even for small sources very far downstream. If σ_S/ℓ increases without bound (as it would in a shear flow), then the plume is eventually torn apart by the turbulence and loses its integrity, violating an essential assumption of our model. An increasing fraction of the variance is contributed by the convergence and divergence of the regions of the plume. In effect the plume becomes eventually N independently wandering plumes, as in the large source model above, and N continues to increase without bound. This suggests that the ratio of centerline r.m.s. to mean concentration will eventually approach zero.

This will, of course, also be true of the large source model: when the individual plumes have spread enough to begin to split, the effective N will begin to increase over the value determined by the source size, and the ratio of centerline r.m.s. to mean concentration will begin to drop.

This raises an interesting question in connection with laboratory data in decaying, homogeneous flow. It is not difficult to demonstrate that $\sigma_s \approx 2\ell$ asymptotically in such a flow. That is, relative to local scales the plume is no longer diffusing. The plume certainly does not remain coherent, but the extent of the splitting is not progressive - the plume scale and the turbulence scale are growing at the same rate, and the effective value of N does not continue to increase. Thus, the ratio of centerline r.m.s. to mean temperature here will approach a constant. This flow is thus fundamentally unlike a shear flow in this respect.

Note that our conclusion regarding the asymptotic behavior of ratios of centerline r.m.s. to mean concentration values in shear flows has far-reaching implications. In long-range transport models for acid rain, it is important to know the asymptotic value of the ratio of standard deviation to mean in the plume. The effective bulk reaction rate for conversion of SO_2 to H_2SO_4, whatever the mechanism, is dependent on the level of concentration fluctuations. In most current models, it is assumed that the ratio in (6) approaches zero asymptotically, and thus that the bulk reaction rate can be determined from data without concentration fluctuations. Our simple reasoning supports this conclusion. The data of Fackrell & Robins (1982) and of Csanady (1967a, b) are ambiguous: they appear to indicate approach to a non-zero constant, but do not have sufficient streamwise extent to exclude the possibility of ultimate approach to zero.

These are the essential qualitative features of plume dispersal that a good model should be able to reproduce. In what follows, we will examine some of the properties and limitations of the second order model, and will present some preliminary calculations of dispersal from a line source in homogeneous, isotropic decaying turbulence.

2. Pope's Problem

Pope (1983) has pointed out that the exact equations for a passive scalar additive are linear in additive concentration, so that fields from different sources are superposable. Because of this linearity, a field of several passive scalars with the same diffusivity may be recombined in arbitrary linear combinations without changing the equations.

The usual second order equations do not have this property, however (Lumley, 1983a, b). The equations for thermal dissipation are not homogeneous of second degree in additive concentration, and hence fields from different sources are not simply superposable, nor can several fields with the same diffusivity be combined arbitrarily. The equations for the second moments used by most workers are linear, since the coefficients used in the pressure-concentration gradient terms are constant. However, these forms do not satisfy realizability (Lumley, 1978, 1984a, b). If the forms are modified so as to satisfy realizability (Lumley & Mansfield, 1984), they are no longer linear, and hence not superposable either. Shih (1984) has suggested a modified form of the realizability condition, which permits superposability, and satisfies realizability weakly, but which does not reproduce system behavior near a limiting state. That is, suppose that u_1 and θ are nearly perfectly correlated. It is possible to solve exactly for the behavior of the system near this condition, and the strong realizability condition reproduces this behavior, while the weak condition does not.

No simple solution for this difficulty presents itself. The difficulty, in all the equations, is the concentration time scale, $\langle\theta^2\rangle/\langle\varepsilon_\theta\rangle$. If this time scale is assumed to be the same for all the fields, then they may be superposed, and different fields can be combined linearly. A basic assumption of the second order modeling is that the field of a particular variable can be complete described by the scales of the energy-containing range, or in other words, that the spectrum is simple. A superposition of two scalar fields having different time scales, or a linear combination of two such fields (which is the same thing) no longer has a simple spectrum, and hence cannot reasonably be described by equations having a single time scale.

While it is possible to construct multiple time scale models, they become extremely complex, and it is not clear that it is worth the effort; probably a large eddy simulation would be more sensible. The practical question is, how much does it matter? That is, many dispersion situations in the atmosphere involve multiple sources of differing sizes, so that the concentration field consists of the superposition of a number of fields of different time scales. In attempting to describe this combined field by equations having a single time scale, necessarily a compromise everywhere, how much error is made, and where is the error worst? In this preliminary study we present (below) calculations for a single source, with and without a decaying isotropic homogeneous background field of different time scale. We will find that, although the model is not superposable, the results of the calculation of the combined fields is not significantly different from the superposition of the two fields. The true time scale of the superposed fields is the energy-weighted average of the time scales of the individual fields, and the model computes something approaching this.

In the second order models, the part of the pressure correlation that is not a transport term is customarily split into a so-called rapid term, and a return-to-isotropy term (Lumley, 1978), corresponding repectively to the part of the pressure linear in the mean velocity gradient (or the buoyancy), and the part quadratic in the fluctuating velocity. We have been discussing the form of the latter, since in a homogeneous turbulence without mean velocity gradients or buoyancy, the rapid term plays no role. In shear flows and atmospheric flows, however, the rapid terms are extremely important. As pointed out in Lumley (1978), the models for the rapid terms in general use do not satisfy realizability, although they do satisfy superposability. Attempts to make these terms satisfy realizability (Lumley, 1978) have resulted in their violating superposability. Recently, however, Shih (1984) has devised forms for the rapid terms which satisfy realizability exactly, and satisfy in addition the requirement of superposability. The key is the fact that, while the time scale of the return-to-isotropy term is the spectral transfer time scale, the time scale of the rapid term is determined by the mean velocity gradient, or the buoyancy. Hence, this time scale does not change when two fields are combined. If the rapid terms are represented as general functions of momentum and heat flux, and all conditions of symmetry, incompressibility, and normalization are applied, as well as realizability relations between the velocity gradient and buoyant rapid terms in the Reynolds stress and heat flux equations, together with the requirement of superposability, it is possible to unambiguously determine all coefficients.

Another interesting situation arises in connection with the cross-dissipation, if we have several scalars. Two scalars are enough to illustrate the point. Consider scalars θ and ϕ. Each has a dissipation $\langle\varepsilon_\theta\rangle = \gamma\langle\theta_{,i}\theta_{,i}\rangle$, $\langle\varepsilon_\phi\rangle = \mu\langle\phi_{,i}\phi_{,i}\rangle$ (where a comma denotes differentiation with respect to the space copordinate, we use the Einstein summation convention, and angle brackets indicate an ensemble average. The cross-dissipation is $\langle\varepsilon_{\theta\phi}\rangle = 0.5(\gamma + \mu)\langle\theta_{,i}\phi_{,i}\rangle$. Realizability requires (Lumley & Mansfield, 1984) that

$$\begin{aligned}&\langle\varepsilon_{\theta\phi}\rangle = 0.5\langle\theta\phi\rangle(\langle\varepsilon_\theta\rangle/\langle\theta^2\rangle + \langle\varepsilon_\phi\rangle/\langle\phi^2\rangle)\\ &\text{if } \langle\theta^2\rangle\langle\phi^2\rangle - \langle\theta\phi\rangle^2 = 0\end{aligned} \tag{7}$$

On the other hand, superposability, or the straightforward application of Schwarz's inequality, requires that

$$\langle\varepsilon_\theta\rangle\langle\varepsilon_\phi\rangle - \langle\varepsilon_{\theta\phi}\rangle^2 \geq 0 \tag{8}$$

at least for the case of equal diffusivities. These two requirements do not appear to be in conflict. Eq. (7) may be satisfied by setting $\langle\varepsilon_{\phi\theta}\rangle$ equal to the right hand side multiplied by $f(\rho^2, r')$ under all circumstances, where $\rho^2 = \langle\phi\theta\rangle^2/\langle\theta^2\rangle\langle\phi^2\rangle$, and $r' = \langle\varepsilon_\theta\rangle\langle\phi^2\rangle/\langle\varepsilon_\phi\rangle\langle\theta^2\rangle$, with $f(1, r') = 1$, $f \geq 1$. Then we obtain to satisfy both requirements $1 \leq f^2 \leq 4/\rho^2(\sqrt{r'} + 1/\sqrt{r'})^2$. Note that we must have $r' = 1$ if $\rho^2 = 1$, and the right side of the inequality must not be less than one; the equations must be arranged to produce this, and no clear way of doing so has presented itself.

3. Equations, Model and Initial Conditions

We will present here calculations of the dispersion of a line source in homogeneous, isotropic turbulence with and without a decaying background thermal field of different time scale. In the case without the background, we will attempt to reproduce the data of Warhaft (1984). The model is essentially the same as that presented in Shih & Lumley (1982). The basic equations are

$$U\partial_x\langle\Theta\rangle + \partial_y\langle\theta v\rangle = \gamma\partial^2{}_y\langle\Theta\rangle \tag{9}$$

$$U\partial_x\langle\theta^2\rangle + 2\langle\theta v\rangle\partial_y\langle\Theta\rangle + \partial_y\langle\theta^2 v\rangle = -2\langle\varepsilon_\theta\rangle + \gamma\partial^2{}_y\langle\theta^2\rangle \tag{10}$$

$$U\partial_x\langle\theta v\rangle + \langle v^2\rangle\partial_y\langle\Theta\rangle + \partial_y\langle\theta v^2\rangle = F + \gamma\partial^2{}_y\langle\theta v\rangle \tag{11}$$

$$U\partial_x\langle q^2\rangle = -2\langle\varepsilon\rangle \tag{12}$$

$$U\partial_x\langle\varepsilon\rangle = -\psi_0\langle\varepsilon\rangle^2/\langle q^2\rangle \tag{13}$$

$$U\partial_x\langle\varepsilon_\theta\rangle + \partial_y\langle\varepsilon_\theta v\rangle = -\psi^\theta\langle\varepsilon_\theta\rangle^2/\langle\theta^2\rangle \tag{14}$$

where U is the mean velocity, and F contains the pressure gradient - temperature correlation. To close this set of equations we need models for $\langle\theta^2 v\rangle$, $\langle\theta v^2\rangle$, F, ψ_0, ψ^θ and $\langle\varepsilon_\theta v\rangle$.

We will use the models for these terms used in Shih & Lumley (1982) with the exception of certain small changes introduced in Lumley & Mansfield (1984). Note that a number of misprints in Shih & Lumley (1982) have been corrected here.

$$F = 0.2\partial_y\langle\theta q^2\rangle - \phi^\theta\langle\theta v\rangle\langle\varepsilon\rangle/\langle q^2\rangle \tag{15}$$

$$\phi^\theta = 1 + r + \{1.1 + 0.55\mathrm{Tanh}[4(r-1)]\}(\beta - 1)F_D{}^{1/2} \tag{16}$$

$$F_D = (1 - \rho^2)/(1 - \rho^2/3)^3 \tag{17}$$

$$\rho = \langle\theta v\rangle/(\langle\theta^2\rangle\langle v^2\rangle)^{1/2} \tag{18}$$

$$r = \langle\varepsilon_\theta\rangle\langle q^2\rangle/\langle\varepsilon\rangle\langle\theta^2\rangle \tag{19}$$

$$\psi_0 = 14/5 + 0.98\exp[-2.83R_\ell{}^{-1/2}] \tag{20}$$

$$\psi^\theta = 2 - (2 - \psi_0)/r + 2.05\langle\theta v\rangle\partial_y\langle\Theta\rangle/\langle\varepsilon_\theta\rangle \tag{21}$$

$$\langle\theta^2 v\rangle = -[\langle v^2\rangle\partial_y\langle\theta^2\rangle + \partial_y\langle\theta v\rangle^2][\langle\theta^2\rangle/\langle\varepsilon_\theta\rangle]/2(1 + \phi^\theta/r) \tag{22}$$

$$\langle\theta v^2\rangle = -[-\langle q^2\theta\rangle(\beta - 2)/3 + 2\langle v^2\rangle(\langle q^2\rangle/\langle\varepsilon\rangle)\partial_y\langle\theta v\rangle]/(\beta + 2\phi^\theta) \tag{23}$$

$$\langle\theta q^2\rangle = -[\langle v^2\rangle\partial_y\langle\theta v\rangle][\langle q^2\rangle/\langle\varepsilon\rangle]/(1 + \phi^\theta) \tag{24}$$

$$\langle\varepsilon_\theta v\rangle = -(\langle\theta^2\rangle/\langle\varepsilon_\theta\rangle)[1/2(1 + \phi^\theta/r)][\langle v^2\rangle\partial_y\langle\varepsilon_\theta\rangle](1 + \rho^2) \tag{25}$$

$$\beta = 2 + (8/R_\ell{}^{1/2})\exp[-7.77/R_\ell{}^{1/2}] \tag{26}$$

It should be emphasized that the expressions for the third moments are derived from first principles and contain no adjustable constants. The coefficients appearing in them are the relaxation coefficients for the momentum flux (zero in this flow) and heat flux, which are determined in homogeneous flows. The way in which these expressions are derived is somewhat like non-equilibrium kinetic theory of mixtures, where the variances and fluxes play the role of the various species. It also resembles in many respects the two-point EDQNM closure (Eddy Damped Quasi-Normal Markovian; Orszag, 1970), which also has only one undetermined coefficient, which must be evaluated from the spectral constant in a homogeneous flow. The various other models are not so firmly based, but are for the most part dictated by the requirements of realizability. This model reproduced virtually within experimental error the data of LaRue et al (1981); all constants were optimized on other flows (Shih & Lumley, 1982).

The mechanical turbulence measured by Warhaft (1984) obeyed the following law:

$$\langle u^2\rangle/U^2 = 0.121(x/M)^{-1.4},\ \langle v^2\rangle/U^2 = 0.076(x/M)^{-1.32} \tag{27}$$

Despite the slight anisotropy, slowly shifting along the tunnel, the energy $\langle q^2\rangle$ decays as $(x/M)^{-1.35}$ with variation only in the fourth significant figure of the power. Since our purpose here is to examine the temperature transport, rather than to improve the interpolation formula for ψ_0, we have arbitrarily taken $\psi_0 = 3.48$, which gives the observed decay. The interpolation formula (20) gives a value slowly shifting from 3.60 to 3.54, corresponding to decay exponents of 1.25 and 1.30. Equation (20) was optimized on the data of Comte-Bellot & Corrsin (1966) (see Lumley & Newman, 1977), which were taken at higher Reynolds number for the most part.

We show calculations for three runs, with the source placed at $x_0/M = 52$, 52 and 20. In the three cases the initial profiles are at values of $x'/M = 0.36$, 8.1 and 17 respectively. x'/M indicates distance from the source. The calculations extend to distances of $x'/M =$ 8, 100 and 133 respectively. For simplicity, we will refer to these as Cases I, II and III respectively. They correspond exactly to the single-source measurements of Warhaft (1984).

From the basic equations, it is clear that the value of $\langle\theta v\rangle$ (specifically, the slope at the centerline) determines the rate of decay of the centerline value of $\langle\Theta\rangle$, and that by adjusting the value of $\langle\varepsilon_\theta\rangle$ we can control the initial rate of decay of the centerline $\langle\theta^2\rangle$ (which is influenced only by this term and the transport). Hence, we can select the initial conditions so as to guarantee that the centerline values of $\langle\Theta\rangle$ and $\langle\theta^2\rangle$ begin with the right value and slope. Whether they will continue to follow the experimental curves depends on whether the parameterization of the various terms is correct.

If it is assumed that the various profiles are self-preserving, which will allow us to obtain a crude estimate for $\langle\theta v\rangle$, we may obtain

$$\langle\theta v\rangle = v'\theta_0 g(\eta),\ \langle\Theta\rangle = \theta_0 F(\eta)$$

$$g(\eta) = -[U\partial_x\theta_0\ell_\theta/v'\theta_0]\eta F,\ \text{or}\ \langle\theta v\rangle = -[U\partial_x \ell n(\theta_0)]y\langle\Theta\rangle \tag{28}$$

where we have used conservation of heat, and θ_0 is the centerline value of $\langle\Theta\rangle$. The requirement of self-preservation also gives

$$\partial_x(1/\theta_0) \propto v' \tag{29}$$

Equations (28) and (29), together with the experimental data, allow the determination of $\langle\theta v\rangle$ at the initial location. This was satisfactory for cases II and III, but for case I the initial value of $\langle\theta v\rangle$ had to be adjusted empirically to give the observed initial slope of $\langle\Theta\rangle$. This is not surprising, since case I is very far from self-preserving. The initial curves for $\langle\Theta\rangle$ and $\langle\theta^2\rangle$ for each run were taken from analytic or spline (for $\langle\theta^2\rangle$ closest to the source) fits to the data. We took r constant across the flow, so that $\langle\varepsilon_\theta\rangle = r\langle\theta^2\rangle\langle\varepsilon\rangle/\langle q^2\rangle$ initially.

For the plume with a decaying background field, we simply added a uniform $\langle\theta^2\rangle_1$ to the initial $\langle\theta^2\rangle_0$ for the plume. This assumes that the plume temperature fluctuations and the background temperature fluctuations are initially uncorrelated. The background temperature fluctuations are taken to have r = 1, so that they might have been produced by a heated grid. From the definition, we can write

$$r = [r_0\langle\theta^2\rangle_0 + r_1\langle\theta^2\rangle_1]/[\langle\theta^2\rangle_0 + \langle\theta^2\rangle_1] \tag{30}$$

which assumes that initially the dissipation scales of the two fields are also not correlated. From this we obtain the initial $\langle\varepsilon_\theta\rangle$ values.

4. Results and Discussion

Following figures 1, 2 and 3 [where, for comparison, we have reproduced the profiles of normalized r.m.s. temperature for Cases I, II and III from Warhaft (1984)], the figures are in pairs, each pair referring to a single case. In each case, the first figure gives the evolution of the centerline values of the mean temperature profile [values from curves faired through Warhaft's (1984) data are open circles], and of the ratio of the centerline value of the r.m.s. temperature profile to that of the mean profile [Warhaft (1984) = eight-pointed stars], and the half-widths (at half the centerline value) of both the mean temperature profile [Warhaft (1984) = five-pointed stars] and the r.m.s. temperature profile [Warhaft (1984) = closed circles]. The second figure of each pair gives the profiles of the r.m.s. temperature. The values of x'/M displayed are the same as the measurements of Warhaft (1984). In all cases the mean temperature profile is approximately Gaussian in both experiment and calculation, so that we have not reproduced them here.

From the first pair of figures (4 & 5), it is clear that case I lies largely in the regime in which the instantaneous plume is below the Kolmogorov microscale in width, so that spreading is by molecular action. Because molecular spreading is so important to this case, we have included the molecular effects in the equations, although they will be negligible in cases II and III. It is evident that so close to the source, the data are not reproduced well. The mean temperature and the width of the mean profile are satisfactory (as they are in all the cases); but the evolution of the ratio of centerline r.m.s to mean, and the width of the r.m.s. profile, are poor; the details of the r.m.s. profiles are not well-reproduced. The off-axis peaks decay much too fast. The reason for this is probably that the transport model used is appropriate for quasi-homogeneous situations, and particularly situations in which intermittency is not vital. Very close to the source, of course, gradients are very large, and intermittency is dominant. Hence, transport is probably underpredicted. Since the initial time scale ratio is adjusted to give the proper initial slope of the evolu-

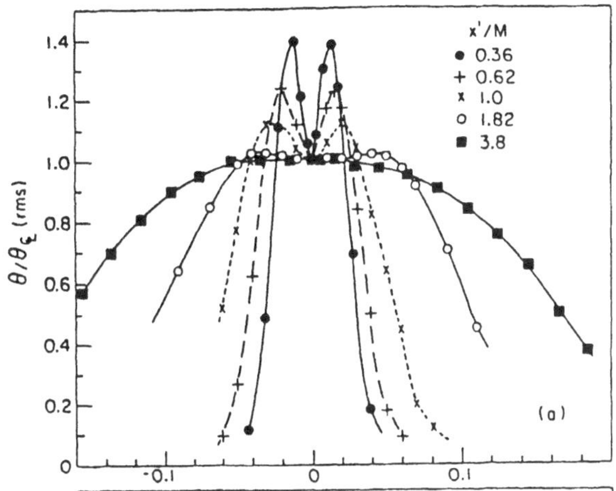

Figure 1. Case I. r.m.s. temperature profiles normalized by the centerline r.m.s. values, from Warhaft (1984). The wire was at $x_0/M = 52$. For these measurements close to the wire, a .025 mm wire was used.

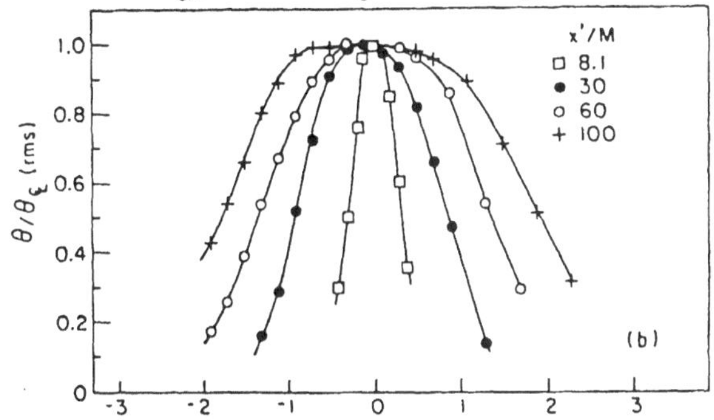

Figure 2. Case II. r.m.s. temperature profiles normalized by the centerline values, from Warhaft (1984). The wire was at $x_0/M = 52$. The wire deameter was 0.127 mm.

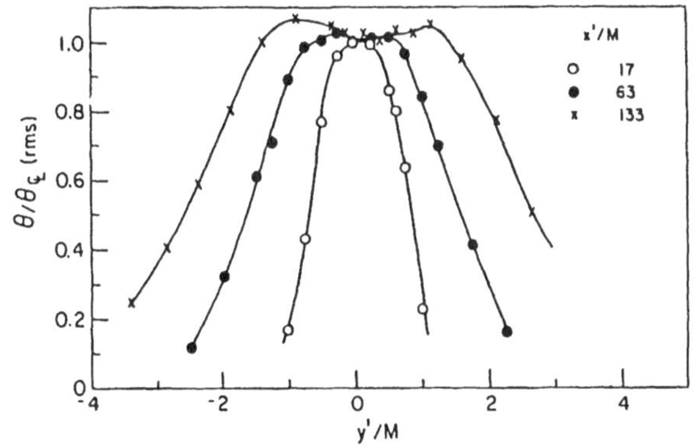

Figure 3. Case III. r.m.s. temperature profiles normalized by the centerline values, from Warhaft (1984). The wire was at $x_0/M= 20$. The wire diameter was 0.127 mm.

tion curve of the r.m.s. value, this means that the initial level of $\langle\varepsilon_\theta\rangle$ is probably too high, resulting in too rapid decay of values of r.m.s. temperature.

In figures 6 & 7 we have case II. It is evident that this case still begins in the regime in which the instantaneous plume is thinner than the Kolmogorov microscale, since the ratio of centerline r.m.s. to mean temperature is growing initially. We can see here, both in the data and in the computations, the decrease to a constant value (following the initial rise), corresponding to the instantaneous plume width

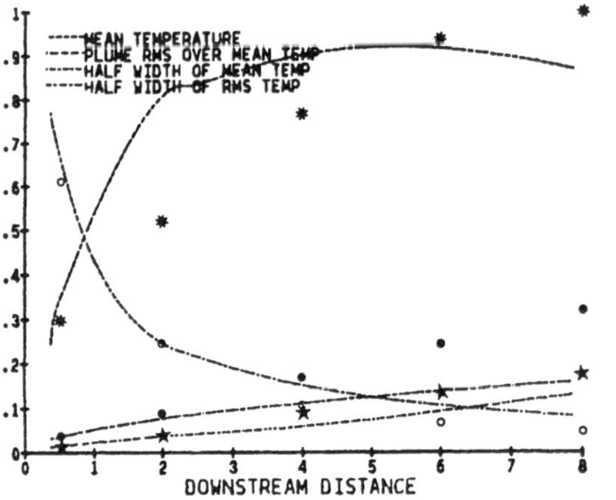

Figure 4. Centerline temperature evolution Case I

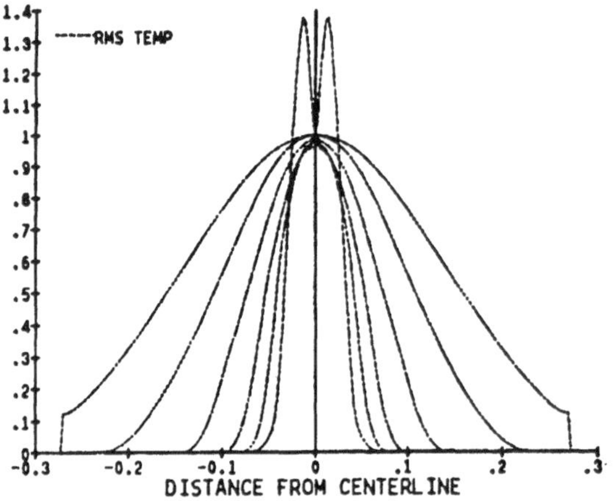

Figure 5. Rms at X/M = 0.36, 0.6, 1.0, 1.82, 3.8, 8.0

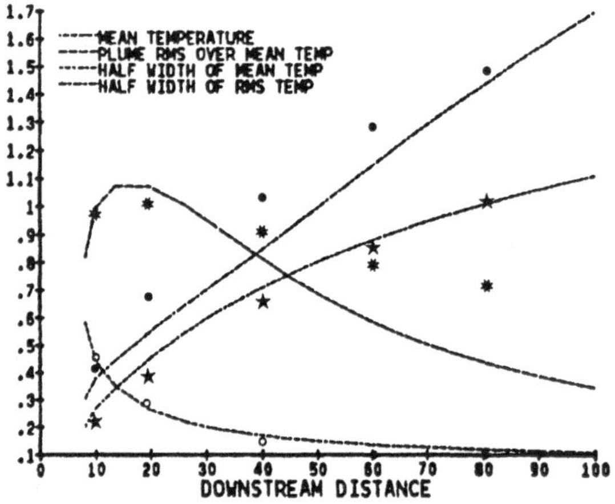

Figure 6. Centerline temperature evolution Case 2

beginning to spread inertially. Again, the centerline mean temperature and width are well reproduced, while the ratio of centerline r.m.s. to mean decreases too fast. This is probably a result of the same mechanism already seen in case I. The curves of normalized r.m.s. tempera-

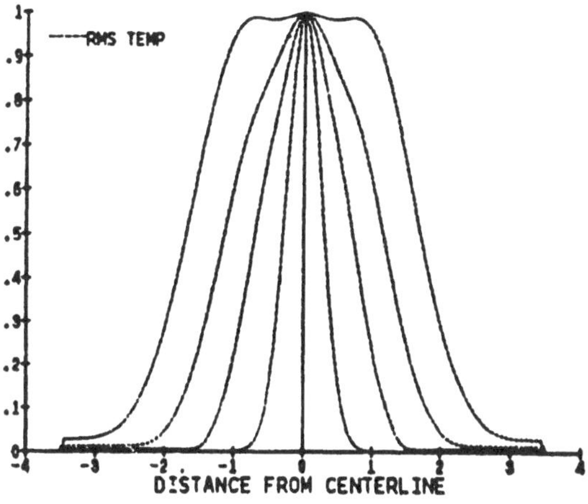

Figure 7. Rms temp profiles at X/M=8, 30, 60, 100

ture are not badly reproduced, although the appearance of off-axis peaks as the instantaneous plume grows to a substantial fraction of the mean plume width is perhaps a bit premature.

To determine whether our hypothesis (above) were correct, we artificially increased all the transport by 10%, decreasing the dissipation by a corresponding amount. The results are shown in figures 8 & 9 (also for case II). It is evident that the trend is in the right direction; the off-axis peaks are suppressed (perhaps too much), and the decay of the centerline value of the r.m.s. temperature is substantially slowed. Of course, this artificial increase in the transport does not produce a transport of the right structure, to correspond to intermittency. We intend in the future to introduce some of the intermittency corrections developed for other purposes (Chen & Lumley, 1984; Dancey, 1984).

In figures 10 & 11 we present case III. Here it is clear that we begin in the inertial range, since the ratio of centerline r.m.s. to

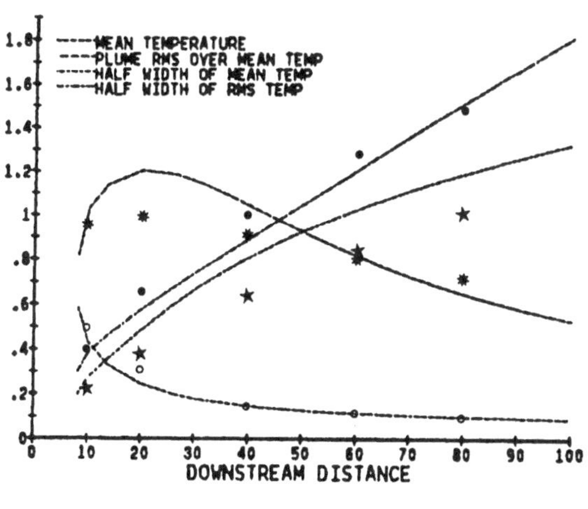

Figure 8. Centerline temperature evolution Case 2

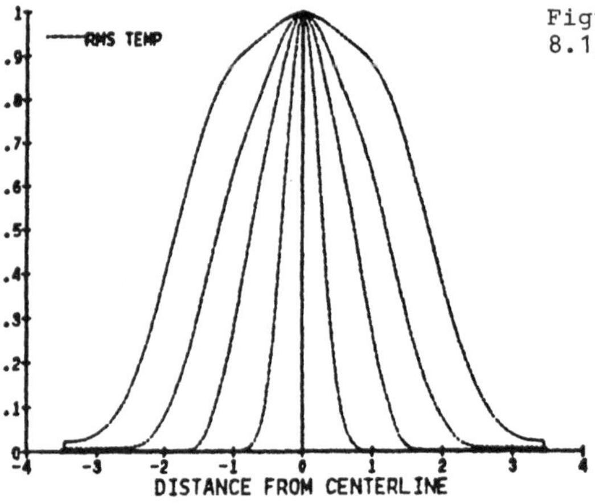

Figure 9. Rms temp profiles at X/M= 8.1, 30, 60, 100

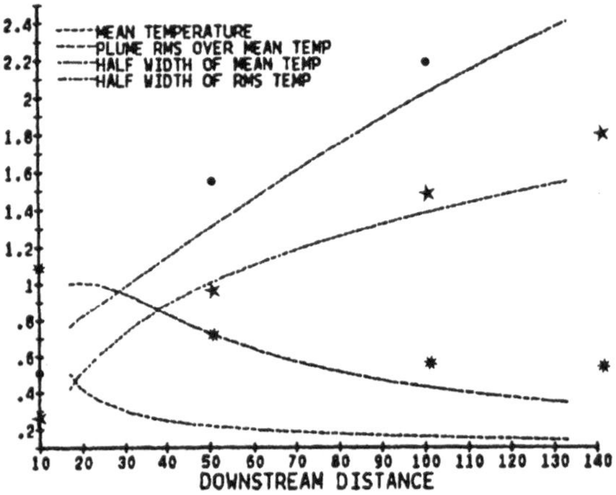

Figure 10. Centerline temperature evolution Case 3

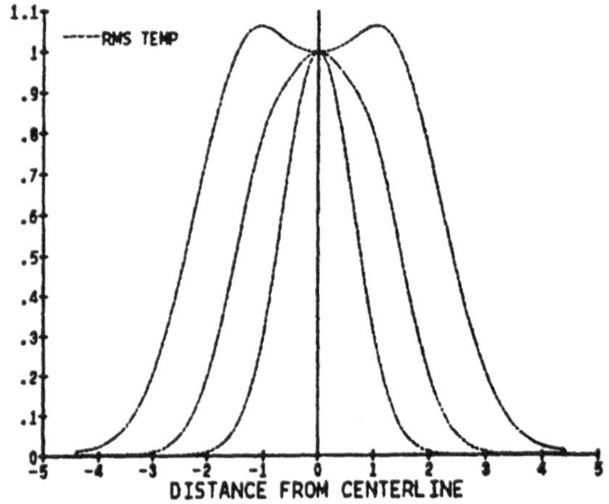

Figure 11. Rms temp profiles at X/M=17, 63, 133

mean temperature is initially decreasing. Now all centerline values and widths are well reproduced, and the curves of r.m.s. temperature are all well reproduced, with the possible exception of the slightly too-prominent off-axis peaks.

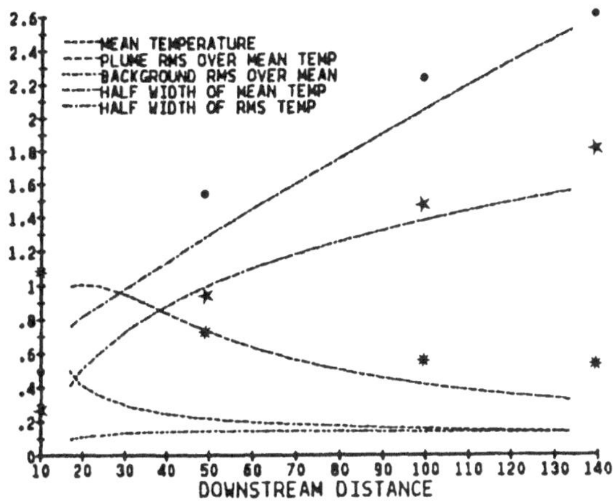

Figure 12. Centerline temperature evolution Case 3

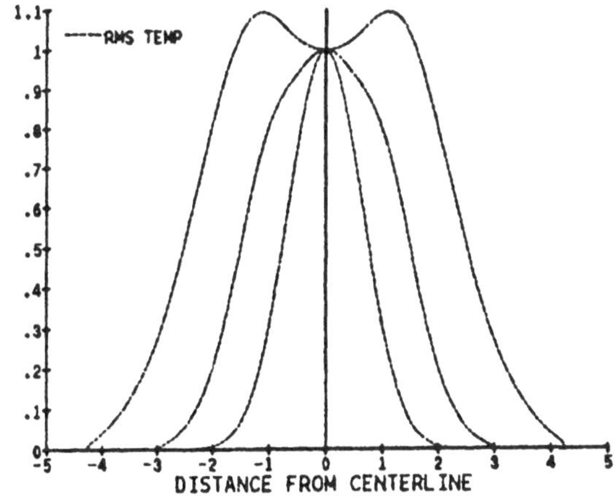

Figure 13. Rms minus background at X/M=17, 63, 133

In figures 12 & 13 we present case III with a background r.m.s. temperature fluctuation level of 10% of the peak value in the plume, and in figures 14 & 15 with a background level of 30%. In both cases, to determine the centerline values and half-widths, we have subtracted the background. It is clear that, although the model is not formally superposable, the difference between the results and a simple superposition is very small at the 10% level. At the 30% level the half-widths and centerline values are worse, but not disastrous; the r.m.s. profiles, however, have a bizarre appearance. This is largely an artifact of the presentation. At x'/M = 133, the peaks of the plume are only 20% above the background, and the centerline value has nearly reached the background level. When the background is subtracted, and the result devided by the centerline value, the curious appearance results. It is undeniable, however, that the peaks are too large relative to the centerline, and that the centerline is decaying too fast.

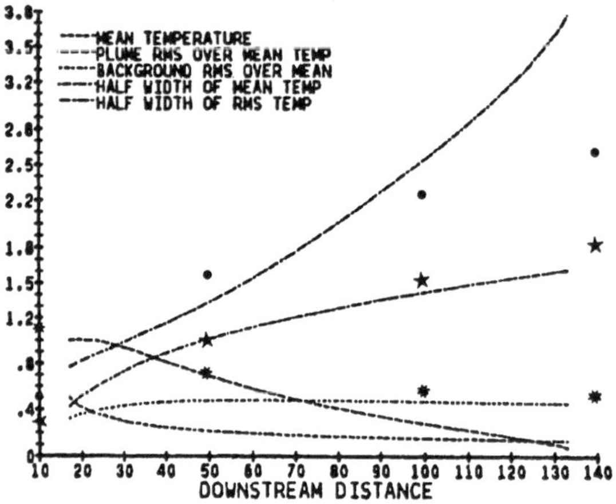

Figure 14. Centerline temperature evolution Case 3

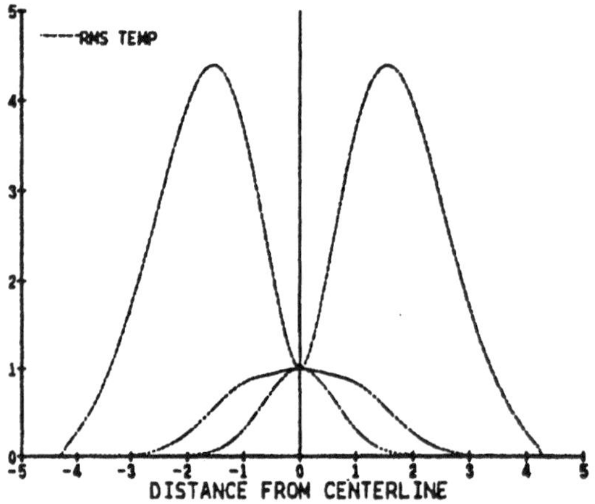

Figure 15. Rms minus background at X/M=17, 63, 133

5. Conclusions

We may conclude that the second order model outlined in Shih & Lumley (1982) is quite satisfactory for describing the evolution of plumes that begin in the inertial range. Modification of the transport will be necessary to describe plumes which are initially thinner than the Kolmogorov microscale; it seems likely that a correction for intermittency will help in this regime. Since the majority of plumes that one is called upon to calculate in geophysical situations begin well into the inertial range, it appears probable that the model as formulated here is more than adequate for the description of geophysical phenomena.

The performance of the model in this regime in the presence of overlapping temperature fields of differing time scale may be regarded as satisfactory as long as the magnitudes are very different. Evidently the lack of formal superposability in the model may not be as serious a concern as had been thought, at least under these circumstances.

Of course, we present here a situation in which the two temperature fields, although their time scales are quite different, are initially uncorrelated, and can never develop a correlation, as one can easily show from the equation for the cross-correlation. The situation is more complicated if the fields are correlated. This is the situation described in Warhaft (1984), in which plumes from adjacent sources interact. In a future paper, we will attempt to calculate this case. We will, in addition, make calculations for finite sources, to explore the effect of source size on temperature variance.

Acknowledgments

We are grateful,for the help and encouragement of our colleagues S. Leibovich, S. Pope, T.-H. Shih, and Z. Warhaft.

References

Chatwin, P. C. & Sullivan, P. J. 1979. The relative diffusion of a cloud of passive contaminant in incompressible turbulent flow. J. Fluid Mech. 91: 337-355.

Chen, J.-Y. & Lumley, J. L. 1984. Second order modeling of the effect of intermittency on scalar mixing. Proceedings, 20th International Symposium on Combustion. In Press.

Comte-Bellot, G. & Corrsin, S. 1966. The use of a contraction to improve the isotropy of grid-generated turbulence. J. Fluid Mech. 25: 657-682.

Csanady, G. T. 1967a. Concentration fluctuations in turbulent diffusion. J. Atmos. Sci. 24: 21-28.

Csanady, G. T. 1967b. Variance of local concentration fluctuations. Physics of Fluids Supplement: S76-S78.

Dancey, C. 1984. The Effect of Intermittency on Scalar Mixing. Ph. D. Thesis. Ithaca, NY: Cornell.

Durbin, P. A. 1980. A stochastic model of two-particle dispersion and concentration fluctuations in homogeneous turbulence. J. Fluid Mech. 100: 279-302.

Fackrell, J. E. & Robins, A. G. 1982. Concentration fluctuations and fluxes in plumes from point sources in a turbulent boundary layer. J. Fluid Mech. 117: 1-26.

LaRue, J. C. , Libby, P. A. & Seshadri, D. V. R. 1981. Further results on thermal mixing layer downstream of a turbulence grid. In Third Symposium on Turbulent Shear Flows, ed. L. J. S. Bradbury et al, pp. 15.1-15.6. Davis: U. of California.

Lumley, J. L. 1978. Computational modeling of turbulent flows. In Advances in Applied Mechanics 18, ed. C.-S. Yih, pp123-176. New York: Academic.

Lumley, J. L. 1983a. Turbulence modeling. J. Applied Mech. 50: 1097-1103.

Lumley, J. L. 1983b. Atmospheric Modeling. The Institution of Engineers, Australia: Mechanical Engineering Transactions. ME8: 153-159.

Lumley, J. L. & Mansfield, P. 1984. Second order modeling of turbulent transport in the surface mixed layer. Boundary Layer Meteorology. In Press.

Lumley, J. L. & Newman, G. R. 1977. The return to isotropy of homogeneous turbulence. J. Fluid Mech. 84: 581-597.

Orszag, S. A. 1970. Analytical theories of turbulence. J. Fluid Mech. 41: 363- .

Pope, S. B. 1983. Consistent modeling of scalars in turbulent flows. Physics of Fluids, 26: 404-408.

Shih, T.-H. 1984. Second Order Modeling of Scalar Turbulent Flow. Ph. D. Thesis. Ithaca NY: Cornell.

Shih, T.-H. & Lumley, J. L. 1982. Modeling heat flux in a thermal mixing layer. In Refined Modeling of Flows, Vol. I (eds. J. P. Benqué et al). Paris: Presses Ponts et Chaussées, pp. 239-250.

Warhaft, Z. 1984. The interference of thermal fields from line sources in grid turbulence. J. Fluid Mech. 144: 363-387.

Acoustic Wave Propagation in Fluids

Timothy S. Margulies
Office of Research, U.S. Nuclear Regulatory Commission, Washington, DC 20555, USA
W.H. Schwarz
Department of Chemical Engineering, The Johns Hopkins University
Baltimore, MD 21218, USA

The classical viscothermal problem of infinitesimal, planar acoustic-wave propagation in a single-component Newtonian fluid is extended to more general multicomponent materials that are diffusive, reacting and viscoelastic. The attenuation and dispersion of the sound wave are determined by solving the linearized (first-order) equations of mass, linear momentum, energy and chemical kinetics. General results are obtained in the form of a biquadratic characteristic equation (called the Kirchhoff-Langevin equation) for the complex propagation coefficient $\varkappa \equiv -(\alpha + i\omega/c)$, where α is the attenuation coefficient, c is the phase speed of the progressive wave and ω is the angular frequency.

First, the case of a single-component Newtonian fluid is examined. Then the problem of a nondiffusive, equilibrium mixture with coupled chemical reactions is considered. Next, the effect of diffusion on acoustic wave propagation in a binary, nonreacting fluid is treated, and results in a sixth-order characteristic equation for $\varkappa$. Then the precise theory for predicting attenuation and dispersion of a plane sound wave propagating in a single-component viscoelastic fluid is developed, where the response of the stress-tensor is represented by Noll's simple-fluid theory which incorporates the memory of the material. Finally, a combined treatment unifying the theories for the effect of viscoelasticity and reactions is presented. Calculated results and experimental measurements are compared for different classes of materials where possible.

1. Introduction

The theoretical prediction of acoustic wave propagation in fluids is important for understanding medical diagnostic and sonar performance characteristics, for measuring physico-chemical and thermostatic properties of fluids and for examining constitutive relations. The purpose of this paper is to present a fluid dynamical theory for examining a wide variety of acoustical classes of materials, including polymer solutions which exhibit non-Newtonian viscous behavior (or memory effects) and chemical reactions. Here we introduce the forced small-amplitude acoustic wave propagation problem.

Given a semi-infinite homogeneous fluid otherwise at rest, the y, z plane of fluid is oscillated harmonically in the x-direction with frequency f as shown pictorially in Fig. 1.1.

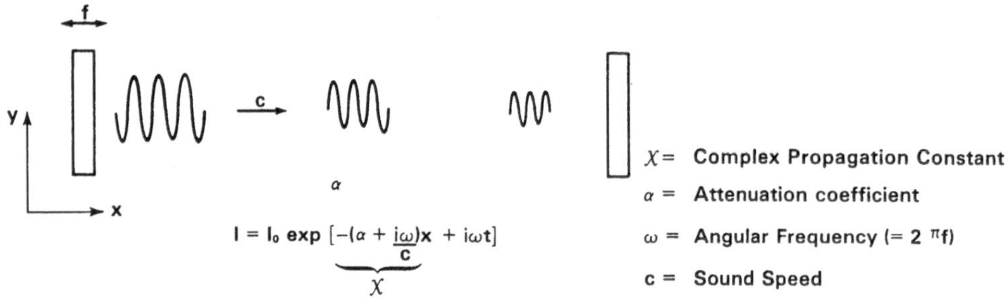

Fig. 1.1 Forced sound wave propagation in a fluid

The displacement of the plane boundary and resulting longitudinal wave motion within the fluid is assumed to be infinitesimal. The wave amplitude progressively diminishes with transmission distance as the mechanical energy which is imparted by the boundary surface is dynamically converted into thermal energy. This analysis assumes that the actual motion is well approximated by the exact solutions to the first-order field equations of the mixture, in which all nonlinear terms are omitted, and moreover, all coefficients are evaluated at their uniform equilibrium reference values.

The total value of each variable, I, in the list of system variables chosen to define this one-dimensional perturbed motion is decomposed into an equilibrium (static) and time-dependent incremental acoustic contribution.

$$I \begin{pmatrix}\text{Total}\\ \text{Variable}\end{pmatrix} = I_o \begin{pmatrix}\text{Static}\\ \text{Variable}\end{pmatrix} + I^a \begin{pmatrix}\text{Acoustic}\\ \text{Variable}\end{pmatrix} \tag{1.1}$$

Furthermore, each acoustic variable, I^a, such as pressure p^a, temperature θ^a, or velocity v^a is represented as a damped sinusoidal progressive wave by

$$I^a = \mathrm{Re}\left[I_o^a \exp(\chi x + i\omega t)\right] \tag{1.2}$$

$\chi = -(\alpha + ik)$ is the complex propagation variable where α is the spatial absorption coefficient and k is the magnitude of the wave number vector perpendicular to surfaces of constant phase. $k = \omega/c$ is the spatial analog of frequency and represents the number of wavecrests per unit length, or simply 2π multiplied by the reciprocal of wavelength in this one-dimensional problem. Absorption and dispersion measures are found by solving the appropriate balance equations and constitutive equations for the fluid (or mixture) under study. Exact solutions are sought, in the analyses in order not to preclude the interactions among the various dissipative mechanisms that contribute to an increase of the entropy, rather than using an approximate approach that relies solely upon the energy equation and omits dispersion estimates.

The paper is arranged as follows. First, we will examine the exact treatment for solving the small-amplitude, sound wave propagation problem for the classical Newtonian viscous and Fourier heat-conducting fluid. Then the acoustic problem will be extended to account for several coupled reactions which occur simultaneously in a nondiffusing mixture. The viscothermal case of diffusion in a binary mixture of fluids (without chemical reaction) is also presented. The next part of the paper will develop the precise theory predicting attenuation and dispersion of a plane sound wave propagating in a single-component, non-Newtonian medium. Here, the response of the fluid may be represented by Noll's simple fluid theory which incorporates memory effects and viscoelastic relaxation spectra. Finally, a combined treatment unifying the above theories for different acoustical classes of materials and describing existing experimental data is proposed for a reacting fluid system with non-Newtonian viscous characteristics.

2. Single-Component Newtonian Fluid

2.1 Continuum Balance Equations

The general continuum field equations for a single component material are obtained from the balances of total mass, linear momentum, moment of momentum and energy. Omitting radiation supply, these balance postulates are written in spatial or Eulerian differential form using rectangular Cartesian coordinates as:

$$\frac{\partial \rho}{\partial t} + \frac{\partial (\rho v_j)}{\partial x_j} = 0 \tag{2.1}$$

$$\rho \frac{d v_i}{d t} = \frac{\partial S_{ij}}{\partial x_j} + \rho b_i \tag{2.2}$$

$$S_{ij} = S_{ji} \tag{2.3}$$

$$\rho \frac{d \varepsilon}{d t} = S_{ij} d_{ji} + \frac{\partial h_i}{\partial x_j} \tag{2.4}$$

where $\frac{d[\,]}{dt}$ denotes the total derivative (and equals

$\frac{\partial[\,]}{\partial t} + v_j \frac{\partial[\,]}{\partial x_j}$). S_{ij} , b_i and h_i denote the components of the total stress tensor $\underline{\underline{S}}$, the external body-force vector per unit mass $\underline{b}$, and $\underline{h}$ the heat-flux vector, respectively. The specific internal energy ε (per unit mass) is defined as the difference of the total and kinetic energy densities, and d_{ij} represents the components of $\underline{\underline{d}}$, the symmetric part of the velocity-gradient tensor called the rate-of-deformation or stretching tensor.

Constitutive relations are required to complete the system of balance equations in order to derive the acoustic equations for a particular class of materials. First, the linear forms of the constitutive relations representing classical fluids are examined. Here, classical means that the response of the material depends only upon the present values of the field variables. Viscoelastic models for materials which exhibit memory effects, for example, are excluded from this class but will be considered in a later section.

The classical constitutive relations considered may be explicitly written for:

1) the stress relation as a homogeneous, linear, isotropic function of the stretching tensor, $\underline{\underline{d}}$, where $2\underline{\underline{d}}$ = grad $\dot{\underline{x}}$ + (grad $\dot{\underline{x}})^T$; that is,

$$S_{ij} = (-p + \lambda_0 d_{kk})\,\delta_{ij} + 2\eta_0 d_{ij} \tag{2.5}$$

where $p = \hat{p}(\theta, v)$ is the pressure; η_0 is the coefficient of shear viscosity or first coefficient of viscosity, λ_0 is the "dilational" or volume coefficient of viscosity or the second coefficient of viscosity, and $v = 1/\rho$ is the specific volume. The stress tensor can be written in the equivalent form:

$$S_{ij} = (-p + K_0 d_{kk})\,\delta_{ij} + 2\eta_0 \langle d_{ij}\rangle \tag{2.6}$$

where $\langle d_{ij}\rangle$ is the traceless part of $\underline{\underline{d}}$; i.e., $\langle d_{ij}\rangle = d_{ij} - \frac{1}{3} d_{kk}\,\delta_{ij}$ and K_0 is called the bulk viscosity of the fluid, $K_0 = (\lambda_0 + \frac{2}{3}\eta_0)$. This constitutive relation for the stress tensor defines a linearly viscous or Newtonian fluid and satisfies the moment-of-momentum equation (2.3).

2) the heat in-flux vector is represented as a linear function of the temperature gradient,

$$h_i = k_\theta \frac{\partial \theta}{\partial x_i} \tag{2.7}$$

where k_θ is the thermal conductivity tensor. This is often referred to as Fourier's law of heat conduction.

3) An equation of state $\varepsilon = \hat{\varepsilon}(\theta, v)$ or $\psi = \psi(\theta, v)$ needs to be specified. ψ is the specific Helmholtz free energy ($\psi = \varepsilon - \theta\eta$)

which is a function of the temperature and specific volume and η is the specific entropy function. Alternatively, the function $\check{\varepsilon}(v,\eta)$, called a caloric equation, may be used for the equation of state. Further, the following relations hold:

$$\theta = \frac{\partial \check{\varepsilon}}{\partial \eta} \qquad p = \check{p}(v,\eta) = -\frac{\partial \check{\varepsilon}}{\partial v} = -\frac{\partial \hat{\Psi}}{\partial v}$$

$$\theta = \frac{\partial \tilde{\eta}(v,\varepsilon)}{\partial \varepsilon} \qquad \eta = \frac{\partial \hat{\Psi}}{\partial \theta} \tag{2.8}$$

These relations are consistent with the basic principles of continuum mechanics which was demonstrated by Coleman and Mizel [1].

Now, using (2.4) and (2.8), obtain

$$\rho \dot{\varepsilon} = \rho \theta \dot{\eta} - \frac{p}{v} \dot{v} = S_{ij} d_{ji} + \frac{\partial h_i}{\partial x_i} \tag{2.9}$$

Since $S_{ij} d_{ij} = -p(\dot{v}/v) + T_{ij} d_{ij}$, (2.9) becomes

$$\rho \theta \dot{\eta} = T_{ij} d_{ji} + \frac{\partial h_i}{\partial x_i} \tag{2.10}$$

which is the entropy equation. This can be written in terms of the temperature field $\theta(\underline{x},t)$, thereby introducing standard thermodynamic parameters such as heat capacities for which experimental data have been tabulated for specific substances.

Consider $\hat{\eta}(\theta,v)$, hence

$$\dot{\eta} = \left(\frac{\partial \hat{\eta}}{\partial \theta}\right)\dot{\theta} + \left(\frac{\partial \hat{\eta}}{\partial v}\right)\dot{v}$$

where

$$\frac{\partial \hat{\eta}}{\partial v} = \frac{\partial \hat{p}}{\partial \theta} = -\frac{c_0^2 \rho \beta_\theta}{\gamma}$$

and

$$\frac{\partial \hat{\eta}}{\partial \theta} = \frac{c_v}{\theta}$$

Furthermore,

$$c_p - c_v = \theta \beta_\theta^2 / \rho \beta_p = \theta \beta_\theta^2 c_0^2 / \gamma$$

$$\gamma = c_p / c_v$$

$$c_0^2 = -v^2 \frac{\partial \check{p}(\eta,v)}{\partial v}$$

where c_0 is called the 'reference speed of sound'. Therefore, using these relations and Fourier's form for the heat-influx vector, (2.10) becomes

$$\rho c_v \dot{\theta} = \frac{(c_p - c_v)}{\beta_\theta} \dot{\rho} + T_{ij} d_{ji} + k_\theta \frac{\partial^2 \theta}{\partial x_i \partial x_i}$$

2.2 The Linearized Acoustics Equations

The balance equations may be systematically linearized by replacing the total variables, say temperature θ , pressure p, velocity v and density ρ by the sum of an equilibrium part and the incremented acoustic contribution (1.1). Products of acoustic variables (written with a superscript "a") which appear explicitly are omitted. For infinitesimal waves propagating into a fluid at rest ($v_0 \equiv 0$) with a uniform equilibrium state, the linearized balance equations for a Newtonian viscous, Fourier heat-conducting fluid become:

$$\frac{\partial \rho^a}{\partial t} + \rho_0 \frac{\partial v^a}{\partial x} = 0 \qquad (2.11)$$

$$\rho_0 \frac{\partial v^a}{\partial t} + \frac{\partial p^a}{\partial x} - \frac{\partial t^a_{xx}}{\partial x} = 0 \qquad (2.12)$$

$$\rho_0 c_v \frac{\partial \theta^a}{\partial t} - \frac{(c_p - c_v)}{\beta_\theta} \frac{\partial \rho^a}{\partial t} - k_\theta \frac{\partial^2 \theta^a}{\partial x^2} = 0 \qquad (2.13)$$

which are usually referred to as the first-order acoustic equations (neglecting radiation supply and body forces). Here $t^a_{xx} = \beta_0 \frac{\partial v^a}{\partial x}$ and we define $\beta_0 = \left(\frac{4}{3}\eta_0 + K_0\right)$ as the "acoustic viscosity".

To solve this system of three linear partial differential equations in terms of ρ^a, v^a and θ^a, the nonlinear dependence of pressure on the other thermodynamic variables is linearly approximated. A small change in ρ and θ in $\hat{p}(\theta, \rho)$ is given by

$$p^a = p - p_0 = \frac{\partial \hat{p}}{\partial \rho} \cdot \left\{ \rho^a + \left[\left(\frac{\partial \hat{p}}{\partial \theta}\right) / \left(\frac{\partial \hat{p}}{\partial \rho}\right) \right] \right\} \theta^a \qquad (2.14)$$

$$= \frac{c_0^2}{\gamma} \left\{ \rho^a + \rho_0 \beta_\theta \theta^a \right\}$$

Also,

$$\frac{\partial \hat{p}}{\partial \rho} = \frac{c_0^2}{\gamma} = -v^2 \frac{\partial \hat{p}}{\partial v} \quad ; \quad \frac{\partial \hat{p}}{\partial \theta} = \frac{\beta_\theta}{\beta_p} \quad ; \quad \frac{\partial \bar{\rho}}{\partial \theta} = -\rho \beta_\theta$$

and

$$c_0^2 = \frac{\partial \check{p}(\rho, \eta)}{\partial \rho}$$

$\alpha_\eta \equiv \frac{1}{v} \frac{\partial v}{\partial p}$ is called the adiabatic compressibility and γ , the ratio of specific heats, can be written as:

$$\gamma = \beta_\gamma / \alpha_\eta$$

where $\beta_\gamma \equiv \frac{1}{\bar\rho}\frac{\partial \bar\rho}{\partial p}$ is the 'isothermal compressibility' and β_θ is the 'isothermal expansion coefficient'. Thus the linearized equation of motion in one-dimension (x) becomes:

$$\rho_0 \frac{\partial v^a}{\partial t} + \frac{c_0^2}{\gamma}\frac{\partial \rho^a}{\partial x} + \frac{c_0^2}{\gamma}\beta_\theta \rho_0 \frac{\partial \theta^a}{\partial x} = \beta_0 \frac{\partial^2 v^a}{\partial x^2} \tag{2.15}$$

2.3 Kirchhoff-Langevin Equation

An exact solution of the complete first-order equations for forced monochromatic plane waves in Newtonian fluids has been given by Truesdell [2]. At x = 0, any acoustic variable, I^a, is represented by the real part of $I^a = I_0^a \exp(i\omega t)$. This disturbance is propagated through the material as a damped harmonic oscillation $I^a = I_0^a \exp(\chi x + i\omega t)$. Assume solutions of the form

$$\begin{aligned}
\theta^a &= \theta_0^a \exp(\chi x + i\omega t)\\
\rho^a &= \rho_0^a \exp(\chi x + i\omega t)\\
v^a &= v_0^a \exp(\chi x + i\omega t)
\end{aligned} \tag{2.16}$$

and substitute into (2.11), (2.13) and (2.15) to obtain:

$$\begin{aligned}
&i\omega \rho_0^a + \rho_0 \chi v_0^a = 0\\
&i\omega \rho_0 v_0^a + \frac{c_0^2}{\gamma}\chi \rho_0^a + \frac{c_0^2}{\gamma}\beta_\theta \theta_0^a \chi - \beta_0 \chi^2 v_0^a = 0\\
&i\omega \rho_0 c_v \theta_0^a - \frac{i\omega (c_p - c_v)}{\beta_\theta}\rho_0^a - k_\theta \chi^2 \theta_0^a = 0
\end{aligned} \tag{2.17}$$

Nontrivial solutions exist if and only if the determinant of the coefficients of these algebraic equations vanishes. The values of χ are found by solving:

$$\begin{vmatrix}
0 & i\omega & \rho_0 \chi \\
\frac{c_0^2}{\gamma}\beta_\theta \rho_0 \chi & \frac{c_0^2}{\gamma}\chi & i\omega\rho_0 - \beta_0 \chi^2 \\
i\omega \rho_0 c_v - k_\theta \chi^2 & -\frac{i\omega (c_p - c_v)}{\beta_\theta} & 0
\end{vmatrix} = 0$$

The characteristic equation formed in this way results in a biquadratic equation in the dimensionless complex propagation variable $\varkappa/k_0$ which can be solved by using the standard "quadratic formula."

Truesdell [2] has also written the characteristic equation in terms of dimensionless parameters which specify the properties of the medium as discussed below. In addition to $\varkappa$ and the three variables (v_0^a , ρ_0^a and θ_0^a), there are nine parameters that define the acoustic behavior of the material at angular frequency ω . However, only eight of these parameters are independent since we have the thermodynamic identity

$$\gamma - 1 = \frac{\Theta \beta_\Theta^2 c_0^2}{c_p}$$

Therefore, only eight parameters are independent: ω , c_0 , ρ_0 , k_θ , K_0 , η_0 , c_p and c_v . Since each of these can be expressed in fundamental units such as mass, length, time and temperature, it follows from the pi-theorem of dimensional analysis [3] four independent dimensionless ratios can be formed. The dimensionless propagation constant may be expressed as:

$$\frac{\varkappa}{k_0} = f\left[\left(\frac{c_p}{c_v}\right), \left(\frac{K_0}{\eta_0}\right), \left(\frac{\eta_0 c_p}{k_\theta}\right), \left(\frac{\omega \eta_0}{\rho_0 c_0^2}\right)\right]$$

The first dimensionless parameter is γ or the ratio of specific heats, and the third ratio is recognized as the Prandtl number (P_N). The second ratio is incorporated into a dimensionless quantity called the viscosity number defined as:

$$V_N = 2 + \lambda_0/\eta_0 = \frac{4}{3} + K_0/\eta_0$$

hence $V_N \geq 4/3$. The fourth group of parameters is called the Stokes number (S_N). Now

$$\left(\frac{\varkappa}{k_0}\right) = g\left[\gamma, V_N, P_N, S_N\right]$$

In dimensionless terms, Truesdell [2] obtained the following equation for the dimensionless complex propagation constant $\varkappa/k_0$

$$\left[(i - \gamma X_N) X_N Y_N\right](\varkappa/k_0)^4 + \left[1 + i X_N (1 + \gamma Y_N)\right](\varkappa/k_0)^2 + 1 = 0 \qquad (2.18)$$

It was found convenient to define $X_N = V_N S_N$ and $Y_N = (P_N V_N)^{-1}$. The result is called the Kirchhoff-Langevin biquadratic equation in recognition of Kirchhoff who formed a secular equation in this way for perfect gases and Langevin who considered a fluid with an arbitrary equation of state [4].

The dimensionless groups are summarized below:

Heat capacity ratio $\gamma = c_p/c_v$

Viscosity number $\mathcal{V}_N = 2 + \lambda_0/\eta_0 = \frac{4}{3} + K_0/\eta_0$

Stokes number $S_N = \omega\eta_0/\rho_0 c_0^2$

Frequency number $X_N = \mathcal{V}_N S_N = \omega\eta_0\mathcal{V}_N/\rho_0 c_0^2$

Thermoviscous number $Y_N = k_\theta/\eta_0\mathcal{V}_N c_p = (P_N\mathcal{V}_N)^{-1}$

Frequency number (K_0 =0) $\bar{X}_N = 8\pi f\eta_0/3\rho_0 c_0^2$

Prandtl number $P_N = (\eta_0/\rho_0)/(k_\theta/\rho_0 c_p) = \eta_0 c_p/k_\theta$

The problem has been reduced to finding the function:

$$\frac{\chi}{k_0} = g\left[\langle\gamma, Y_N\rangle, X_N\right]$$

where the inner brackets are dimensionless groups representing the fluid, and only X_N contains the frequency of the sound.

Each Newtonian fluid is represented as a point in the physical property space $\langle\gamma, Y_N\rangle$. The real part of χ/k_0 corresponds to α/k_0 and the imaginary part corresponds to $k/k_0 = c_0/c$, which for each fixed value of $\langle\gamma, Y_N\rangle$ are continuous functions of X_N, the frequency number. Therefore, the absorption coefficient and sound speed can be precisely calculated from the Kirchhoff-Langevin equation and tabulated for each fluid with frequency or X_N as the variable.

The range of the physical property space is given by: $1 \leq \gamma < \infty$ and $Y_N \geq 0$. For most fluids, $\gamma < 2$. Liquid argon is an exception [5].

The standard quadratic formula can be used to solve for $(k_0/\chi)^2$:

$$\left(\frac{k_0}{\chi}\right)^2 = \frac{-\left[1 + iX_N(1+\gamma Y_N)\right] \mp \left\{\left[1 + iX_N(1+\gamma Y_N)\right]^2 - 4(i\gamma X_N)X_N Y_N\right\}^{1/2}}{2} \tag{2.19}$$

which gives an expression relating α and k to the three independent parameters X_N, γ, and Y_N. Each parameter in (2.19) occurs in one or more product combinations, and hence it cannot be generally assumed that the effects of viscosity and heat conduction are linearly additive. Therefore, the practice of determining contributions from these effects separately, and superimposing the results must be justified explicitly and quantitatively.

Equation (2.19) has two pairs of non-coincident complex roots:

$$\left(\frac{k_o}{\varkappa}\right)^2 = \frac{-[1 + iX_N(1+\gamma Y_N)] - \{1 - X_N^2(1-\gamma Y_N)^2 + 2iX_N[1-(2-\gamma)Y_N]\}^{1/2}}{2} \tag{2.20}$$

and

$$\left(\frac{k_o}{\varkappa}\right)^2 = \frac{-[1 + iX_N(1+\gamma Y_N)] + \{1 - X_N^2(1-\gamma Y_N)^2 + 2iX_N[1-(2-\gamma)X_N]\}^{1/2}}{2} \tag{2.21}$$

Here the curly brackets $\{\}$ means the principal determination of the square root. Now, only those solutions which yield positive values for α, corresponding to real attenuation, were retained. The two solutions (2.20) and (2.21) comprise the two branches of a complex square root; one branch pertains to typical compressional sound waves identified as type I, and obtained from (2.20); the other, the so-called thermal waves referred to as type II are obtained from (2.21). One is not completely justified, however, in describing these simply as 'pressure' and 'thermal' waves, since all field variables are simultaneously perturbed and propagated by each wave type; and both types of waves are always propagated by any source. On the other hand, the absorption and dispersion characteristics of types I and II waves will, in general, be quite different. Type II waves are more rapidly attenuated than type I waves and have not been observed in any experiment to date.

Clearly the calculated results for α and c are not easily interpreted, because of the algebraic complexity of the equations. However, power series expansions do provide explicit formulae that are valid over certain ranges and can be readily examined for physical content. However, to establish the range of validity of any series expansion, as well as the analyticity of $(k_o/\varkappa)^2$ with respect to $\langle\gamma, Y_N\rangle$ and X_N, Truesdell [2] examined carefully the branches of (2.18) and (2.19).

The amplitude attenuation per wave-length A, was defined as:

$$A = 2\pi\alpha/k = \alpha\lambda$$

and in a similar manner the amplitude attenuation per reference wave-length, A_o was defined as:

$$A_o = 2\pi\alpha/k_o = \alpha\lambda_o$$

These are standard measures of absorption, while $\delta \equiv n^2 \equiv (c/c_o)^2$, the square of the ratio of the actual speed to the reference speed, is a measure of the dispersion.

For fluids such that $Y_N < \frac{1}{2-\gamma}$ in the low-frequency range ($X_N \ll 1$), one may approximate (2.21) by

$$c = c_0$$

$$\left(\frac{\alpha}{k_0}\right) = \frac{1}{2}\left[1 + (\gamma-1)\, Y_N\right] = \left(\frac{\alpha}{k}\right)$$

Thus, dispersion is predicted to be negligible and the absorption coefficient is given by the 'Kirchhoff' value:

$$\left(\frac{\alpha_K}{k_0}\right) = \frac{1}{2}\, X_N \left[1 + (\gamma-1)\, Y_N\right]$$

Substituting for k_0, X_N, Y_N, and γ, this takes the familiar form:

$$\left(\frac{\alpha_K}{f^2}\right) = \frac{2\pi^2 \left[\left(\frac{4}{3}\eta_0 + K_0\right) + k_0\left(c_v^{-1} + c_p^{-1}\right)\right]}{\rho_0\, c_0^3} \qquad (2.22)$$

Using the 'Stokes assumption', viz., $K_0 = 0$ $\left(\lambda_0' = -\frac{2}{3}\eta_0\right)$, (2.22) becomes:

$$\left(\frac{\alpha_S}{f^2}\right) = \frac{2\pi^2 \left[\frac{4}{3}\eta_0 + k_0\left(c_v^{-1} - c_p^{-1}\right)\right]}{\rho_0\, c_0^3}$$

This form for the absorption coefficient is referred to as the 'classical' or the Stokes relation. Due to the large number of assumptions and approximations inherent in its determination, it lacks, in general, the ability to correctly predict the absorption characteristics of all fluids over a large range of frequencies.

Some of the results of the viscothermal Newtonian theory are presented and then compared to experimental results to indicate the inadequacy of that model. The asymptotic limits of the measures of absorption and dispersion are given in Table 2.1.

TABLE 2.1

Asymptotic Limits of Absorption and Dispersion Measures from Newtonian Viscothermal Theory

	$X_N \to 0$	$X_N \to \infty$
$A_0 = \alpha\lambda_0$	0	0
$A = \alpha\lambda$	0	2π
$\delta = (c/c_0)^2$	1	∞

Since the limits of A_0 are zero as $X_N \to \infty$ and $X_N \to 0$, the absorption measure has at least one maximum, hence indicates a viscous relaxation time. Also a consequence of the Newtonian viscothermal theory is that the dispersion increases indefinitely and is determined by the viscosity only, since for $X_N \to \infty$,

$$c^2 \sim \frac{2(\eta_0 + \lambda_0)\,\omega}{\rho_0}$$

Using the asymptotic results listed above, Truesdell [2] has noted the following behavior for the absorption and dispersion measures. 1) A_0 experiences exactly one, usually broad maximum that characterizes visco-thermal resonance and this lies in the frequency range $1 \leq X_N|_{A_{0,max}} \leq 3$. The maximum value of A_0 will always have the same magnitude $(A_{0,max} = 2.2)$; 2) A is a monotone increasing function of X_N. 3) The curve of δ versus X_N is curved positively upward at all points. These results are exhibited by the curves shown in Fig. 2.1.

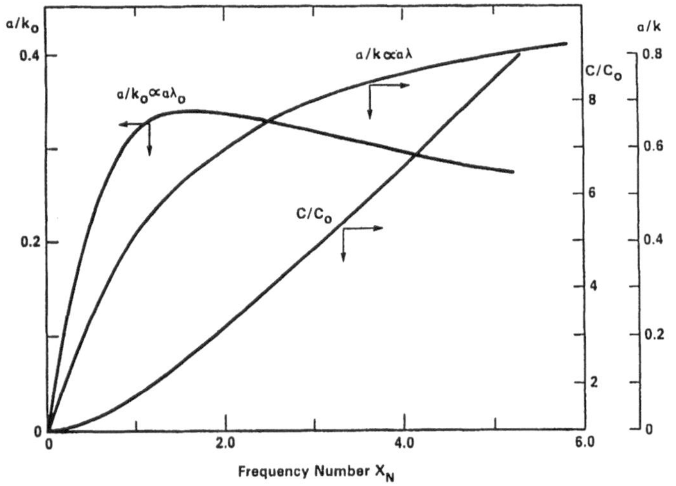

Figure 2.1 Graph of absorption and dispersion for classical, linearly viscous fluid $\langle \gamma, Y_N \rangle = \langle 1.40, 0.25 \rangle$.

Now, in the range of frequencies such that $X_N \ll 1$, the Stokes' value for the absorption coefficient is the lowest value that can be calculated by the viscothermal theory, i.e., $\left(\frac{\alpha_s}{f^2}\right) \leq \left(\frac{\alpha}{f^2}\right)$ since the bulk viscosity $K_0 \geq 0$. There are experimental data for several fluids where $X_N \ll 1$ and $\left(\frac{\alpha_s}{f^2}\right) > \left(\frac{\alpha}{f^2}\right)$ which must be analyzed using alternative constitutive relation assumptions.

2.4 Approximate Formula

For the case of a non-conducting fluid ($k_\theta = 0$), $Y_N = 0$, the Kirchhoff-Langevin equation obtains the simple quadratic form:

$$\left(\frac{x}{k_0}\right)^2 + \frac{1}{1 + i X_N} = 0 \tag{2.23}$$

or

$$\left[\left(\frac{\alpha}{k_0}\right) - \left(\frac{k}{k_0}\right)^2\right] + i\,\frac{2\alpha k}{k_0^2} = -\,\frac{1 - i X_N}{1 + X_N^2} \tag{2.24}$$

If we equate the real parts, obtain

$$\left(\frac{k}{k_0}\right)^2 = \frac{1}{1 + X_N^2} + O\left[(\alpha/k_0)^2\right] \tag{2.25}$$

The order of magnitude of the correction term has been included here to examine the validity of the approximation that uses the relation: $\frac{\alpha}{k_0} \ll \frac{k}{k_0}$ or $\alpha\lambda \ll 1$ to obtain the dispersion relation:

$$\left(\frac{c}{c_0}\right)^2 = 1 + X_N^2$$

from (2.25) by neglecting the correction term. On comparing the imaginary parts of (2.24), obtain

$$\left(\frac{\alpha}{k}\right) = \frac{1}{2}\left(\frac{k_0}{k}\right)^2 \frac{X_N}{1 + X_N^2} \tag{2.26}$$

For $X_N \ll 1$

$$\left(\frac{\alpha}{k_0}\right) = \frac{1}{2} X_N + O\left(X_N^2\right) \tag{2.27}$$

and using (2.25).

$$\left(\frac{k}{k_0}\right)^2 = \frac{1}{1 + X_N^2} + O\left(X_N^2\right)$$

For small X_N, the (α/k_0) term cannot be neglected and

$$\left(\frac{k}{k_0}\right)^2 = \frac{1}{1 + X_N^2} + \frac{X_N}{1 + X_N^2}\left(\frac{1}{1 + X_N^2} + \frac{1}{2} X_N + O\left(X_N^2\right)\right)^{1/2} \tag{2.28}$$

or

$$\left(\frac{k}{k_0}\right) = \frac{1 + X_N\left[\frac{1}{2} X_N + O\left(X_N^2\right)\right]^{1/2}}{1 + X_N^2}$$

Therefore

$$\left(\frac{k_0}{k}\right)^2 = \left(\frac{c}{c_0}\right)^2 = 1 + \frac{3}{4} X_N^2 + O\left(X_N^4\right)$$

Now using (2.27)

$$\left(\frac{\alpha}{k}\right) = \frac{1}{2} X_N + O\left(X_N^2\right)$$

3. Newtonian Fluid Mixture with Coupled Reactions (Nondiffusive System)

3.1 Continuum Balance Equations

Ultrasonic relaxation techniques have contributed significanty to our understanding of fast chemical reaction mechanisms in fluids and to the measurement of reaction parameters (e.g., rate constants) [6]. The aim of this section is to present a hydroacoustic theory which accounts for sound absorption and dispersion in a mixture of reacting fluids in equilibrium, such that several reactions can occur simultaneously. The theory is applied to several chemical systems using physico-chemical data available in the literature. The theoretical results reduce to the Kirchhoff-Langevin biquadratic equation obtained for the classical, viscothermal non-reacting case.

Here we present the continuum theory involving a perturbation of a multistep sequence of reactions omitting diffusion and radiation supply. The previously considered field equations for the balance of total mass, linear momentum, moment of momentum and energy are used in the derivation. Also, a partial mass balance equation for each constituent of the mixture must be included. For constituents which have a common absolute temperature θ and which can occupy any place $\underline{x}$ at time t in the mixture of (α = 1,2,...k) species the balance equations are written in rectangular Cartesian coordinates (2.1-2.4), and

$$\frac{d\omega_\alpha}{dt} = \overset{+}{\omega}_\alpha \tag{3.1}$$

where ω_α is the molar concentration of the αth constituent. ω_α equals the moles of α per total mass (mol g^{-1}) and $\overset{+}{\omega}_\alpha$ represents the production due to chemical reactions.

The form of the equation for the mass balance of a chemical constituent can be rewritten by a change of variables to any other suitable concentration measure. However, additional terms may arise due to a chain rule time differentiation of ω_α whose neglect or retention in the equation must be consistent with the analysis. For instance, in the collisional rate theory of chemical kinetics for gaseous systems one usually considers $\dot{m}_\alpha = \overset{+}{m}_\alpha$, where this form of the equation of balance in terms of the molar density (mol cm^{-3}) is valid only for constant volume systems.

3.2 Chemical Reactions and Their Description

The acoustic theory is appropriate for a general reaction scheme composed of j = 1,2,...r coupled reactions. This may be expressed in terms of the <u>signed</u> stoichiometric matrix, $\overset{*}{S}_{\alpha j}$, as follows:

$$\sum_{\alpha=1}^{k} S^{*}_{\alpha j} C_{\alpha} = 0 \qquad \begin{array}{l} \alpha = 1, 2, \ldots k \\ j = 1, 2, \ldots r \end{array}$$

where the positive values of the stoichiometric coefficients correspond to the reaction products while negative values correspond to reactants. Also, by convention, a neutral species such as an inert solvent is assigned a stoichiometric number of zero. The rank of the stoichiometric matrix determines the maximum number of independent reactions or calculated relaxation times. Let rank ($S^{*}_{\alpha j}$) = $R \leq r$. Then the total collection of reaction steps may be completely represented by an arbitrarily selected subset of R reactions.

In order to express the composition changes during the progress of a reaction in a nondiffusing mixture, it is convenient to define a degree-of-advancement variable for each reaction by

$$\omega_{\alpha} = \omega^{\circ}_{\alpha} + S^{*}_{\alpha j} \zeta_{j}$$

where ω°_{α} is a reference molar density, say at equilibrium. If ζ_j is the degree of advancement of the jth chemical reaction, it has contributed $S^{*}_{\alpha j} \zeta_j$ (moles g^{-1}) to the total change in the molar density of the αth component of the mixture. All composition changes may be expressed in terms of these degree-of-advancement variables corresponding to the chosen R independent reactions.

The equation of partial mass balance may now be written alternatively as

$$\dot{\omega}_{\alpha} = \overset{+}{\omega}_{\alpha} = S^{*}_{\alpha j} \overset{+}{\zeta}_{j}$$

and $\dot{\zeta}_j = \overset{+}{\zeta}_j$

where $\dot{\zeta}_j$, the time rate of change of ζ_j, is called the reaction velocity and $\overset{+}{\zeta}_j$ denotes the production term that must be specified.

3.3 Constitutive Relations

A compatible set of constitutive relations is required to complete the system of balance equations in order to derive the acoustic equations for Newtonian viscous, Fourier heat-conducting fluids with chemical reactions. The linear forms for the stress tensor and heat-flux vector previously described in (2.6) and (2.7), as well as the following relations define the material under study:

1) $\varepsilon = \hat{\varepsilon}(\theta, v, \underset{\sim}{\zeta})$ for the specific internal energy; and

2) the linear reaction rate by a continuum theory of thermodynamics for a nondiffusive mixture of fluids is:

$$\overset{+}{\zeta}_{\sigma} = \overline{\overset{+}{\zeta}}_{\sigma}(\theta, p, \underset{\sim}{\zeta}) + \overline{R}_{\sigma} \, \mathrm{tr} \, \underset{\approx}{d}$$

The first term is the classical mass-action form for the chemical kinetics $\overset{+}{\overline{\zeta}}$ and is generally represented as

$$\overset{+}{\zeta}_\sigma = \overset{+}{\overline{\zeta}}_\sigma(\theta, p, \underline{\zeta}) = k_\sigma^F(\theta, p) \prod_{\alpha=1}^{k} a_\alpha^{S_{\alpha\sigma}^-} - k_\sigma^R(\theta, p) \prod_{\alpha=1}^{k} a_\alpha^{S_{\alpha\sigma}^+}$$

where $S_{\sigma\alpha}^-$ and $S_{\sigma\alpha}^+$ denote the positive stoichiometric constants for the reactants and products, respectively ($S_{j\alpha}^* = S_{j\alpha}^- - S_{j\alpha}^+$). k_σ^F and k_σ^R are the forward and reverse reaction rate coefficients which are assumed independent of the activities, although $\overline{a}_\alpha(\theta, p, \underline{\zeta})$. This is consistent with the equilibrium condition

$$\frac{k_\sigma^F}{k_\sigma^R} = K_\sigma(\theta, p) = \prod_{\alpha=1}^{k} a_\alpha^{S_{\sigma\alpha}^*} = \exp(\overset{\ominus}{A}/R\theta)$$

where $\overset{\ominus}{A}_\sigma(\theta, p)$ is the standard affinity of the σth reaction.

The trace term indicates a dependence on the dilational parts of the mean velocity gradient. This relation for the reaction rate is similar to that derived by the Onsager theory of irreversible thermodynamics when considering a state near equililbrium such that the affinity is zero [7]. This term indicating a mechanical contribution to the reaction velocity has been included in both single [8] and multistep forced plane wave acoustic analyses [9]. For simplicity, the chemical production due to mechanical effects will not be addressed in the following derivation.

3.4 The Linearized Equations of Acoustics

Next, we will derive the first-order equations of acoustics using the balance laws and constitutive relations.

The energy equation is expressed in terms of θ by first transferring it into the form of an entropy equation as before. Let the internal energy be written as $\varepsilon = \check{\varepsilon}(\eta, v, \underline{\zeta})$. Then

$$\dot{\varepsilon} = \left(\frac{\partial \check{\varepsilon}}{\partial v}\right)\dot{v} + \left(\frac{\partial \check{\varepsilon}}{\partial \eta}\right)\dot{\eta} + \left(\frac{\partial \check{\varepsilon}}{\partial \zeta_\sigma}\right)\dot{\zeta}_\sigma$$

Using the following relations

$$p = -\frac{\partial \check{\varepsilon}}{\partial v} \; ; \quad \theta = \frac{\partial \check{\varepsilon}}{\partial \eta} \; ; \quad A_\sigma = \frac{\partial \check{\varepsilon}}{\partial \zeta_\sigma} \; ;$$

one can rewrite (2.4) as

$$\rho\dot{\varepsilon} = \rho p \dot{v} + \rho\theta\dot{\eta} - \rho A_\sigma \dot{\zeta}_\sigma = S_{ij} d_{ji} + \frac{\partial h_i}{\partial x_i}$$

Assuming Fourier heat conduction and using the relationship $S_{ij} = -p\dot{v}/v + T_{ij} d_{ij}$ one obtains for the local change of entropy

$$\rho\theta\dot{\eta} = \rho A_\sigma \dot{\zeta}_\sigma + T_{ij} d_{ji} + k_\theta \frac{\partial^2\theta}{\partial x_k \partial x_k}$$

Also, $\bar{\eta}(\theta, p, \underset{\sim}{\zeta})$ gives

$$\dot{\eta} = \left(\frac{\partial\bar{\eta}}{\partial\theta}\right)\dot{\theta} + \left(\frac{\partial\bar{\eta}}{\partial p}\right)\dot{p} + \left(\frac{\partial\eta}{\partial\zeta_\sigma}\right)\dot{\zeta}_\sigma$$

so that

$$\rho c_{p,\zeta}\dot{\theta} - \theta\beta_{\theta,\zeta}\dot{p} + h_\sigma\dot{\zeta}_\sigma = T_{ij} d_{ji} + k_\theta \frac{\partial^2\theta}{\partial x_k \partial x_k}$$

with $\frac{c_{p,\zeta}}{\theta} = \frac{\partial\bar{\eta}}{\partial\theta}$; $\frac{\partial\bar{\eta}}{\partial p} = -\frac{\partial\bar{v}}{\partial\theta} \equiv -v\beta_{\theta,\zeta}$; $\frac{\partial\bar{\eta}}{\partial\zeta_\sigma} = (A_\sigma + h_\sigma)/\theta$

Now the equilibrium heat capacity which is measured by calorimetry ($\underset{\sim}{\zeta}$ at equilibrium) is defined by

$$c_p^e = c_{p,\zeta} + \frac{h_\sigma^2}{\theta}\left(\frac{\partial A_\sigma}{\partial\zeta_\sigma}\right)^{-1}$$

$c_{p,\zeta}$ is called the instantaneous, or "frozen" specific heat at constant pressure (measured at fixed $\underset{\sim}{\zeta}$). The equilibrium isobaric and isothermal coefficients of compressibility are

$$\beta_\theta^e = \beta_{\theta,\zeta} - \frac{v_\sigma h_\sigma}{v\,\theta}\left(\frac{\partial\bar{A}_\sigma}{\partial\zeta_\sigma}\right)^{-1}$$

$$\beta_p^e = \beta_{p,\zeta} - \frac{v_\sigma^2}{v}\left(\frac{\partial\bar{A}_\sigma}{\partial\zeta_\sigma}\right)^{-1}$$

$$\rho c_p^e(\gamma^e - 1)\beta_p^e = \theta\gamma^e(\beta_\theta^e)^2$$

$$\gamma^e = c_p^e / c_v^e$$

Linearizing the hydrodynamic equations (total mass, linear momentum and energy) gives

$$\frac{\partial\rho^a}{\partial t} + \rho_0\frac{\partial v^a}{\partial x} = 0 \tag{3.2}$$

$$\rho_0\frac{\partial v^a}{\partial t} = -\frac{\partial p}{\partial x} + \beta_0\frac{\partial^2 v^a}{\partial x^2} \tag{3.3}$$

$$\rho_0 c_{p,\zeta}\frac{\partial\theta^a}{\partial t} - \theta_0\beta_{\theta,\zeta}\frac{\partial p^a}{\partial t} + \rho_0 h_\sigma\frac{\partial\zeta_\sigma}{\partial t} = k_\theta\frac{\partial^2\theta^a}{\partial x^2} \tag{3.4}$$

Further, the density relation $\bar{\rho}(\theta, p, \underline{\xi})$ can be linearly approximated

$$\rho^a = \rho - \rho_0 = \left(\frac{\partial\bar{\rho}}{\partial\theta}\right)\theta^a + \left(\frac{\partial\bar{\rho}}{\partial p}\right)p^a + \left(\frac{\partial\bar{\rho}}{\partial\xi_\sigma}\right)\xi_\sigma^a$$

$$= (\beta_{\theta,\xi}/v_0)\,\theta^a + (\beta_{p,\xi}/v_0)\,p^a - (1/v_0^2)\,\xi_\sigma^a$$

where $\beta_{p,\xi} \equiv \frac{1}{\rho_0}\frac{\partial\bar{\rho}}{\partial p}$

Equation (3.2) obtains

$$-\frac{\beta_{\theta,\xi}}{v_0}\frac{\partial\theta^a}{\partial t} + \frac{\beta_{p,\xi}}{v_0}\frac{\partial p^a}{\partial t} - \frac{1}{v_0^2}\frac{\partial\xi_\sigma^a}{\partial t} + \rho_0\frac{\partial v^a}{\partial x} = 0 \quad (3.5)$$

Next, consider the chemical kinetic equations and their linearization. It is convenient to decompose the reaction velocity $\overset{+}{\underline{\xi}}$ into velocities for the forward and reverse directed reactions respectively.

$$\overset{+}{\xi}_j = \overset{+}{\xi}{}_j^F - \overset{+}{\xi}{}_j^R \qquad j = 1, 2, \ldots R$$

For small departures from the strong equilibrium point $(\theta_0, p_0, \underline{\xi}^e)$, where $\bar{A}_\sigma \equiv 0$,

$$\ln \overset{+}{\xi}{}_\sigma^F \simeq \ln(\overset{+}{\xi}{}_\sigma^F)^e + \frac{\partial}{\partial\theta}\ln(\overset{+}{\xi}{}_\sigma^F)^e\,\theta^a$$

$$+ \frac{\partial}{\partial p}\ln(\overset{+}{\xi}{}_\sigma^F)^e\,p^a + \frac{\partial}{\partial\xi_j}(\ln\overset{+}{\xi}{}_\sigma^F)^e\,\xi_j^a$$

and $\xi_\sigma^a = \xi_\sigma - \xi_\sigma^e$

Let $\overset{+}{\xi}{}_\sigma^e = \overset{+}{\bar{\xi}}{}_\sigma^F(\theta_0, p_0, \underline{\xi}) = \overset{+}{\bar{\xi}}{}_\sigma^R(\theta_0, p_0, \underline{\xi}^e)$

Now $\ln\dfrac{\overset{+}{\xi}{}_\sigma^F}{\overset{+}{\xi}{}_\sigma^R} = \ln\left[1 + \dfrac{\overset{+}{\xi}{}_\sigma^F - \overset{+}{\xi}{}_\sigma^R}{\overset{+}{\xi}{}_\sigma^e}\right] \simeq \dfrac{\overset{+}{\xi}{}_\sigma^F - \overset{+}{\xi}{}_\sigma^e}{\overset{+}{\xi}{}_\sigma^e}$

for $\left(\dfrac{\overset{+}{\xi}{}_\sigma^F - \overset{+}{\xi}{}_\sigma^R}{\overset{+}{\xi}{}_\sigma^e}\right) \ll 1$. Combining with a similar expression obtained for $\overset{+}{\xi}{}_\sigma^R$ yields

$$\frac{\overset{+}{\xi}_\sigma}{\overset{+}{\xi}{}_\sigma^e} = \frac{\partial}{\partial\theta}\ln\left(\frac{\overset{+}{\xi}{}_\sigma^F}{\overset{+}{\xi}{}_\sigma^R}\right)\theta^a + \frac{\partial}{\partial p}\ln\left(\frac{\overset{+}{\xi}{}_\sigma^F}{\overset{+}{\xi}{}_\sigma^R}\right)p^a + \frac{\partial}{\partial\xi_j}\ln\left(\frac{\overset{+}{\xi}{}_\sigma^F}{\overset{+}{\xi}{}_\sigma^R}\right)\xi_j^a$$

Since $\dfrac{\partial}{\partial\theta}\left(\dfrac{\bar{A}_\sigma}{R\theta}\right) = \dfrac{h_\sigma}{R\theta^2}$

and $$\frac{\partial}{\partial p}\left(\frac{\bar{A}_\sigma}{R\theta}\right) = -\frac{v_\sigma}{R\theta}$$

the linearized kinetics equations are written

$$\frac{\overset{+}{\zeta}_\sigma}{\overset{+}{\zeta}^e_\sigma} = \left(\frac{h_\sigma}{R\theta^2}\right)\theta^a + \left(\frac{v_\sigma}{R\theta}\right)p^a + \left(\frac{\partial(\bar{A}_\sigma/R\theta)}{\partial \zeta_\rho}\right)\zeta^a_\rho = \frac{\dot{\zeta}_\sigma}{\overset{+}{\zeta}^e_\sigma} \tag{3.6}$$

$\frac{\partial}{\partial \zeta_\rho}(\bar{A}_\sigma/R\theta)$ may be evaluated by writing the chemical affinities in terms of a linear stoichiometric combination of chemical potentials (partial molar Gibbs free energies). Let

$$\bar{A}_\sigma = -S^*_{\sigma\rho}\mu_\rho$$

where $$\mu_\rho = \bar{\mu}^\theta_\rho(\theta,p) + R\theta \ln(\gamma_\rho x_\rho)$$

is the usual representation for the chemical potential. $\bar{\mu}^\theta_\rho(\theta,p)$ is the standard function, γ_ρ is the activity coefficient and $x_\alpha = \omega_\alpha / \sum_{\alpha=1}^{k}\omega_\alpha$ is the mole fraction of component α. Then calculate

$$\frac{\partial \bar{A}_\sigma}{\partial \zeta_\rho} = -\left\{\frac{S^*_{\sigma j}S^*_{j\rho}}{\omega^\circ_j} - \frac{S_\sigma S_\rho}{\omega^\circ} + S_{\sigma j}\left(\frac{\partial \ln\gamma_j}{\partial \zeta_\rho}\right)\right\} \equiv a_{\sigma\rho}$$

where $$S_\rho \equiv \sum_{\alpha=1}^{k} S^*_{\rho\alpha}$$

3.5 Chemical Relaxation Times

Equation (3.6) is a linear system of coupled inhomogeneous differential equations of first order in ζ^a_σ. For $\theta^a = p^a = 0$ and (i.e., a single elementary reaction), a chemical relaxation time towards equilibrium can be defined as

$$\frac{1}{\tau_1} = \frac{\overset{+}{\zeta}^e_1}{R\theta_0}\left(\frac{\partial \bar{A}_1}{\partial \zeta_1}\right)^e$$

so that $$\dot{\zeta}^a_1 = -{}^1\!/\!_{\tau_1}\,\zeta^a_1$$

Different chemical relaxation times may be defined in a similar way according to the particular set of independent variables used for ζ_1. That is, relaxation times may be defined with $\{\theta,\eta\}$ held fixed, for example, where η is the specific entropy rather than at fixed θ and p. τ^*_σ for fixed $\{\theta,\eta\}$ is related to τ_σ by $\tau_\sigma = C_{p,\sigma}\tau^*_\sigma$ where

$$C_{p,\sigma} \equiv 1 + \sum_{\sigma=1}^{R}\frac{H_\sigma \tau_\sigma}{c_p R\theta_0^2}$$

When considering multi-step systems, the kinetics equations at constant $\{\theta, p\}$ become

$$\dot{\zeta}_\sigma^a = - G_{\sigma j} \zeta_j^a$$

where $$G_{\sigma j} = \overset{+}{\zeta}{}_{\underline{\sigma}}^e \, a_{\underline{\sigma} j}$$

The chemical relaxation times can be calculated from the reciprocals of the eigenvalues of $\underline{\underline{G}}$ or $\underline{\underline{J}}$ (defined below to separate the linearized kinetic equations into independent, disjoint equations). The determinantal equation for $\underline{\underline{G}}$ can be converted into that for $\underline{\underline{J}}$ by a symmetrization procedure of Jost [10]. Castellan [11] has discussed eigenvalue determinations by direct and approximate methods for various chemical systems. $\lambda_\sigma = \tau_\sigma^{-1}$ are obtained from the equation

$$\det (\lambda_\sigma \delta_{\sigma j} - J_{\sigma j}) = 0$$

with

$$J_{\sigma \ell} = (\overset{+}{\zeta}{}_{\underline{\sigma}}^e)^{1/2} a_{\underline{\sigma}\underline{\ell}} (\overset{+}{\zeta}{}_{\underline{\ell}}^e)^{1/2} = J_{\ell\sigma}$$

The kinetics equations may now be expressed in normal form. Let $\underline{\underline{M}}$ be a matrix constructed from the σ eigenvectors corresponding to the eigenvalues of the matrix $\underline{\underline{J}}$ and define $\zeta_j^a = (\overset{+}{\zeta}{}_{\underline{j}}^e)^{1/2} M_{\underline{j}\sigma} Z_\sigma$ then

$$\dot{Z}_\ell + \left(\frac{H_\ell}{R\theta^2}\right) \theta^a + \left(\frac{V_\ell}{R\theta}\right) p^a + \lambda_{\underline{\ell}} Z_{\underline{\ell}} = 0 \qquad \ell = 1, 2, \ldots R \tag{3.7}$$

3.6 Modified Kirchhoff-Langevin Equation

Treating equations (3.3), (3.4), (3.5) and (3.7) as a system of linear partial differential equations for infinitesimal plane waves and introducing damped harmonic wave solutions for each acoustic variable results in a modified Kirchhoff-Langevin equation. That is, assuming

$$\begin{pmatrix} \theta^a \\ p^a \\ v^a \\ Z_\ell^a \end{pmatrix} = \begin{pmatrix} \theta_0^a \\ p_0^a \\ v_0^a \\ (Z_\ell)_0 \end{pmatrix} \exp (\chi x + i\omega t)$$

results in a homogeneous system of algebraic equations in terms of the acoustic amplitude variables $\{ \theta_0^a, p_0^a, v_0^a, (Z_\ell^a)_0 \}$. Nontrivial solutions exist if

$$\det \begin{pmatrix} \underline{\underline{A}} & \underline{\underline{B}} \\ \underline{\underline{C}} & \underline{\underline{D}} \end{pmatrix} = \det (\underline{\underline{A}} - \underline{\underline{B}}\,\underline{\underline{D}}^{-1}\underline{\underline{C}}) \det \underline{\underline{D}} = 0 \tag{3.8}$$

where the submatrices are given by

$$\underline{\underline{A}} = \begin{pmatrix} -i\omega\beta_{\theta,\xi} & i\omega\beta_{p,\xi} & \chi \\ 0 & \chi & i\rho_0\omega-\beta_0\chi^2 \\ i\rho_0\omega c_{p,\xi}-k_\theta\chi^2 & -i\omega\Theta_0\beta_{\theta,\xi} & 0 \end{pmatrix} \quad \underline{\underline{B}} = \begin{pmatrix} -i\omega\rho_0 V_1 & -i\omega\rho_0 V_2 & \cdots & -i\omega\rho_0 V_R \\ 0 & 0 & \cdots & 0 \\ i\omega\rho_0 H_1 & i\omega\rho_0 H_2 & \cdots & i\omega\rho_0 H_R \end{pmatrix}$$

$$\underline{\underline{C}} = \begin{pmatrix} H_1/R\Theta^2 & V_1/R\Theta & \cdots 0 \\ \vdots & \vdots & \vdots \\ H_R/R\Theta^2 & V_R/R\Theta & \cdots 0 \end{pmatrix}$$

$$\underline{\underline{D}} = \begin{pmatrix} \frac{1+i\omega\tau_1}{\tau_1} & 0 & \cdots & 0 \\ 0 & \frac{1+i\omega\tau_2}{\tau_2} & \cdots & 0 \\ \vdots & \vdots & & \vdots \\ 0 & 0 & \cdots & \frac{1+i\omega\tau_R}{\tau_R} \end{pmatrix}$$

In general, a complex-valued biquadratic equation for (χ/k_0) is obtained [9]. This may be written in terms of dimensionless parameters as follows:

$$\left(\frac{\chi}{k_0}\right)^4 \left[i X_N Y_N (i\gamma X_N B_p^\omega + 1) \right] + \left(\frac{\chi}{k_0}\right)^2 \left[C_p^\omega + i\gamma X_N \left(C_p^\omega B_p^\omega - \frac{\gamma-1}{\gamma} (B_\theta^\omega)^2 \right) + Y_N B_p^\omega \right] + \gamma \left[C_p^\omega B_p^\omega - \frac{\gamma-1}{\gamma} (B_\theta^\omega)^2 \right] = 0 \qquad (3.9)$$

where

$$\Omega_\sigma \equiv \tau_\sigma \omega \qquad \text{(Reaction frequency number)}$$

and

$$\gamma_\xi = c_{p,\xi} / c_{v,\xi} \qquad \text{(Ratio of specific heats)}$$

Also,

$$C_p^\omega = 1 + \sum_{\sigma=1}^{R} \frac{\Delta C_{p\sigma}}{1+i\omega\tau_\sigma} \;; \qquad \Delta C_{p\sigma} = \frac{H_\sigma^2 \tau_\sigma}{c_p R \Theta_0^2}$$

$$B_\theta^\omega = 1 + \sum_{\sigma=1}^{R} \frac{\Delta B_{\theta\sigma}}{1+i\omega\tau_\sigma} \;; \qquad \Delta B_{\theta\sigma} = \frac{H_\sigma V_\sigma \tau_\sigma}{v \beta_\theta R \Theta_0^2}$$

$$B_p^\omega = 1 + \sum_{\sigma=1}^{R} \frac{\Delta B_{p\sigma}}{1+i\omega\tau_\sigma} \;; \qquad \Delta B_{p\sigma} = \frac{V_\sigma^2 \tau_\sigma}{v \beta_p R \Theta_0}$$

$$H_\rho = (\overset{+}{\zeta}{}^e_\sigma)^{1/2} M_{\sigma\rho} h_\sigma \;; \qquad V_\rho = (\overset{+}{\zeta}{}^e_\sigma)^{1/2} M_{\sigma\rho} v_\sigma$$

For the cases of 1) a single component non-reacting fluid and 2) a fluid mixture with a single elementary reaction, the algebraic equations correspond to the results of Truesdell [2] and Mazo [12], respectively. In the case of a non-diffusive, non-reacting system $(B_p^\omega \to 1, C_p^\omega \to 1, B_\theta^\omega \to 1)$ the classical Kirchhoff-Langevin equation (2.18) is obtained. The propagation constant χ can be determined by solving the complex algebraic equation numerically. Two of the four values of χ are unphysical (negative values of α) and the other two correspond to type I and type II waves respectively. The type I waves are chosen to correspond to the classical theories. The type II waves are found to have very large attenuation and have not been observed experimentally.

3.7 Approximations and Calculated Results

The calculation of χ from the general biquadratic equation is algebraically complicated and it is not apparent that the viscous heat conduction and reaction effects can be separated. We should note that for a nonconducting fluid $k_\theta = Y_N = 0$, the biquadratic equation reduces to the quadratic equation

$$-\left(\frac{\chi}{k_0}\right)^2 = \frac{\gamma\left[C_p^\omega B_p^\omega - \frac{\gamma-1}{\gamma}(B_\theta^\omega)^2\right]}{C_p^\omega + i\gamma X_N\left[C_p^\omega B_p^\omega - \frac{\gamma-1}{\gamma}(B_\theta^\omega)^2\right]}$$

or

$$-\left(\frac{\chi}{k_0}\right)^2 \sim \frac{\gamma\left[C_p^\omega B_p^\omega - \frac{\gamma-1}{\gamma}(B_\theta^\omega)^2\right]}{C_p^\omega}\left\{1 - \frac{i\gamma\left[C_p^\omega B_p^\omega - \frac{\gamma-1}{\gamma}(B_\theta^\omega)^2\right]}{C_p^\omega}\right\}$$

From the above equation it is seen that the viscous contribution to the absorption and dispersion is linearly additive to the reactive contribution only for $\omega\tau_\sigma^* >> 1$. Other approximations and power series expansions have been considered; in particular, those needed to obtain the approximate formula typically used to interpret experimental measurements. Simple and direct relationships are sought for fitting data from experimental measurements as shown in Table 3.1.

The assumptions needed to obtain these simple formulae from the exact biquadratic solution are: 1) an ideal solution, 2) viscous, thermal and diffusive effects are negligible, 3) either a dilute solution or $S_\sigma = 0$, 4) $\alpha << k$, and a reaction sequence orthonormalization. We wrote a computer program to calculate the absorption and speed of sound by the "direct" method, which requires all the physical property data. For the single-reaction case, however, the approximate equation is precisely that obtained by others. For

TABLE 3.1

Typical Formulas Used to Compute Chemical Reaction Contribution to the Acoustic Absorption (Neglecting Viscothermal Effects)

Single Reaction Case:	EXAMPLES
$\alpha\lambda = \dfrac{\pi(\gamma-1)\tau_1\left[h_1-(c_p/v\,\beta_\theta)V_1\right]^2\omega\tau_1}{c_p R\Theta^2\left[1+(\omega\tau_1)^2\right]}$	$A \rightleftharpoons B$
Uncoupled Reaction Case:	
$\alpha\lambda = \sum_{\rho=1}^{R} \dfrac{\pi(\gamma-1)\tau_\rho\left[h_\rho-(c_p\, v\,\beta_\theta)V_\rho\right]^2\omega\tau_\rho}{c_p R\Theta^2\left[1+(\omega\tau_\rho)^2\right]}$	$A \rightleftharpoons B$ $C \rightleftharpoons D$
Multiple Coupled Independent Reaction Case:	
$\alpha\lambda = \sum_{\rho=1}^{R} \dfrac{\pi(\gamma-1)\tau_\rho\left[H_\rho-(c_p/v\,\beta_\theta)V_\rho\right]^2\omega\tau_\rho}{c_p R\Theta^2\left[1+(\omega\tau_\rho)^2\right]}$	$A \rightleftharpoons B$ $B \rightleftharpoons C$

illustrative purposes, Fig. 3.1 shows the sound absorption in acetic acid. The chemical system and sound absorption calculations were made assuming a monomer-dimer reaction ($2A \rightleftharpoons A_2\,(D)$) in equilibrium which exhibits a single relaxation.

Figures 3.2 and 3.3 present calculations performed on a cobalt polyphosphate (Co-PP) system where the ultrasonic absorption due to chemical reactions is considered analogous to Eigen's ion-pair formation mechanism in simple electrolyte solutions [6]. Here the chemical model considers ion-pairs exhibiting three states of hydration that are in equilibrium relative to the two reaction sequence $A \rightleftharpoons B \rightleftharpoons C$. State A corresponds to the hydration shells of the polyion site and the counterion in contact without overlapping; state B consists of relatively unaffected counterions and partially dehydrated polyion sites and state C consists of the hydration shells of both the counterions and the polyion sites being modified. It was assumed that the excess absorption due to the above described chemical process can be separated from other effects by subtracting the absorption using TMA^+ (tetramethyl-ammonium) as the counterion. We remark that approximate calculations of $\frac{\alpha(\text{Co-PP})}{f^2}$ minus $\frac{\alpha(\text{TMA}^+\text{-PP})}{f^2}$ differ from the exact by less than one percent which is well within experimental error. The plot of α/f^2 for a 0.068 N Co-PP is shown in Fig. 3.2 and the relative contributions to the $\alpha\lambda$ or absorption per

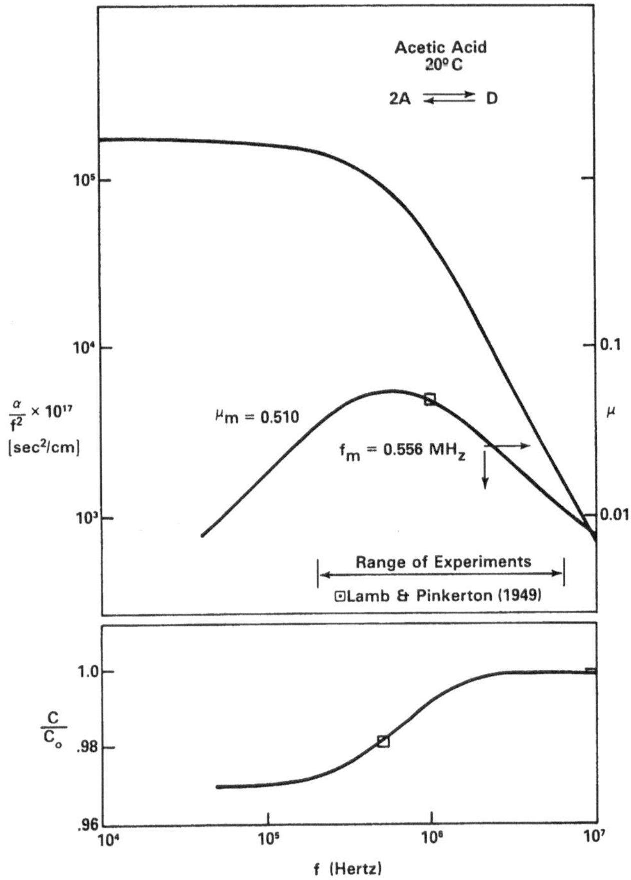

Figure 3.1: Graph of α/f^2 and c/c_o versus frequency for acetic acid (J. Lamb and J. M. M. Pinkerton, Proc. Roy. Soc. A199, 14, 1949).

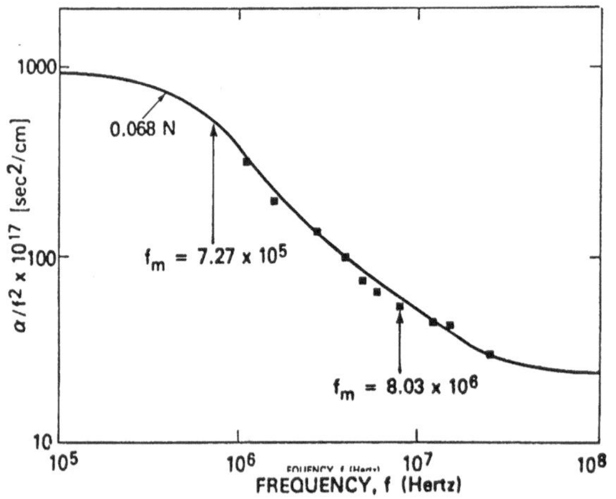

Figure 3.2: Plot of α/f^2 versus frequency for 0.068N Cobalt-Polyphosphate Solution

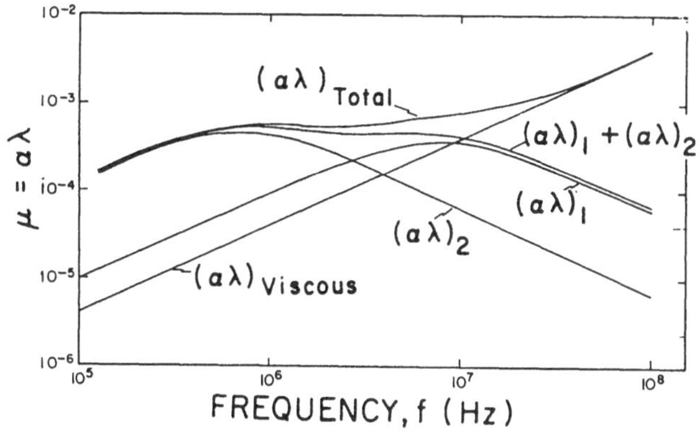

Figure 3.3: Absorption per wavelength versus frequency for 0.068N Cobalt-Polyphosphate Solution

wavelength curve are shown in Fig. 3.3. The relaxation frequencies are the same as those for a 0.125 N solution (13) and provide quantitative information about the reaction rates. The amplitude of the spectra for the reactions, $(\alpha\lambda)_1$, and $(\alpha\lambda)_2$, are interpreted as mainly volume changes associated with the reactions.

Calculations have also been performed for other chemical systems, each having two reaction steps, including aqueous solutions of glycine and water-p-dioxane mixtures [13]. The calculated values using the exact and approximate formulae for the multiple-reaction cases examined were well within experimental error. However, it was shown that the viscous term in the case of p-dioxane-water is not small compared to the reactive terms over the entire range of frequencies. Therefore, assuming that the viscous term can be decoupled from the kinetics by a linear subtraction is not necessarily valid and must be explicitly justified. From the calculations it is not possible to determine if the difference between the exact and the approximate values is due to the viscous approximation or the chemical approximation. Because the criteria for the chemical approximation are satisfied and the error decreases at high frequencies where the viscous term dominates, we attribute the difference to the nonlinear viscous contribution at low frequencies. Furthermore, the exact and approximate calculation and interpretation of sound absorption spectra for quantifying thermodynamic parameters (e.g., heat or volumetric changes) associated with the reactions may differ. In general, the exact results depend upon $\underline{\underline{M}}$, the transformation used to diagonalize the kinetics equations. The chemical systems previously studied indicate that the approximate formula provides estimates for the first reaction parameters which closely agree with those derived from the exact calculation. However, the second reaction (heat or volume change) calculation is affected by the transformation used even in dilute cases where the chemical relaxation times are well separated.

3.8 Dependent Reactions

A general reaction scheme may be composed of r-coupled reaction steps containing dependent reactions. As a consequence, the total collection of reaction sequences may be completely represented by a subset (say R) of independent reactions that span the reaction space. Chemical mixtures with dependent reactions have been considered in the acoustic theory for a classical Newtonian viscothermal fluid. There the reaction sequences were rewritten according to a procedure similar to Heilman [14]. Let

$$\sum_{\alpha=1}^{k} S^{*}_{\sigma\alpha} C_{\alpha} = \sum_{\alpha=1}^{k} \sum_{\beta=1}^{R} B_{\sigma\beta} S^{**}_{\beta\alpha} C_{\alpha} = 0 \qquad \sigma = 1,2,.. r$$

$B_{\sigma\beta}$ is a matrix of dimension (rxR) whose first rows are identical with the identity matrix and $S^{**}_{\beta\alpha}$ is the matrix of dimension (Rxk) consisting of the first R rows of $S^{*}_{\beta\alpha}$ (i.e., the chosen independent set of reactions).

Let's consider a simple example of a dependent reaction, a triangular reaction scheme,

$$\begin{matrix} \sigma = 1 & A \rightleftharpoons B \\ \sigma = 2 & B \rightleftharpoons C \\ \sigma = 3 & C \rightleftharpoons A \end{matrix} \qquad \begin{pmatrix} -1 & 1 & 0 \\ 0 & -1 & 1 \\ 1 & 0 & -1 \end{pmatrix} \equiv S^{*}_{\sigma\alpha}$$

where Rank ($S^{*}_{\sigma\alpha}$) = 2. If we choose the first two reactions (or rows in the stoichiometric matrix) to be independent, calling them $S^{**}_{\beta\alpha}$ $(\beta = 1,2 \; ; \; \alpha = 1,3)$

$$S^{*}_{\sigma\alpha} = B_{\sigma j} S^{**}_{j\alpha} = \begin{pmatrix} 1 & 0 \\ 0 & 1 \\ -1 & -1 \end{pmatrix} \begin{pmatrix} -1 & 1 & 0 \\ 0 & -1 & 1 \end{pmatrix}$$

Now the acoustics theory must be appropriately modified by transforming the various thermodynamic and kinetic parameters

$$h_{\sigma} = B_{\sigma j} h^{*}_{j}$$

$$v_{\sigma} = B_{\sigma j} v^{*}_{j}$$

and

$$a_{\rho\sigma} = B_{\rho u} a^{*}_{uj} B^{T}_{j\sigma} \qquad \sigma, \rho = 1,2,.. r \;\; ; \;\; u,j = 1,2,.. R \tag{3.10}$$

where nonideal terms are neglected. For our triangular reaction example,

$$a^*_{\rho\sigma} = M_e \begin{pmatrix} \frac{1}{x^e_A} + \frac{1}{x^e_B} & -\frac{1}{x_B} \\ -\frac{1}{x^e_B} & \frac{1}{x^e_B} + \frac{1}{x^e_C} \end{pmatrix}$$

and

$$a_{\rho\sigma} = M_e \left(\begin{array}{cc:c} \frac{1}{x^e_A} + \frac{1}{x^e_B} & -\frac{1}{x^e_B} & -\frac{1}{x^e_A} \\ -\frac{1}{x^e_B} & \frac{1}{x^e_B} + \frac{1}{x^e_C} & -\frac{1}{x^e_C} \\ \hdashline -\frac{1}{x^e_A} & -\frac{1}{x^e_C} & \frac{1}{x^e_A} + \frac{1}{x^e_C} \end{array}\right)$$

According to (3.10), this equals

$$M_e \begin{pmatrix} 1 & 0 \\ 0 & 1 \\ -1 & -1 \end{pmatrix} \begin{pmatrix} \frac{1}{x^e_A} + \frac{1}{x^e_B} & -\frac{1}{x^e_B} \\ -\frac{1}{x^e_B} & \frac{1}{x^e_B} + \frac{1}{x^e_C} \end{pmatrix} \begin{pmatrix} 1 & 0 & -1 \\ 0 & 1 & -1 \end{pmatrix}$$

Also, it can be shown that the eigenvalues (or reciprocal relaxation times) are computed from $J^*_{\rho\sigma}$.

$$J^*_{\rho\sigma} = P_{\rho s}\, a^*_{sm}\, P_{m\sigma}$$

where $\quad P^2_{su} = B^T_{s\sigma}\, \overset{+}{\xi}{}^e_{\underline{\sigma}}\, B_{\underline{\sigma} u}$

That is, the matrix square root in the example becomes

$$P_{su} = \left(B^T_{s\sigma}\, \overset{+}{\xi}{}^e_{\underline{\sigma}}\, B_{\sigma u}\right)^{1/2} = \begin{pmatrix} \overset{+}{\xi}{}^e_1 + \overset{+}{\xi}{}^e_3 & \overset{+}{\xi}{}^e_3 \\ \overset{+}{\xi}{}^e_3 & \overset{+}{\xi}{}^e_2 + \overset{+}{\xi}{}^e_3 \end{pmatrix}^{1/2}$$

For $\quad \overset{+}{\xi}{}^e_3 \to 0 \; ; \quad P_{su} \to \begin{bmatrix} (\overset{+}{\xi}{}^e_1)^{1/2} & 0 \\ 0 & (\overset{+}{\xi}{}^e_2)^{1/2} \end{bmatrix}$

and the two independent reaction case is obtained (e.g., $A \rightleftharpoons B$ and $B \rightleftharpoons C$).

In general, for the dependent reaction case one obtains a biquadratic equation for χ similar to the independent reaction case; however, the thermodynamic quantities and coefficients are redefined [9].

4. Binary Mixture of Newtonian Fluids with a Generalized Fick's Law of Diffusion

4.1 Continuum Balance Equations and Constitutive Relations

The classical problem of the propagation of forced acoustic waves in a binary mixture of nonreactive fluids, particularly monatomic gases, has been previously examined in considerable detail using hydrodynamic [15-17] and kinetic theories [18-20]. Here, the hydrodynamic approach will be presented using appropriate constitutive equations including a generalized form of Fick's law of diffusion [21] to obtain a set of linear partial differential equations from the balance postulates. The constitutive equations are deduced from a thermodynamic theory that uses the entropy inequality and also certain invariance principles [22]. Specifically, for a mixture of two constituents (nonreacting) A and B for the heat flux (Fourier's law)

$$q_i = -k_\theta \theta_{,i} - k_D u_i \tag{4.1}$$

for the caloric equation of state

$$\epsilon = \hat{\epsilon}(\theta, p, \omega_A) \tag{4.2}$$

for the inner part of the total stress tensor;

$$S_{ij} = -p\delta_{ij} + 2\eta_0 d_{ij} + 2\eta_D D_{ij} + \left[(K_0 - \tfrac{2}{3}\eta_0) d_{kk} + (K_D - \tfrac{2}{3}\eta_D) D_{kk} \right] \delta_{ij} \tag{4.3}$$

for the symmetric diffusive tensor

$$\hat{\tau}_{ij} = -p\delta_{ij} + 2\eta_0^\tau d_{ij} + 2\eta_D^\tau D_{ij} + \left[(K_0^\tau - \tfrac{2}{3}\eta_0^\tau) d_{kk} + (K_D^\tau - \tfrac{2}{3}\eta_D^\tau) D_{kk} \right] \delta_{ij} \tag{4.4}$$

and for a generalized Fick's law

$$\begin{aligned} u_i + \frac{D_{AB} M_A}{c_A c_B \mu_{AA}^\dagger} \dot{u}_i = & \frac{-D_{AB}}{c_A c_B} \left\{ \left[\frac{K_T M_A M_B}{\theta M^2} + \frac{\beta_\omega c_0^2 \beta_\theta M_A}{\gamma \mu_{AA}^\dagger} \right] \theta_{,i} \right. \\ & \left. + \left[\frac{\beta_\omega c_0^2 M_A}{\rho \gamma \mu_{AA}^\dagger} \right] \rho_{,i} + M_A \left[1 + \frac{\beta_\omega^2 c_0^2}{\gamma \mu_{AA}^\dagger} \right] \omega_{A,i} \right\} + \\ & \frac{D_{AB}}{\rho(\omega_A c_A)\mu_{AA}^\dagger} \left[(K_0^\tau - \tfrac{2}{3}\eta_0) d_{jj,i} + (K_D^\tau - \tfrac{2}{3}\eta_D^\tau) D_{jj,i} \right. \\ & \left. + 2\eta_0^\tau d_{ij,j} + 2\eta_D^\tau D_{ij,j} + \mu^\tau \dot{\Omega}_{ij,j} \right] \end{aligned} \tag{4.5}$$

The coefficients of the constitutive relations were arranged to be consistent with the classical results of irreversible thermodynamics [6,7] and the results obtained by Goldman and Sirovich [23]. η_D, K_D, η_0^τ, K_0^τ, η_D^τ, K_D^τ and μ^τ are diffusive shear and bulk viscosity parameters that define the material.

In terms of the variables $\{\theta, p, \omega_A\}$, the generalized form of Fick's law (4.5) can be written as:

$$u_i + \frac{D_{AB} M_A}{\omega_A c_B \mu^{\dagger}_{AA}} \dot{u}_i = \frac{-D_{AB}}{x_A x_B} x_{A,i} - \frac{D_{AB} \beta_\omega}{\rho \omega_A c_B \mu^{\dagger}_{AA}} p_{,i} - \frac{D_{AB} K_T}{x_A x_B \theta} \theta_{,i} + \frac{D_{AB} M_A}{\omega_A c_B \mu^{\dagger}_{AA}} (b_{A i} - b_{B i}) + \frac{D_{AB}}{\rho (\omega_A c_B)^2 \mu^{\dagger}_{AA}} \Big\{ K^{\tau}_{o} d_{kk,i} + K^{\tau}_{D} D_{kk,i} + 2 \eta^{\tau}_{o} \langle d_{ij,j} \rangle + 2 \eta^{\tau}_{D} \langle D_{ij} \rangle_{,j} + \mu^{\tau} \Omega_{ij,j} \Big\} \tag{4.6}$$

where the brackets <> refer to the traceless part of the tensor. On the right hand side, the terms correspond to diffusive transport by a concentration gradient, the pressure-diffusive effect, the thermal-diffusive or Soret effect, diffusive effect due to a difference in body forces, and a diffusive-viscous drag. In addition, an inertial term appears on the left hand side, and has the satisfying effect of changing the parabolic diffusion equation to a hyperbolic form hence making propagating diffusion waves travel with finite speed.

4.2 Linearized Equations for a Binary Nonreactive Fluid

The linearized balance equations for mass, composition of component A, linear momentum, and energy, as well as Fick's law are given by:

$$\partial_t \rho^a + \rho v^a_{k,k} = 0 \tag{4.7}$$

$$\partial_t \omega^a_A + \omega_A c_B u_{k,k} = 0 \tag{4.8}$$

$$\rho \partial_t v^a_j = -\left[(\partial_\theta \hat{p}) \theta^a_{,j} + (\partial_\rho p^a) \rho_{,j} + (\partial_{\omega_A} \hat{p}) \omega^a_{A,j} \right] + (K_o + \tfrac{4}{3} \eta_o) v^a_{k,kj} + (K_D + \tfrac{4}{3} \eta_D) u^a_{k,kj} \tag{4.9}$$

$$\rho c_v \partial_t \theta^a - \frac{(c_p - c_v)}{\beta_\theta} \partial_t \rho^a + \rho \partial_{\omega_A} \hat{\varepsilon} \partial_t \omega^a_A = k_\theta \theta_{,kk} + k_D u^a_{k,k} \tag{4.10}$$

$$u^a_j + \frac{D_{AB} M_A}{\omega_A c_B \mu^{\dagger}_{AA}} \partial_t u^a_j = \frac{-D_{AB}}{\omega_A c_B} \Big\{ \left[\frac{K_T M_B}{\theta M^2} + \frac{\beta_\omega c_o^2 \beta_\theta M_A}{\mu^{\dagger}_{AA}} \right] \theta^a_{,j} + \left[\frac{\beta_\omega c_o^2}{\rho \gamma \mu^{\dagger}_{AA}} \right] \rho^a_{,j} + \left[1 + \frac{\beta^2_\omega c_o^2}{\gamma \mu^{\dagger}_{AA}} \right] \omega^a_{A,j} \Big\} + \frac{D_{AB}}{\rho (\omega_A c_B)^2 \mu^{\dagger}_{AA}} \left[(K^{\tau}_{o} + \tfrac{4}{3} \eta^{\tau}_{o}) v^a_{k,kj} + (K^{\tau}_{D} + \tfrac{4}{3} \eta^{\tau}_{D}) u^a_{k,kj} \right], \tag{4.11}$$

where the thermodiffusion coefficient k_D is given by

$$k_D \equiv -\rho\omega_A c_B \left[\frac{K_T M_B \mu^{\dagger}_{AA}}{M^2} + \partial_{c_A} \bar{h} \right]$$

where

$$h = \varepsilon + p v = \bar{h}(\theta, p, \omega_A)$$

is the specific enthalpy and all coefficients; e.g., $c_p, c_v, k_\theta, \ldots$ are evaluated at equilibrium $\{\theta^e, \rho^e, \omega_A^e\}$.

4.3 Exact Solution - Sixth Order Polynomial for χ

Forced plane waves propagating in the x-direction are assumed to be admitted by (4.7-4.11) where $I = I_0 + I^a \exp(\chi x + i\omega t)$ for $I = \{\theta, p, \omega_A, v, u\}$. These equations obtain

$$i\omega\rho_0^a + \rho_0 \chi v_0^a = 0 \tag{4.12}$$

$$i\omega\,\omega_{A0}^a + \omega_A c_B \chi u_0^a = 0 \tag{4.13}$$

$$\begin{aligned} &i\rho\omega v_0^a + \frac{\rho c_0^2 \beta_\theta}{\gamma}\chi\theta_0^a + \frac{c_0^2}{\gamma}\chi\rho_0^a + \frac{\rho_0 c_0^2 \beta_\omega}{\gamma}\chi\omega_{A0}^a \\ &- (K_0 + \tfrac{4}{3}\eta_0)\chi^2 v_0^a - (K_D + \tfrac{4}{3}\eta_D)\chi^2 u_0^a = 0 \end{aligned} \tag{4.14}$$

$$\begin{aligned} &i\rho c_v \omega\theta_0^a + \frac{c_p - c_v}{\beta_\theta} i\omega\rho_0^a + i\omega \frac{\partial\hat{\varepsilon}}{\partial\omega_A}\omega_{A0}^a \\ &- k_\theta \chi^2 \theta_0^a - k_D \chi u_0^a = 0 \end{aligned} \tag{4.15}$$

$$\begin{aligned} &u_0^a + \frac{D_{AB} M_A}{\omega_A c_B \mu^{\dagger}_{AA}} i\omega u_0^a + \frac{D_{AB}}{\omega_A c_B}\left\{ \left[\frac{K_T M_B}{\theta M^2} + \frac{\beta_\omega c_0^2 \beta_\theta}{\mu^{\dagger}_{AA}\gamma} \right]\chi\theta_0^a \right. \\ &\left. + \left[\frac{\beta_\omega c_0^2}{\rho\gamma\mu^{\dagger}_{AA}} \right]\chi\rho_0^a + \left[1 + \frac{\beta_\omega^2 c_0^2}{\gamma\mu^{\dagger}_{AA}} \right]\chi\omega_{A0}^a \right\} \\ &- \frac{D_{AB}}{\rho(\omega_A c_B)^2 \mu^{\dagger}_{AA}} \left[(K_0^\tau + \tfrac{4}{3}\eta_0^\tau)\chi^2 v_0^a + \right. \\ &\left. (K_D^\tau + \tfrac{4}{3}\eta_D^\tau)\chi^2 u_0^a \right] = 0 \end{aligned} \tag{4.16}$$

Therefore, for nontrivial solutions $\det[\underline{\underline{\Gamma}}] = 0$, where

$$\Gamma_{11} = \Gamma_{13} = \Gamma_{15} = \Gamma_{21} = \Gamma_{22} = \Gamma_{24} = \Gamma_{54} = 0 ,$$

$$\Gamma_{12} = i\omega , \quad \Gamma_{14} = \rho_0\chi , \quad \Gamma_{23} = i\omega , \quad \Gamma_{25} = \omega_A c_B \chi ,$$

$$\Gamma_{31} = \frac{\rho c_0^2 \beta_\theta \chi}{\gamma} , \quad \Gamma_{32} = \frac{c_0^2\chi}{\gamma} , \quad \Gamma_{33} = \frac{\rho c_0^2 \beta_\omega \chi}{\gamma} ,$$

$$\Gamma_{34} = i\rho\omega - (K_o + 4/3\,\eta_o)\,\chi^2 , \quad \Gamma_{35} = -(K_D + 4/3\,\eta_D)\,\chi^2$$

$$\Gamma_{41} = \frac{D_{AB}}{\omega_A c_B}\left[\frac{K_T M_B}{\theta M^2} + \frac{\beta_\omega c_o^2 \beta_\theta}{\mu_{AA}^{\dagger}\gamma}\right]\chi ,$$

$$\Gamma_{42} = \frac{D_{AB}}{\omega_A c_B}\left[\frac{\beta_\omega c_o^2}{\rho\gamma\mu_{AA}^{\dagger}}\right]\chi , \quad \Gamma_{43} = \frac{D_{AB}}{\omega_A c_B}\left[1 + \frac{\beta_\omega^2 c_o^2}{\gamma\mu_{AA}^{\dagger}}\right]\chi ,$$

$$\Gamma_{44} = \frac{-D_{AB}}{\rho(\omega_A c_B)^2\mu_{AA}^{\dagger}}\left(K_o^{\tau} + \tfrac{4}{3}\eta_o^{\tau}\right)\chi^2 ,$$

$$\Gamma_{45} = 1 + \frac{D_{AB}}{\rho(\omega_A c_B)^2\mu_{AA}^{\dagger}}\left[i\omega\rho c_A c_B - \left(K_D^{\tau} + \tfrac{4}{3}\eta_D^{\tau}\right)\chi^2\right] ,$$

$$\Gamma_{51} = i\rho c_v\omega - k_\theta\chi^2 , \quad \Gamma_{52} = -(c_p - c_v)\,i\omega/\beta_\theta ,$$

$$\Gamma_{53} = i\omega\frac{\partial\hat{\varepsilon}}{\partial\omega_A} , \quad \Gamma_{55} = -k_D\chi$$

Also, note the thermodynamic relations,

$$\frac{\partial\bar{h}}{\partial\omega_A} = \frac{\partial\hat{\varepsilon}}{\partial\omega_A} + \frac{(c_p - c_v)\beta_\omega}{\beta_\theta}$$

and

$$\gamma - 1 = \theta\beta_\theta^2 c_o^2 / c_p$$

Now, a sixth-order polynomial is obtained for the complex propagation coefficient χ which is analogous to the Kirchhoff-Langevin biquadratic equation for a single component.

$$\chi^6[\alpha_2\omega^2 + i\alpha_1\omega + \alpha_o] + \chi^4[i\beta_3\omega^3 + \beta_2\omega^2 + i\beta_1\omega]$$

$$+ \chi^2[\gamma_4\omega^4 + i\gamma_3\omega^3 + \gamma_2\omega^2] + [i\delta_5\omega^5 + \delta_4\omega^4] = 0 \tag{4.17}$$

where

$$\alpha_2 = \frac{D_{AB}k_\theta}{\mu_{AA}^{\dagger}(\omega_A c_B)^2}\left[(K_o + \tfrac{4}{3}\eta_o)(K_D^{\tau} + \tfrac{4}{3}\eta_D^{\tau}) - (K_D + \tfrac{4}{3}\eta_D)(K_o^{\tau} + \tfrac{4}{3}\eta_o^{\tau})\right] ,$$

$$\alpha_1 = k_\theta D_{AB}\left\{\frac{c_o^2\beta_\omega}{\gamma\mu_{AA}^{\dagger}(\omega_A c_B)}\left[(K_o^{\tau} + \tfrac{4}{3}\eta_o^{\tau}) + (K_D + \tfrac{4}{3}\eta_D) - \frac{(K_D^{\tau} + \tfrac{4}{3}\eta_D^{\tau})}{\beta_\omega\omega_A c_B}\right]\right.$$

$$\left. - (K_o + 4/3\,\eta_o)\left(1 + \frac{c_o^2\beta_\omega^2}{\gamma\mu_{AA}^{\dagger}}\right)\right\} ,$$

$$\alpha_o = \frac{-\rho c_o^2 k_\theta D_{AB}}{\gamma} ,$$

$$\beta_3 = \frac{D_{AB}}{\mu^{\dagger}_{AA}(\omega_A c_B)^2}\Big\{-\underline{k_\theta\big[c_A c_B(K_0+\tfrac{4}{3}\eta_0)} + \underline{(K_D^\tau + \tfrac{4}{3}\eta_D^\tau)}\big]$$

$$+\, c_v\big[\underline{(K_D + \tfrac{4}{3}\eta_D)(K_0^\tau + \tfrac{4}{3}\eta_0^\tau) - (K_0 + 4/3\,\eta_0)(K_D^\tau + \tfrac{4}{3}\eta_D^\tau)}\big]$$

$$\beta_2 = k_\theta D_{AB}\rho\Big(1 + \frac{c_0^2\beta_\omega^2}{\gamma\mu^{\dagger}_{AA}} + \underline{\frac{c_0^2 M_A^2}{\gamma c_A c_B \mu^{\dagger}_{AA}}}\Big) - (K_0 + \tfrac{4}{3}\eta_0)\Big[k_\theta +$$

$$D_{AB}\rho\mu^{\dagger}_{AA}\theta\Big(\frac{K_T M_B}{\theta M^2} + \frac{c_0^2\beta_\theta\beta_\omega}{\gamma\mu^{\dagger}_{AA}}\Big)^2\Big] + \underline{\frac{\rho D_{AB} c_v(K_0^\tau + \tfrac{4}{3}\eta_0^\tau)c_0^2}{\mu^{\dagger}_{AA}\omega_A c_B\gamma}\Big(\beta_\omega +}$$

$$\underline{\frac{K_T M_B \mu^{\dagger}_{AA}\beta_\theta}{c_v M^2}\Big)} - D_{AB}\rho c_v(K_0 + \tfrac{4}{3}\eta_0)\Big(1 + \frac{c_0^2\beta_\omega^2}{\gamma\mu^{\dagger}_{AA}}\Big)$$

$$-\underline{\frac{D_{AB}\rho c_0^2(K_D^\tau + \tfrac{4}{3}\eta_D^\tau)c_v}{(\omega_A c_B)^2\mu^{\dagger}_{AA}}},$$

$$\beta_1 = \frac{\rho^2 c_0^2}{\gamma}\Big\{D_{AB}\Big[c_p + \frac{K_T^2 M_B^2 \mu^{\dagger}_{AA}}{\theta M^4}\Big] + \frac{k_\theta}{\rho}\Big\},$$

$$\gamma_4 = \frac{-\rho D_{AB}}{\mu^{\dagger}_{AA}(\omega_A c_B)^2}\Big\{\underline{c_A c_B\big[k_\theta + (K_0 + \tfrac{4}{3}\eta_0)c_v\big]} + \underline{(K_D^\tau + \tfrac{4}{3}\eta_D^\tau)c_v}\Big\},$$

$$\gamma_3 = \underline{\frac{\rho^2 c_0^2 c_p D_{AB} M_A^2}{\gamma\mu^{\dagger}_{AA} c_A c_B}} + \rho\big[k_\theta + c_v(K_0 + \tfrac{4}{3}\eta_0)\big] + \rho^2 D_{AB}\Big[\frac{K_T M_B}{M^2}\Big(\frac{K_T M_B \mu^{\dagger}_{AA}}{\theta M^2}$$

$$+ \frac{2c_0^2\beta_\theta\beta_\omega}{\gamma}\Big) + c_v + \frac{c_0^2\beta_\omega c_p}{\gamma\mu^{\dagger}_{AA}}\Big],$$

$$\gamma_2 = \rho^2 c_0^2 c_v, \qquad \delta_5 = \underline{\rho^2 c_v D_{AB} M_A^2/\mu^{\dagger}_{AA} c_A c_B},$$

Equation (4.17) includes terms that arise from the generalized form of Fick's law. The dashed-underscored terms are due to the viscous-diffusive terms and the solid-underscored terms arise from the inertial-diffusive term.

4.4 Approximations

Explicit results obtained by solving the complex sixth-order polynomial (4.17) cannot be interpreted in a simple manner. However, a low frequency expansion does obtain relatively simple representations. Set

$$Z = \frac{x}{\omega} = Z_0 + \omega Z_1 + \omega^2 Z_2 + \ldots = -\Big(\frac{\alpha}{\omega} + \frac{i}{c}\Big)$$

Then $$\alpha c_0/\omega = \alpha/k_0 = \omega/2c_0^3 \left[\frac{K_0 + \frac{4}{3}\eta_0}{\rho} + \frac{(\gamma-1)}{\gamma}\frac{k_\theta}{\rho c_v} + \frac{D_{AB} c_0^2}{\mu_{AA}^{\dagger}} \left(\beta_\omega + \frac{K_T M_B \mu_{AA}^{\dagger} \beta_\theta}{c_p M^2} \right) \right] \tag{4.18}$$

and $$\frac{c_0}{c} = 1 + \frac{\omega^2}{2c_0\gamma_2}\left[Z_1^2 \gamma_2 c_0^2 - \frac{\alpha_0}{c_0^4} + \frac{\beta_2}{c_0^2} - \gamma_4 - 2Z_1\left(\frac{2\beta_1}{c_0} + \gamma_3 c_0 \right) \right] \tag{4.19}$$

Therefore, the results are similar to those obtained by Kirchhoff for the Newtonian single component fluid, viz., the sound propagation is nondispersive and the absorption is proportional to the square of the frequency and depends on viscous, thermal and diffusive effects. For the special case of a mixture of perfect gases, (4.18) becomes

$$\frac{\alpha}{f^2} = \frac{2\pi^2}{c_0^3}\left[\frac{K_0 + 4/3\eta_0}{\rho} + \frac{(\gamma-1)}{\gamma}\frac{k_\theta}{\rho c_v} + D_{AB} x_A x_B \gamma \left(\frac{M_B - M_A}{M_B} + \frac{K_T(\gamma-1)}{x_A x_B \gamma} \right)^2 \right]$$

which is precisely the result obtained by Kohler [15] and Goldman [17]. Kohler, however, used the classical form for Fick's law, which does not include the inertial-diffusive term nor the viscous-diffusive terms. Therefore, at low frequencies, these latter effects are negligible.

4.5 Calculated Results

Numerical computations of the roots of (4.17) were obtained for gaseous mixtures of helium and argon with compositions that corresponded to the experiments of Prangsma, Jonkman and Beenakker [24]. The fluids were considered to be ideal gases, and the viscometric coefficients that appear in the form of Fick's law were computed using the results of Goldman and Sirovich [23,25]. Their relations were given in terms of $\eta_{\alpha\beta}$ which are the forms that appear in the constitutive equations for the individual stresses $\underline{\underline{S}}_\alpha^{\wedge}$. It was more convenient to use the total stress tensor and the symmetric diffusive stress tensor $\overset{\wedge}{\tau} = c_B \underline{\underline{S}}_A^{\wedge} - c_A \underline{\underline{S}}_B^{\wedge}$. The viscometric coefficients are related by:

$$\eta_0 = \eta_{AA} + \eta_{AB} + \eta_{BA} + \eta_{BB},$$

$$\eta_D = c_B(\eta_{AA} + \eta_{BA}) - c_A(\eta_{AB} + \eta_{BB}),$$

$$\eta_D^\tau = 2\{\eta_{AA} c_B^2 - c_A c_B(\eta_{AB} + \eta_{BA}) + \eta_{BB} c_A^2\}$$

and $$\eta_0^\tau = 2\{c_B(\eta_{AA} + \eta_{AB}) - c_A(\eta_{BA} + \eta_{BB})\}$$

All the bulk viscosity coefficients, viz., K_0 , K_D , K_D^τ and K_0^τ were assumed to be identically zero, consistent with the kinetic theory. Although the Maxwell model (inverse fifth-power law) is known to be less suitable than other models, such as the Lennard-Jones 6-12 model, it is not considered that the numerical values will be substantially different so as to alter the results appreciably.

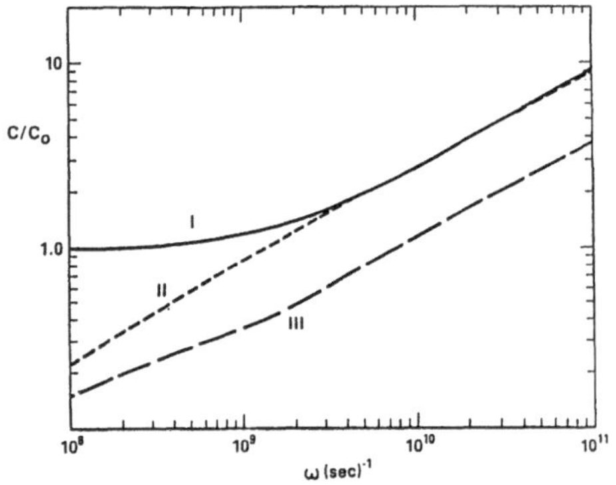

Fig. 4.1a Plot of dispersion versus frequency for a helium-argon mixture

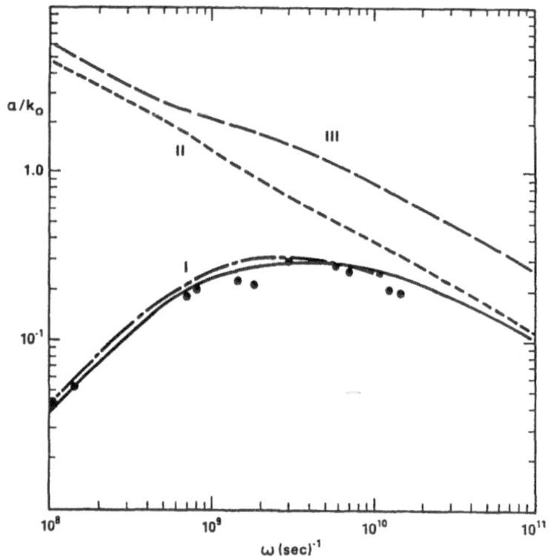

Fig. 4.1b Plot of absorption α/k_0 versus frequency ω for a helium-argon mixture (x_{AR} = 0.10, p = 1 atm, Θ_0 = 293.16°K). Data points ● from Prangsma, Jonkman and Beenakker [24]; ——— -- ——— Theory of Goldman [17]

Calculated curves for the absorption and dispersion for all three wave types are shown in Figs. (4.1-4.3) for X_{AR} =0.1, 0.5 and 0.75. For some of the experiments, the agreement with the theoretical values for the type-I wave is remarkable. However, it is apparent that type-II and type-III waves must be considered in the high-frequency range. The results of Goldman's two-fluid theory for the type-I wave are also shown in these figures. These results are similar; the differences are probably due to the different choice of values of the physical properties of the helium-argon mixture and the more general form for the energy equation.

The absorption data of Greenspan [26,27] for argon, helium, krypton, xenon and neon which are pure gases are shown in Fig. 4.4, where α/k_0 is plotted versus X_N .

All the data obtained at f = 11MHz differ from the results obtained with the Navier-Stokes theory for $X_N \geq 0.1$, except the helium data obtained at f = 1MHz. The 11 MHz data were confirmed by Meyer and

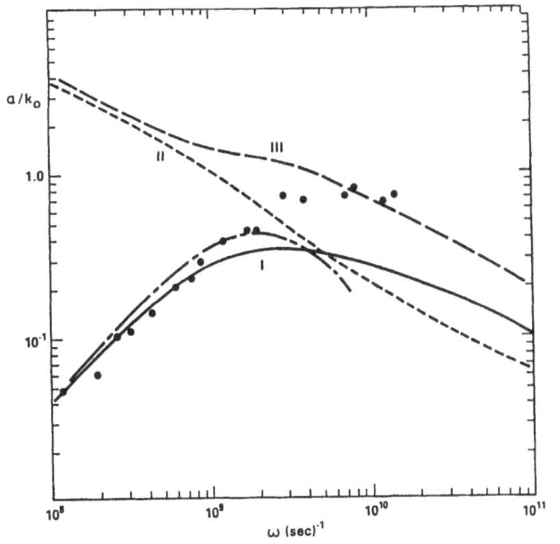

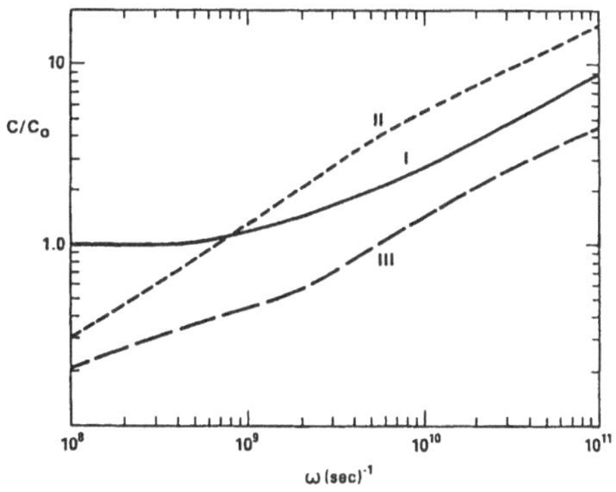

Fig. 4.2a Plot of absorption α/k_o versus frequency ω for a helium-argon mixture (x_{AR} = 0.50, p = 1 atm, Θ_o = 293.16°K). Data points ● from Prangsma, Jonkman and Beenakker [24]. ——— – – ——— Theory of Goldman [17].

Fig. 4.2b Plot of dispersion versus frequency for a helium-argon mixture (x_{AR} = 0.50, p = 1 atm, Θ_o = 293.16°K).

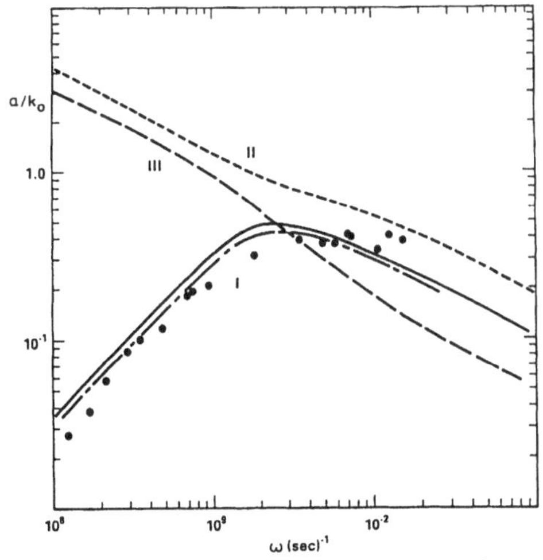

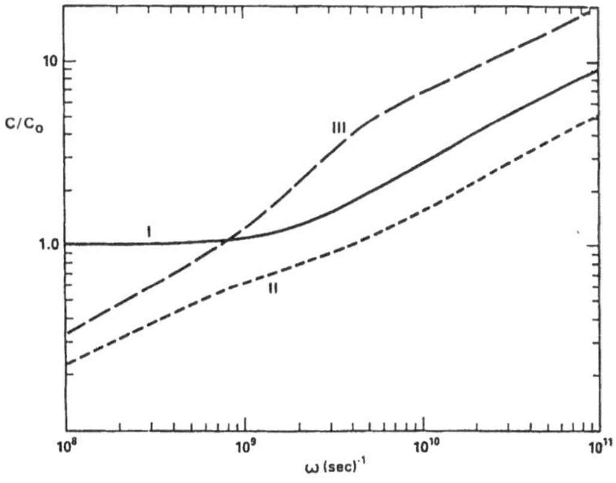

Fig. 4.3a Plot of absorption α/k_o versus frequency ω for a helium-argon mixture (x_{AR} = 0.75, p = 1 atm, Θ_o = 293.16°K). Data points ● from Prangsma, Jonkman and Beenakker [24]. ——— – – ——— Theory of Goldman [17].

Fig. 4.3b Plot of dispersion versus frequency for a helium-argon mixture (x_{AR} = 0.75, p = 1 atm, Θ_o = 293.16°K).

Sessler [28] for argon. The theoretical curve for the type-II wave is also shown in Fig. 4.4. At low frequencies, the attenuation is relatively large and the type-II wave is probably not detectable for $X_N \leq 1.0$.

On the basis of Greenspan's measurements for pure gases, the range of applicabililty of the hydrodynamical theory to the prediction of absorption and dispersion of sound in gases is rather limited. At $X_N \approx 0.1$, there is an apparent deviation of the data from the theoretical curve, i.e., the deviation is greater than the experimental scatter. Therefore, the theory appears to be valid only for $X_N \lesssim 0.1$, for which the absorption is given by the Kirchhoff relation (4.18) and the sound velocity is non-dispersive.

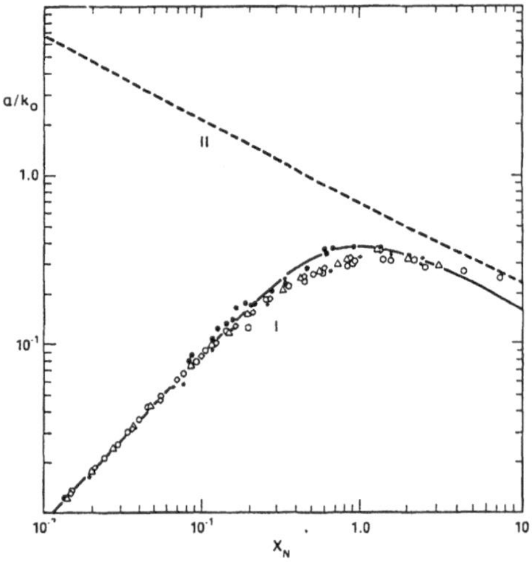

Fig. 4.4 Absorption of sound in pure gases plotted as α/k_o versus the viscous number X_N. Data of Greenspan: 0 argon, ◊ krypton, ⬡ neon, △ xenon, O helium, 11MHz, ● helium 1MHz. The solid lines are the results of the hydrodynamical theory.

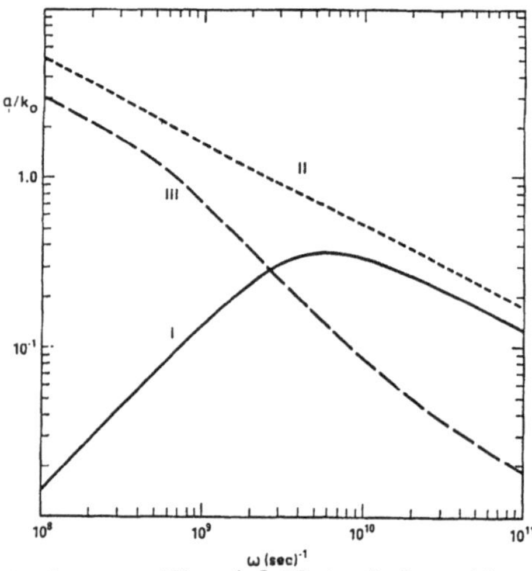

Fig. 4.5 Plot of absorption α/k_o versus frequency ω for a helium-argon mixture ($x_{AR} = 0.999$, $p = 1$ atm, $\Theta_o = 293,16°K$).

On the other hand, there is the possibility that the discrepancy between the experimental work of Greenspan and the results of the hydrodynamical theory was affected by impurities in the gases. To provide some perspective on this possibility, calculations for a mixture of argon and helium ($X_{AR} = 0.999$) were made and shown in Fig. 4.5. The type-I and type-II waves are essentially the same as the type-I and type-II waves of pure argon. However, the type-III wave can occur, and in the range of the measurements it has a lower attenuation than the type-I wave which apparently dominates the measurements. The wave speed of this type-III wave is about ten times faster than the type-I wave. It should be remarked that other kinds of impurities can produce different types of relaxational effects. Further experiments that focus on the different wave types would provide useful information.

5. Viscoelastic Fluids

There is a large class of fluids that do not satisfy the linearly viscous constitutive equation (Newton's law) for the stress tensor. Examples are synthetic polymer melts and solutions, biopolymers, emulsions and other multiphase fluids. During the last two decades, there has been considerable work done by fluid mechanicians to understand some of the flow phenomena exhibited by non-Newtonian materials. The particular problem of determining the characteristics of the propagation of infinitesimal acoustic waves through viscoelastic media is considered here in order to better understand non-Newtonian effects on the mechanics of fluid motions, and also because it provides a method for determining some viscometric parameters of non-Newtonian fluids that are not otherwise accessible.

5.1 Viscoelastic Constitutive Equation

A number of constitutive equations have been proposed to represent the stress tensor in terms of the motion of the fluid. These forms were developed phenomenologically by using hydrodynamical and/or molecular models, or by using the methods of continuum mechanics. A constitutive relation for the stress tensor that has been shown to represent a large class of nonlinear fluids was developed for the model of a <u>simple fluid</u> [29] for which

$$\underline{\underline{S}}(\mathcal{P},t) = \underset{\sigma=0}{\overset{\infty}{\underline{\underline{\mathfrak{S}}}}}\left[\underline{\underline{C}}_t^t(\sigma);\ \rho(t)\right] \tag{5.1}$$

The stress tensor associated with a fluid particle $\mathcal{P}$ at the time t depends on the entire history of the relative right Cauchy-Green tensor $\underline{\underline{C}}_t^t(\sigma)$, where past time $\sigma \equiv t-\tau$ $(-\infty < \tau \leq t)$. The tensor $\underline{\underline{C}}_t^t(\sigma) \equiv \underline{\underline{F}}_t(\sigma)\underline{\underline{F}}_t^T(\sigma)$, where $\underline{\underline{F}}_t(\sigma)$ is the relative deformation gradient. The model of a <u>simple fluid</u> accounts for the memory and also the local deformation of the material by the functional dependence of the stress tensor on the entire history of a measure of the local deformation gradient. To obtain explicit analytical forms for the functional $\underline{\underline{\mathfrak{S}}}[\cdot]$ and to satisfy the intuitive belief that recent events (small σ) influence the stress tensor (associated with $\mathcal{P}$) more than those occurring in the distant past, the concept of "material fading memory" was introduced. This in turn brings in the idea of a relaxation time.

Non-Newtonian fluids that have a long memory have a large viscous-relaxation time.

Using the mathematical representation for "fading memory" and the isotropy condition for a fluid, Coleman and Noll [30] obtained:

$$S_{ij}(P,t) = \left\{ -p\left[\rho(t)\right] + \int_0^\infty \phi(\sigma,\rho)\, \frac{\partial \xi^m}{\partial x^k} \frac{\partial \xi^n}{\partial x^k}\, d_{mn}(\xi, t-\sigma)\, d\sigma \right\} \delta_{ij} + 2\int_0^\infty \psi(\sigma,\rho)\, \frac{\partial \xi^m}{\partial x^i} \frac{\partial \xi^n}{\partial x^j}\, d_{mn}(\xi, t-\sigma)\, d\sigma \tag{5.2}$$

which is the finite first-order linear form for a viscoelastic fluid where $\underline{\xi}$ is the position of P at past time τ. This representation corresponds to small displacements superposed on the rest state.

5.2 Sound Absorption and Dispersion Theory

Having developed a general constitutive equation for the stress tensor to represent viscoelastic fluids, the problem of sound wave absorption and dispersion is considered. If this viscometric model is adequate, analytic results are expected to describe the behavior of small-amplitude acoustic waves in a variety of viscoelastic fluids.

For planar sound wave propagation, it is appropriate to approximate (5.2) by the linear infinitesimal expression:

$$t_{ik}(\underline{x},t) = 2\int_{-\infty}^{t} \psi(t-\tau,\rho)\, d_{ik}(\underline{x},\tau)\, d\tau + \delta_{ik}\int_{-\infty}^{t} \phi(t-\tau,\rho)\, d_{jj}(\underline{x},t)\, d\tau \tag{5.3}$$

We remark that this relation is exactly that obtained by the phenomenological approach of using a superposition of "Maxwell elements" to obtain an expression for a linear viscoelastic fluid [30]. Most other models for isotropic non-Newtonian materials can be represented by (5.3) in the limit of small-displacement motions.

For sound propagation along the x-axis, only the xx component of the extra-stress tensor is required, viz.,

$$t^a_{xx}(x,t) = \int_{-\infty}^{t} \left[\phi(t-\tau;\rho) + 2\psi(t-\tau;\rho)\right] \frac{\partial v^a(x,t)}{\partial x}\, d\tau \tag{5.4}$$

It should be noted that this continuum mechanics model is essentially mechanical; it accounts for the temperature (the thermodynamics) in the classical way. The quantities such as the internal energy (ε), the stress tensor ($\underline{\underline{S}}$), etc., are assumed to depend on the present value of the temperature only, and not on its entire history.

The linearized equations of change (2.11), (2.12) and (2.13) and the equation of state are applicable to this problem, and together with (5.4) form a deterministic system for the four unknowns ρ^a, V^a, θ^a, and t_{xx}^{a}. Damped wave solutions, given by (2.16), are then assumed for the acoustic variables.

The expression for t_{xx}^{a} becomes

$$t_{xx}^{a} = \chi\, v_0^{a} \exp(\chi x) \int_0^{\infty} \left[\phi(t-\tau;\rho) + 2\psi(t-\tau;\rho)\right] \exp(i\omega\tau)\, d\tau \tag{5.5}$$

It is convenient to define the Fourier transforms of the shear-relaxation and compressional-relaxation functions as "complex shear and second-coefficient viscosities" respectively, viz.,

$$\eta^*(\omega) \equiv \int_0^{\infty} \psi(\sigma) e^{-i\omega\sigma} d\sigma \quad ; \quad \lambda^*(\omega) \equiv \int_0^{\infty} \phi(\sigma) e^{-i\omega\sigma} d\sigma \tag{5.6}$$

Also define the bulk compressional relaxation function by

$$K(\sigma) \equiv \phi(\sigma) + \frac{2}{3}\psi(\sigma) \tag{5.7}$$

and the "complex bulk viscosity" by

$$K^*(\omega) \equiv \lambda^*(\omega) + \frac{2}{3}\eta^*(\omega) = \int_0^{\infty} K(\sigma) e^{-i\omega\sigma} d\sigma \tag{5.8}$$

The complex-valued shear viscosity function $\eta^*(\omega)$ has been experimentally determined for a variety of materials [31]. Also there are a number of theoretical models to predict values of the function [32]. These studies will be discussed briefly in a later section. We also remark that there are other representations for the "viscosities" that have physical interpretations, specifically the relaxation spectra defined by:

$$\psi(\sigma) = \int_0^{\infty} \frac{H(\lambda)}{\lambda} e^{-\sigma/\lambda} d\lambda \quad ; \quad \phi(\sigma) \equiv \int_0^{\infty} \frac{L(\lambda)}{\lambda} e^{-\sigma/\lambda} d\lambda$$

and

$$K(\sigma) \equiv \int_0^{\infty} \frac{K(\lambda)}{\lambda} e^{-\sigma/\lambda} d\lambda \tag{5.9}$$

The functions $H(\lambda)$, $L(\lambda)$ and $K(\lambda)$ are the shear-relaxation, the second-coefficient-relaxation and the bulk-relaxation spectra respectively. These forms are obtained in a natural way by extending the concept of a Maxwell viscoelastic fluid having a single relaxation time to the concept of a material having a continuous distribution of relaxation times, such that

$H(\tau)\, d\tau$ = contribution to total viscosity for Maxwell elements with times between τ and $\tau + d\tau$

Further, "molecular" theories for dilute polymer solutions obtain the shear relaxation spectrum directly in terms of "molecular" parameters [33-35].

Now the acoustic stress tensor can be written in a form similar to the Newtonian case,

$$t^a_{xx} \equiv \beta^*(\omega)\, \varkappa\, v^a \tag{5.10}$$

where the acoustic viscoelastic complex viscosity is defined by

$$\beta^*(\omega) = \beta'(\omega) - i\,\beta''(\omega) \equiv \frac{4}{3}\eta^*(\omega) + K^*(\omega) \tag{5.11}$$

$$\beta^*(\omega) \equiv \int_0^\infty \left[\frac{4}{3}\psi(\sigma) + K(\sigma)\right] e^{-i\omega\sigma}\, d\sigma \tag{5.12}$$

or

$$\beta^*(\omega) = \int_0^\infty \frac{\frac{4}{3}H(\lambda) + K(\lambda)}{1 + i\,\omega\lambda}\, d\lambda \tag{5.13}$$

Note that in the limit of $\omega \to 0$,

$$\begin{aligned}\beta^*(\omega) &\sim \frac{4}{3}\eta^*(0) + K^*(0) = \frac{4}{3}\eta'(0) + K'(0)\\ &= \frac{4}{3}\eta_0 + K_0 = \frac{4}{3}\int_0^\infty \psi(\sigma)\, d\sigma + \int_0^\infty K(\sigma)\, d\sigma\\ &= \beta_0\end{aligned} \tag{5.14}$$

which is the acoustic viscosity for a Newtonian fluid. This is consistent with the intuitive belief and also the experimental observations that all isotropic non-Newtonian fluids act Newtonian (linear) for small deformations.

The linearized acoustics balance equations for mass, linear momentum and energy applied to a viscoelastic fluid having classical thermodynamic response obtain the same expressions as those for a Newtonian fluid except that the acoustic viscosity β_0 is replaced by the frequency-dependent acoustic viscosity $\beta^*(\omega)$.

For a linear viscoelastic fluid, the problem of establishing measures of the absorption and dispersion of a forced, plane, monochromatic sound wave propagating into an unbounded fluid otherwise at rest as a function of the physical and thermodynamical properties of the fluid and the frequency of the sound is posed as finding the function:

$$\varkappa = F\left[\omega; \rho, k_0, \langle\bullet\rangle; c_0, c_p, c_v; \beta_0; \theta\right] \tag{5.15}$$

where $\langle\bullet\rangle$ represents those rheological properties of the viscoelastic fluid that are relevant to the acoustics problem. The two relaxation functions $\psi(t)$ and $\phi(t)$, or equivalent functions such as the relaxation spectra $H(\lambda)$ and $L(\lambda)$ must be specified. Here we consider the second-viscosity relaxation spectrum $L(\lambda)$ explicitly rather than the bulk viscosity relaxation spectrum $K(\lambda)$ since it appears in the theory as an independent coefficient, whereas $K(\lambda)$ is defined as the combination $L(\lambda)+\frac{2}{3}H(\lambda)$. We do remark that the bulk viscosity is often convenient to use. Recall that $K_0=0$ for monatomic gases at low pressure (the "dilute" range). There is no loss of generality for either choice.

The relaxation spectra appear to be more convenient to represent the viscometric properties of viscoelastic fluids because there are "molecular/hydrodynamical" theories for polymer solutions that obtain these quantities in terms of molecular parameters explicitly. Using the spectral representation, the rheological properties <•> that are relevant to the acoustic problem may be represented by:

$$\langle\cdot\rangle = \langle \eta_0, \tau_0, \tau_1, \ldots ; m_0, m_1, \ldots ; \lambda_0, \bar{\tau}_0, \bar{\tau}_1, \ldots ; \bar{m}_0, \bar{m}_1, \ldots \rangle \tag{5.16}$$

The $\{\tau_j\}$ and $\{\bar{\tau}_j\}$ are sets of characteristics times, and $\{m_j\}$ and $\{\bar{m}_j\}$ are sets of dimensionless constants that represent the shear and compressional relaxation spectra respectively. Several "molecular" models for viscoelastic fluids [32] obtain finite sets of values. In general $H(\lambda)$ and $L(\lambda)$ are continuous functions, and it is necessary to specify an analytic representation and the parameters that appear in the function. However, these parameters can be described by (5.16). Some examples of discrete and continuous spectral representations are given in Table 5.1.

By using convenient dimensionless parameters, the characteristic equation can be rewritten as:

$$\left(\frac{k_0}{\chi}\right)^4 + \left[i\, X(1+\gamma Y) + (1+PX)\right]\left(\frac{k_0}{\chi}\right)^2 + XY\left[i(1+\gamma PX - \gamma X\right] = 0 \tag{5.17}$$

where X is the frequency number, Y is the thermo-viscous number and P is the non-Newtonian viscosity number. These quantities are defined more completely in Table 5.2.

The problem is then reformulated as finding the function:

$$\frac{\chi}{k_0} = G\Big(X_N, \big[\gamma, Y_N, \mathcal{V}_N ; T_0, T_1, \ldots ; \bar{T}_0, \bar{T}_1, \ldots ; m_0, m_1, \ldots ; \bar{m}_0, \bar{m}_1, \ldots\big]\Big) \tag{5.18}$$

where the inner brackets [•] represent the physical and thermodynamical properties of the fluid. The space needed to represent the physical properties of a Newtonian fluid is the $\langle\gamma, Y_N\rangle$ plane; however, a

Table 5.1

Examples of Relaxation Spectra

Fluid	Spectra: Shear Relaxation $H(\lambda)$	Spectra: Compressional Relaxation $L(\lambda)$	Parameters
Newtonian	$\eta_0 \delta(\lambda)$	$\lambda_0 \delta(\lambda)$	η_0, λ_0
Single Relaxation Time	$\eta_0 \delta(\lambda - \tau_0)$	$\lambda_0 \delta(\lambda - \bar{\tau}_0)$	$\eta_0, \lambda_0 ; \tau_0, \bar{\tau}_0$
Finite Number Relaxation Times	$\sum_{j=0}^{N} \eta^{(j)} \delta(\lambda - \tau_j)$	$\sum_{j=0}^{\bar{N}} \lambda_0^{(j)} \delta(\lambda - \bar{\tau}_j)$	$\eta_0, \lambda_0 ; \tau_0, \dots, \tau_N ;$ $\bar{\tau}_0, \dots, \bar{\tau}_{\bar{N}} ;$ $\eta^{(0)}/\eta_0, \dots, \eta^{(N-1)}/\eta_0 ;$ $\lambda^{(0)}/\lambda_0, \dots, \lambda^{(\bar{N}-1)}/\lambda_0 .$
Gaussian	$\frac{\eta_0}{f \tau_1 \pi^{1/2}} \exp\left\{-\left[\frac{\lambda - \tau_0}{\tau_1}\right]^2\right\}$	$\frac{\lambda_0}{\bar{f} \bar{\tau}_1 \pi^{1/2}} \exp\left\{-\left[\frac{\lambda - \bar{\tau}_0}{\bar{\tau}_1}\right]^2\right\}$	$\eta_0, \lambda_0 ; \tau_0, \tau_1 ; \bar{\tau}_0, \bar{\tau}_1$
Oldroyd +	$\eta_0 \frac{\tau_2}{\tau_1} \delta(\lambda) + \eta_0 \frac{(\tau_1 - \tau_2)}{\tau_1} \delta(\lambda - \tau_1)$		η_0, τ_1, τ_2
Normalization:	$\eta_0 = \int_0^\infty H(\lambda)\, d\lambda$	$\lambda_0 = \int_0^\infty L(\lambda)\, d\lambda$	

+ Yields the constitutive equation $1 + \tau_1 \frac{d_c}{dt} t_{ik} = 2\eta_0 \left(1 + \tau_2 \frac{d_c}{dt}\right) d_{ik}$

TABLE 5.2
Dimensionless Groups for a Linear Viscoelastic Fluid

Parameter	Symbol	
	General Fluid	Newtonian Fluid
Non-Newtonian Viscosity Parameter 1	$P' \equiv \beta'/\beta_0$	$P' = 1$
Non-Newtonian Viscosity Parameter 2	$P'' \equiv \beta''/\beta_0$	$P'' = 0$
Frequency Number	$X \equiv \beta'\omega/\rho_0 c_0^2$	$X = X_N = \beta_0 \omega/\rho_0 c_0^2$
Thermoviscous Number	$Y \equiv k_\theta/\beta' c_p = Y_N/P'$	$Y = Y_N = k_\theta/\eta_0 \mathcal{V}_N c_p$
Non-Newtonian Viscosity Number	$P \equiv \beta''/\beta' = P''/P'$	$P = 0$
Viscosity Number	$\mathcal{V} \equiv \beta'/\eta_0$	$\mathcal{V} = \mathcal{V}_N = \beta_0/\eta_0$
Stress Relaxation Time	$T_i \equiv \tau_i \rho_0 c_0^2/\beta_0$	$T_i = 0$
Compressional Relaxation Time	$\bar{T}_i \equiv \bar{\tau}_i \rho_0 c_0^2/\beta_0$	$\bar{T}_i = 0$

$$\beta' \equiv \int_0^\infty \frac{[2H(\lambda) + L(\lambda)]}{1 + \omega^2\lambda^2}\, d\lambda \quad ; \quad \beta'' \equiv \int_0^\infty \frac{\omega\lambda[2H(\lambda) + L(\lambda)]}{1 + \omega^2\lambda^2}\, d\lambda$$

$$\beta' = \frac{4}{3}\eta'(\omega) + K'(\omega) \quad ; \quad \beta'' = \frac{4}{3}\eta''(\omega) + K''(\omega)$$

$$K(\lambda) = L(\lambda) + \frac{2}{3}H(\lambda)$$

multi-dimensional space (P-space, P>2) is necessary for a general fluid. The dimensions of this space depend on the number of parameters used to specify the compressional and shear relaxation spectra. It is clear that for practical purposes, the space should be a minimum, consistent with adequate correlation of experimental data. Currently, there are few molecular or phenomenological theories available for which it is possible to relate the relaxation functions to physical or molecular properties of the fluid. Consequently, the present methods of choosing a relaxation spectrum are empirical or phenomenological, and are tempered with the desire to retain mathematical simplicity. There is, however, a considerable literature for stress relaxation functions for solutions of large molecular weight polymers in otherwise Newtonian solvents [31].

For a fluid which may be represented by a single relaxation time for compressional relaxation, and a single relaxation time for shear relaxation, (5.18) becomes:

$$\left(\frac{\chi}{k_0}\right) = G\left[X_N\,,\ \langle \gamma, Y_N, \mathcal{V}_N, T_0, \bar{T}_0 \rangle\right] \tag{5.19}$$

If the relaxation times are the same, then $T_0 = \bar{T}_0$ and $\mathcal{V}_N$ does not appear explicitly in the equations, hence

$$\left(\frac{\chi}{k_0}\right) = G\left[X_N\,,\ \langle \gamma, Y_N, T_0 \rangle\right] \tag{5.20}$$

and the physical property space is 3-dimensional. If the relaxation time is zero, then $T_0 = 0$ and (5.20) is the same form as the Newtonian case.

The calculation procedure for the biquadratic characteristic equation is straight-forward, once forms for the compressional and shear relaxation spectra are specified.

Similar to the Newtonian case, there are two possible wave-solutions (Type I and Type II) for viscoelastic fluids. Generally, the Type I wave is considered to be that observed in experiments since the absorption of the Type II wave is much greater, hence those waves would be expected to decay very rapidly. There is very little information available about Type II waves in the literature. Our main effort is concentrated on the Type I waves.

5.3 Power Series Expansions

Power series expansion for the measures of absorption and dispersion may be established in terms of the variable X. This is a convenient representation having results comparable with current theories. For the low-frequency range, we find:

$$\left(\frac{\alpha}{k_0}\right) = \frac{1}{2}\left[1 + (\gamma-1)Y_N\right] X + O\left[X^3\right] \tag{5.21}$$

and

$$n = 1 + \frac{X^2}{4}\left[4\,\mathcal{E}_0 + 3 + 10(\gamma-1)Y_N + (\gamma-1)(3\gamma-7)Y_N^2\right] + O\left[X^4\right] \tag{5.22}$$

where

$$\mathcal{E}_0 \equiv \frac{\rho_0 c_0^2}{\beta_0^2}\int_0^{\infty} \lambda\left[2H(\lambda;\rho) + L(\lambda;\rho)\right] d\lambda \tag{5.23}$$

Now

$$\left(\frac{\alpha}{k_0}\right) = \frac{X_N}{2}\left[1 + (\gamma-1)Y_N\right]\frac{\beta'(\omega)}{\beta_0} + O\left[X^3\right] \tag{5.24}$$

or

$$\left(\frac{\alpha}{f^2}\right) = \left(\frac{\alpha_s}{f^2}\right) \frac{3\,\mathcal{V}_N}{4} \frac{\beta'(\omega)}{\beta_0} \tag{5.25}$$

Both $\eta'(\omega)$ and $K'(\omega)$ are monotonic non-increasing functions with $\eta'(0) = \eta_0$ and $K'(0) = K_0$ and $\mathcal{V}_N \geq \frac{4}{3}$. In the Newtonian theory, $\left(\frac{\alpha_s}{f^2}\right)$ is the smallest possible value. Therefore, if $\left(\frac{\alpha}{f^2}\right) < \left(\frac{\alpha_s}{f^2}\right)$ is experimentally observed, a viscoelastic effect is indicated.

The quantity $K'(\omega)$ may be computed from absorption data by

$$\frac{K'(\omega)}{\eta_0} = \frac{4}{3}\left\{\left[\left(\frac{\alpha}{f^2}\right)\Big/\left(\frac{\alpha_s}{f^2}\right)\right] - \frac{\eta'(\omega)}{\eta_0}\right\} \tag{5.26}$$

and $K(\lambda)$ may be determined by inverting

$$K'(\omega) = \int_0^\infty \frac{K(\lambda)}{1 + \omega^2\lambda^2}\, d\lambda$$

It is of interest to write the form of the dispersion relation for the limiting case where $Y_N \ll 1$. Now,

$$n = 1 + \frac{1}{4} X_N^2 \left[\frac{\beta'(\omega)}{\beta_0}\right]^2 (4\,\varepsilon_0 + 3) + O\left[X^4\right]$$

and note that

$$\lim_{\omega \to 0}\left[\frac{\eta''(\omega)}{\omega}\right] = \int_0^\infty \lambda\, H(\lambda)\, d\lambda \equiv \eta\, \beta_0^2 / \rho_0 c_0^2$$

There is a similar form for the compressional part, call it m. Then

$$\varepsilon_0 = (2\eta + m)$$

Therefore

$$n = 1 + \frac{1}{4} X_N^2 \left[\frac{\beta'(\omega)}{\beta_0}\right]^2 \left[4\,(2\eta + m) + 3\right]$$

which gives an estimate of the dispersion in a viscoelastic fluid for "small" values of X.

In order to obtain some idea of the possible results of this theoretical approach, fluids represented by discrete relaxation spectra and continuous relaxation spectra have been examined. It should be stated that the representation of the spectrum for a real fluid by a single relaxation time or several relaxation times is a simplistic method used to characterize fluids. If there is a complete lack of data for a substance, then this approach is often necessary.

Discrete Relaxation Spectra

For a single relaxation time for shear, and a single relaxation time for compression, the relaxation spectra are given by the analytic forms

$$H(\lambda) = \eta_0 \, \delta(\lambda - \tau_0) \qquad ; \qquad L(\lambda) = \lambda_0 \, \delta(\lambda - \bar{\tau}_0)$$

The parameters which determine the absorption and dispersion measures are given by the set

$$\left[X_N \, ; \, \langle \gamma, Y_N, \mathcal{V}_N, T_0, \bar{T}_0 \rangle \right]$$

where the inner bracket represents the physical properties of the fluid and X_N is the frequency variable. Each fluid is then represented by a point in a 5-dimensional space. There is a special case when $T_0 = \bar{T}_0$, or when the relaxation times for both the compressional and shear modes are the same. Then $\mathcal{V}_N$ does not appear explicitly and the P-space reduces to 3-dimensions.

If there are two relaxation times for shear and two different relaxation times for compression, then the spectra are represented by:

$$H(\lambda) = \eta^{(0)} \, \delta(\lambda - \tau_0) + \eta^{(1)} \, \delta(\lambda - \bar{\tau}_1)$$

and

$$L(\lambda) = \lambda^{(0)} \, \delta(\lambda - \tau_0) + \lambda^{(1)} \, \delta(\lambda - \bar{\tau}_1)$$

The coefficients $\eta^{(0)}$ and $\eta^{(1)}$ are related to η_0, and $\lambda^{(0)}$ and $\lambda^{(1)}$ are related to λ_0 by the normalization relations given in Table 5.1. Define m_0 and $\bar{m}_0$ by

$$m_0 \equiv \eta^{(1)}/\eta_0 \qquad ; \qquad m_0 = \lambda^{(1)}/\lambda_0$$

Therefore

$$m^{(0)} = \eta_0 (1 - m_0) \qquad ; \qquad \lambda^{(0)} = \lambda_0 (1 - \bar{m}_0)$$

The measures of absorption and dispersion are therefore determined by the parameters

$$\left[X_N \, ; \, \langle \gamma, Y_N, \mathcal{V}_N, T_0, T_1, T_0, T_1 \, ; \, m_0, \bar{m}_0 \rangle \right]$$

The P-space, whose coordinates are given by the inner brackets is 9-dimensional. For each additional pair of relaxation times, the P-space is increased by four dimensions. The dimension (d) of the space is given by d = 3 + 2$(m+\bar{m})$ where m and $\bar{m}$ are the number of shear and compressional relaxation times respectively. It does not seem desirable to consider more than two relaxation times since the number of adjustable constants would be sufficiently large to adequately correlate any experimental data.

For the one-time model, the measures of absorption α/k and dispersion n were computed and are shown plotted in Figs. 5.1 and 5.2 respectively for various values of the parameters T_o and $\bar{T}_o$. The values of γ, Y_N and V_N were chosen to be 1.0, 3.0×10^{-4}, and 2.34 respectively which are typical for a liquid. It is seen that the absorption curve may have a single maximum, or a maximum and an inflexion point depending on the ratio of the characteristic times T_o and $\bar{T}_o$. Now, we comment that even with a simple 2-relaxation time model, it is possible to obtain results which represent many available data. However, we must note that the theory neglected any changes in the thermodynamical properties of the fluid caused by the passage of the acoustic wave, other than those which occur through the usual caloric equation of state. The molecular theorists have noted that when a disturbance is applied to a gas which is in thermostatic equilibrium, a finite time is required for the energy of the system to be distributed among the possible energy levels. During the transition from one equilibrium state to another, it is possible to consider the fluid as a mixture of two or more substances with a reaction or sequence of reactions occurring among them. Other relaxation theories have been proposed, based on different models, and in general do not

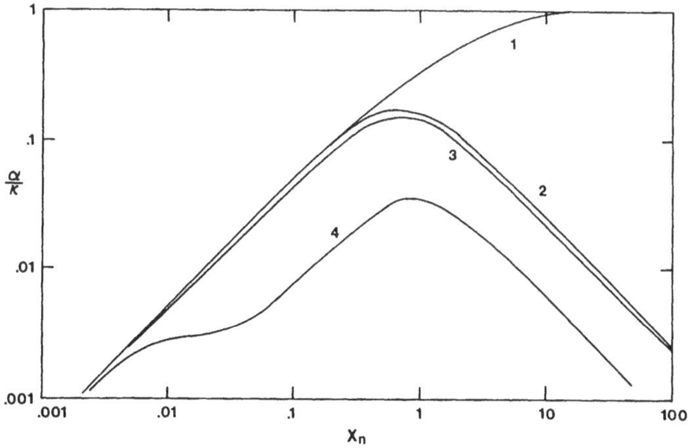

Figure 5.1: Graph of α/k versus X_N for a single-relaxation viscoelastic fluid

Case	T_o	$\bar{T}_o$
2	1	1
3	1	100
4	100	1
1	Newtonian	

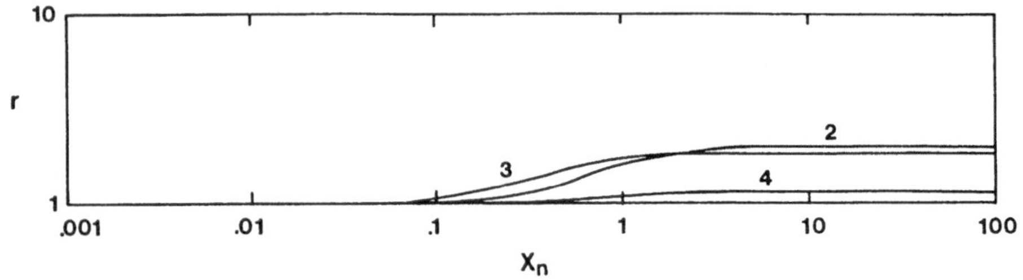

Figure 5.2: Graph of Dispersion r versus X_N for the one-time model. The numbers correspond to those of Fig. 5.1.

account for the viscosity of the system, other than superposing the classical results.

Power-series expansions for the measures of absorption and dispersion for the low-frequency range obtain

$$\left(\frac{\alpha}{k_0}\right) = \frac{X_N}{2}\left[1 + (\gamma-1)Y_N\right]\left\{\frac{2}{\mathcal{V}_N}\,\frac{1}{1+X_N^2 T_0^2} + \frac{\mathcal{V}_N-2}{\mathcal{V}_N}\,\frac{1}{1+X_N^2\bar{T}_0^2}\right\} + O\left[X^3\right]$$

and

$$\left(\frac{\alpha}{f^2}\right) = \frac{2\pi}{C_0 f}\left(\frac{\alpha}{k_0}\right) = \frac{\pi}{2C_0 f}\left[1+(\gamma-1)Y_N\right]\left\{\frac{2}{\mathcal{V}_N}\,\frac{X_N}{1+X_N^2 T_0^2} + \frac{\mathcal{V}_N-2}{\mathcal{V}_N}\,\frac{X_N}{1+X_N^2\bar{T}_0^2}\right\}$$

where

$$\frac{\eta'(\omega)}{\eta_0} = \frac{1}{1+X_N^2 T_0^2} \quad ; \quad \frac{\lambda'(\omega)}{\lambda_0} = \frac{1}{1+X_N^2\bar{T}_0^2}$$

If $T_0 = \bar{T}_0$,

$$\left(\frac{\alpha}{k_0}\right) = \left[1+(\gamma-1)Y_N\right]\frac{X_N}{2\left(1+X_N^2 T_0^2\right)} + O\left[X^3\right]$$

Continuous Spectrum

The results for a continuous spectrum will be examined by considering the Gaussian model:

$$H(\lambda) \equiv \frac{\eta_0}{f\tau_1\pi^{1/2}}\exp\left\{-\left[\frac{\lambda-\tau_0}{\tau_1}\right]^2\right\}$$

The limiting case of $\tau_1 \to 0$ is the Dirac delta function $\eta_0\delta(\lambda-\tau_0)$. The factor f is a normalization constant and is fixed once τ_0 and τ_1 are specified.

The Gaussian model contains two arbitrary times: τ_0 and τ_1. If a similar form is applicable for the compressional relaxation model, then two additional times are needed. Therefore, the dimensionless parameters which govern the measures of absorption and dispersion are given by

$$\left[X_N\ ;\ \langle\gamma, Y_N, \mathcal{V}_N, T_0, T_1, \bar{T}_0, \bar{T}_1\rangle\right] \tag{5.27}$$

and each fluid for this class is represented by a point in the 7-dimensional P-space whose coordinates are given by the bracketed terms in (5.27).

A physical interpretation can be ascribed to the viscosity spectra for the Gaussian case:

$$\xi \equiv (\lambda - \tau_0)\,\tau_1$$

then $H(\xi)d\xi$ = amount of viscosity in the range of values between ξ and $\xi + d\xi$.

and the normalization is given by:

$$\eta_0 = \int_0^\infty H(\lambda)\,d\lambda = \int_0^\infty H(\xi)\,d\xi$$

Now

$$\eta_0 = \frac{\eta_0}{f\sqrt{\pi}} \int_{-\tau_0/\tau_1}^{\infty} \exp(-\xi^2)\,d\xi$$

or

$$f = \frac{1}{2}\left[1 + \operatorname{erf}(\tau_0/\tau_1)\right]$$

where

$$\operatorname{erf} u = \frac{2}{\sqrt{\pi}} \int_0^u \exp(-\xi^2)\,d\xi$$

Now $\frac{1}{2} \le f \le 1$ and also $\tau_0/\tau_1 = T_0/T_1$

The Gaussian form for the relaxation spectrum does not obtain simple results for the absorption and dispersion. In general, numerical computations are required. For the case $\tau_0 = 0$, results can be obtained in terms of tabulated functions although the calculations are not apparently reduced.

5.4 Comparison to Experimental Data

Available experimental data for viscoelastic fluids are generally incomplete since absorption and dispersion measurements have not been obtained over a sufficiently large range of frequencies. It is usually not possible to estimate the asymptotic limits at large or small frequencies. Further, in order to use the theoretical results to obtain values for the frequency-dependent bulk viscosity and subsequently the compressional relaxation function, independent information on the frequency-dependent shear viscosity is required. There are very few cases for which both sets of data have been obtained.

Barlow, Harrison and Lamb [36] have presented data on the response to an oscillatory shear wave for various silicone oils. The experiments covered an extensive range of frequencies and temperatures, and both $\eta'(\omega)$ and $\eta''(\omega)$ were reported. An interesting feature of some of these fluids is that the value of $\overline{X}_N$ (modified thermoviscous number, obtained from X_N with $K_0 \equiv 0$) is small enough so that the power series expansions in X can be used. For the special case that $K'(\omega) = 0$ and $n = 1$

$$\left(\frac{\alpha}{f^2}\right) = \left(\frac{\alpha}{f^2}\right) \frac{\eta'(\omega)}{\eta_0} \tag{5.28}$$

Values of X_N (100 MHz) were calculated for several silicone oils of different molecular weights. It was found that the simplified expression (5.28) is only applicable for a limited range of materials and temperatures.

Hunter and Derdul [37] have reported sound absorption data for three of the silicone oils used by Barlow et al. [36] for shear measurements, i.e., the 100, 350, and 1000 centistokes (cS) materials for θ = -10 , 10 and 30 °C. The absorption data are shown plotted in Fig. 5.3. The authors report that the absorption data for the 1000 cS silicone were about 10-15% higher than the 350 cS fluid. The theoretical predictions for the silicone oils, obtained by using the shear data of Barlow et al. [36] and assuming that $K'(\omega)=0$, are also graphed in Fig. 5.3. The calculations for the 1000 cS material were computed by the complete expressions. However, as described earlier, provided that $X_N \ll 1$, then (5.28) is valid. This relation is the same as that used by Hunter and Derdul [37].

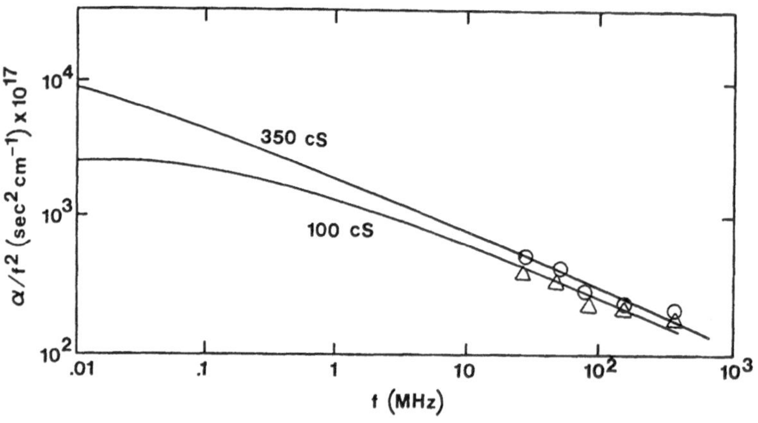

Figure 5.3: Graph of the absorption coefficient versus frequency for 100 cS and 350 cS Silicone Oil (DC 200 Series).

The experimental data for the 100 cS and 350 cS fluids are accurately determined by the theory, assuming that X_N is negligible small in the range of frequencies examined. Also, the values of the absorption computed by the exact formulae are somewhat larger (10-15%) than for the 350cS material which is in agreement with the experimental measurements.

Hunter and Derdul [37] also reported that the dispersion was negligible for the range of frequencies investigated ($30 \leq f \leq 350$MHz) which is in agreement with the theory.

It is concluded that for silicone oils the bulk viscosity $K'(\omega)$ is very small in the range of frequencies examined. Absorption measurements at lower frequencies would be desirable.

5.4 Application to Solutions of Polymeric Materials

Provided that $X_N \ll 1$, then the power series expansion for the absorption measure is applicable and reduces to the rather simple form:

$$\left(\frac{\alpha}{f^2}\right) = \left(\frac{\alpha_s}{f^2}\right)\left\{\left[1 + \eta_{sp}\,\phi'_\eta(\omega)\right] + \frac{3}{4}\frac{K_s}{\eta_s}\left[1 + K_{sp}\,\phi'_K(\omega)\right]\right\} \tag{5.29}$$

where

$$\phi'_\eta(\omega) \equiv \left[\frac{\eta'(\omega) - \eta_s}{\eta_0 - \eta_s}\right] \quad ; \quad \phi'_K(\omega) \equiv \left[\frac{K'(\omega) - K_s}{K_0 - K_s}\right]$$

The subscript s refers to the solvent, and K_0 and η_0 refer to the bulk and shear viscosity of the solution respectively, and $\eta_{sp} \equiv (\eta_0 - \eta_s)/\eta_s$ and $K_{sp} \equiv (K_0 - K_s)/K_s$ refer to the specific shear and bulk viscosities respectively. For very dilute solutions,

$$\eta_0 = \eta_s\left(1 + [\eta]\rho_p\right)$$

where $[\eta]$ is the intrinsic viscosity $[\eta] = \lim_{\rho_p \to 0}\left(\frac{\eta_0 - \eta_s}{\eta_s \rho_p}\right)$ and ρ_p is the concentration (g of polymer/cm^3 of solution). It is also convenient to define an intrinsic bulk viscosity $[K]$, viz.,

$$[K] \equiv \lim_{\rho_p \to 0}\left(\frac{K_0 - K_s}{K_s \rho_p}\right) \equiv \lim_{\rho_p \to 0} K_{red.}$$

hence the specific bulk viscosity becomes:

$$K_{sp} \doteq [K]\,\rho_p$$

Adequate data are not available for this quantity in order to determine its properties. Also, note that we must require $K_s \neq 0$ for an intrinsic bulk viscosity to be defined.

For very dilute solutions of polymeric materials, (5.29) may be written

$$\left(\frac{\alpha}{f^2}\right)_{red} \equiv \frac{(\alpha/f^2)_{sp}}{\rho_p} \equiv \left[\frac{\left(\frac{\alpha}{f^2}\right) - \left(\frac{\alpha_K}{f^2}\right)}{\left(\frac{\alpha_K}{f^2}\right)\rho_p}\right] = \frac{[\eta]\,\phi'_\eta(\omega) + [K]\,\phi'_K(\omega)\,\frac{3}{4}\frac{K_s}{\eta_s}}{\left(1+\frac{3}{4}\frac{K_s}{\eta_s}\right)}$$

The function $\phi'_\eta(\omega)$ may be obtained by experimental methods or by use of a molecular theory such as that proposed by Zimm [35] or Rouse [34] for monodisperse polymers. In Fig. 5.4, calculated from Zimm's theory, $\phi'_\eta(\omega)$ has been plotted versus frequency using molecular weight as a parameter. In the limit of $\omega \to 0$, both $\phi'_\eta(\omega)$ and $\phi'_K(\omega)$ are unity. Now, define

$$\phi'_\eta(\infty) = \lim_{\omega\to\infty} \phi'_\eta(\omega) \quad ; \quad \phi'_K(\infty) \equiv \lim_{\omega\to\infty} \phi'_K(\omega)$$

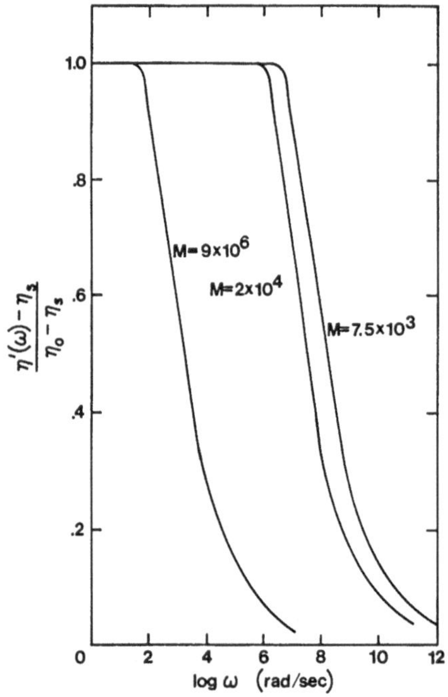

Figure 5.4: Graph of $\phi'_\eta(\omega)$ versus ω for polymer solutions of various molecular weights calculated by the theory of Zimm.

Both the theories of Zimm and Rouse predict that $\phi'_\eta(\infty) = 0$, which appears to be an unacceptable result since experimental data indicate otherwise. Theoretical attempts [38] have been made to correct this deficiency, and also to account for polydispersity of the polymer. Unfortunately, the theories have been restricted to dilute solutions where inter-molecular interactions or entanglements are unimportant. Many of the reported data on sound absorption are for concentrated solutions.

Now for small enough ω, such that $X \ll 1$, $\phi'(\omega) \sim 1$, and $\phi'_K(\omega) \sim 1$, obtain

$$\left(\frac{\alpha}{f^2}\right)_{red.} \equiv \frac{(\alpha/f^2)_{sp}}{\rho_p} \sim \frac{[\eta] + [K]}{\left(1 + \frac{3}{4}\frac{K_s}{\eta_s}\right)}$$

For a particular polymer-solvent system it is generally possible to write an expression of the form [39]:

$$[\eta] = A\ M_w^m$$

where A and m are constants for a fixed temperature ($0 \lesssim m \lesssim 2$). Further $[\eta] \geq 0$ (typical values are of order unity) which states that the shear viscosity of the solution is greater than the pure solvent.

At the present time, a discussion of the properties of the intrinsic bulk viscosity is speculative since there is a complete lack of theory and experiment on this subject. However, (5.30) does provide a relation from which $[K]$ can be determined by experiments.

For large enough ω, where $X_N \ll 1$, (5.29) becomes:

$$\left(\frac{\alpha}{f^2}\right)_{red} \equiv \frac{(\alpha/f^2)_{sp}}{\rho_p} \sim \frac{[\eta]\ \phi_\eta(\infty) + [K]\ \phi_K(\infty)\frac{3}{4}\frac{K_s}{\eta_s}}{\left(1 + \frac{3}{4}\frac{K_s}{\eta_s}\right)}$$

In summary, (5.29) is the fundamental relation representing the absorption coefficient as a function of frequency for polymer solutions. If the solutions are dilute, the data that are needed to compute the absorption of acoustic waves are: 1) the intrinsic viscosity, 2) the concentration, 3) the molecular weight, or the relation between intrinsic viscosity and molecular weight, 4) experimental data, or a molecular theory, for $\phi'_\eta(\omega)$ 5) the value of the bulk viscosity of the solution, or the intrinsic bulk viscosity and 6) the function $\phi'_K(\omega)$. In fact, sound absorption data will provide us with a knowledge of 5) and 6). For concentrated solutions, there are no guidelines currently available to estimate the absorption and dispersion measures.

Equation (5.29) may also be written as:

$$\frac{\Delta\alpha}{f^2\rho_p} = S_0 + C_0$$

where

$$S_0 \equiv \left(\frac{\alpha_K}{f^2}\right)_s \left\{\frac{[\eta]\ \phi'_\eta(\omega)}{\left(1 + \frac{3}{4}\frac{K_s}{\eta_s}\right)}\right\} = S_0(\omega; M_w) \quad \text{and}$$

$$C_0 \equiv \left(\frac{\alpha_K}{f^2}\right)_s \left\{\frac{[K]\ \phi'_K(\omega)\frac{3}{4}\frac{K_s}{\eta_s}}{\left(1 + \frac{3}{4}\frac{K_s}{\eta_s}\right)}\right\} = C_0(\omega; M_w)$$

In the limit: $\omega \to 0$, $\phi'(\omega) \to 1$ therefore

$$S_0(0;M_w)=\left(\frac{\alpha_K}{f^2}\right)_s\left\{\frac{[\eta]}{1+\frac{3}{4}\frac{K_s}{\eta_s}}\right\}=\left(\frac{\alpha_s}{f^2}\right)_s[\eta]\ ;\quad C_0(0;M_w)=\left(\frac{\alpha_K}{f^2}\right)_s\left\{\frac{[K]\frac{3}{4}\frac{K_s}{\eta_s}}{1+\frac{3}{4}\frac{K_s}{\eta_s}}\right\}$$

Now

$$S_0(\omega;M_w)=S_0(0;M_w)\,\phi'_\eta(\omega)$$

and

$$C_0(\omega;M_w)=C_0(0;M_w)\,\phi'_K(\omega)$$

In the limit: $\omega\to$ large, but such that $X_N \ll 1$, obtain

$$S_0(\infty;M_w)=\left(\frac{\alpha_K}{f^2}\right)_s\left\{\frac{[\eta]\,\phi'_\eta(\infty)}{1+\frac{3}{4}\frac{K_s}{\eta_s}}\right\}=\left(\frac{\alpha_s}{f^2}\right)_s\left\{[\eta]\,\phi'_\eta(\infty)\right\}$$

and

$$C_0(\infty;M_w)=\left(\frac{\alpha_K}{f^2}\right)_s\left\{\frac{[K]\,\phi'_K(\infty)\,\frac{3}{4}\frac{K_s}{\eta_s}}{1+\frac{3}{4}\frac{K_s}{\eta_s}}\right\}$$

Now it is accepted that $\phi'_\eta(\infty)\neq 0$ from experimental observations and the theoretical work of Peterlin [38], which predicts that

$$\phi'_\eta(\infty)=g(M_w)\neq 0$$

In fact, $\phi'_\eta(\infty)$ decreases as the molecular weight increases.

For solutions which cannot be considered as dilute:

$$\left(\frac{\alpha}{f^2}\right)=\left(\frac{\alpha_s}{f^2}\right)\left[1+\frac{3}{4}\left(\frac{K_s}{\eta_s}\right)\right]+\left(\frac{\alpha_s}{f^2}\right)_s\left[\eta_{sp}\,\phi'_\eta(\omega)+\frac{3}{4}\frac{K_s}{\eta_s}K_{sp}\,\phi'_K(\omega)\right]$$

where the conduction term was neglected. However, if conduction is non-negligible, then write

$$\left(\frac{\alpha}{f^2}\right)=\left(\frac{\alpha_s}{f^2}\right)\left\{1+\frac{3}{4}\frac{K_0}{\eta_0}\left[1+\frac{3}{4}\frac{k_\theta}{\eta_0}\left(c_v^{-1}+c_p^{-1}\right)\right]^{-1}\right\}$$

Now,

$$\Delta\left(\frac{\alpha}{f^2}\right)\equiv\left(\frac{\alpha}{f^2}\right)-\left(\frac{\alpha_K}{f^2}\right)_s=\left(\frac{\alpha}{f^2}\right)_s\left[\eta_{sp}\,\phi'_\eta(\omega)+\frac{3}{4}\frac{K_s}{\eta_s}K_{sp}\,\phi'_K(\omega)\right]$$

The data which are available indicates that the sign of K_{sp} can be positive or negative, and the bulk viscosity of a solvent can be increased or decreased by the addition of a sòlute.

6. Viscoelastic Fluids with Coupled Reactions

A simple set of first-order acoustic equations for a reacting system with non-Newtonian viscosity may be written as

$$-i\omega\beta_\theta\theta^a+i\omega\beta_p p^a-\rho_0 v_\alpha(\xi^e_\alpha)^{1/2}M_{\alpha\sigma}\,i\omega Z^a_\sigma+\chi V^a=0$$

$$i\omega\rho_0 V^a+\chi p^a-V^a\chi^2\int_0^\infty\frac{2H(\lambda,p)+L(\lambda,p)}{1+i\omega\lambda}\,d\lambda$$

$$i\omega\rho_0 c_p \Theta^a - i\omega\Theta_0\beta_\theta\, p^a + i\rho_0\omega h_\sigma (\overset{+e}{\xi}_\sigma)^{1/2} M_{\sigma\delta} Z^a_\delta - k_\theta \chi^2\Theta^a = 0$$

$$i\omega Z^a_\rho - M^{-1}_{\rho\sigma}(\overset{+e}{\xi}_\sigma)^{1/2}\left(\frac{h_\sigma}{R\Theta_0^2}\right)\Theta^a + M^{-1}_{\rho\sigma}(\overset{+e}{\xi}_\sigma)^{1/2}\left(\frac{v_\sigma}{R\Theta_0}\right)p^a$$
$$+ \left(M^{-1}_{\rho\sigma}\, J_{\rho\tau}\, M_{\tau\upsilon}\right)Z^a_\upsilon - M^{-1}_{\rho\sigma}\,\bar{R}_\sigma(\xi^e_\sigma)^{-1/2}\chi V^a = 0$$

where damped plane wave solutions have been assumed. Here the equations are similar to (3.3), (3.4), and (3.7) except for the viscoelastic response model of Noll [29] now subsuming the Newtonian viscous term previously used in the linear momentum equation and a visco-reactive term being introduced. In general, a biquadratic equation for the complex propagation variable χ is found:

$$a\chi^4 + b\chi^2 + c = 0$$

where

$$a \equiv k_\theta\left[1 + \beta^* i\omega\beta_p B_p^\omega\right] + a_r$$

$$b \equiv -i\rho_0 c_p\omega C_p^\omega + \rho_0 k_\theta\beta_p\omega^2 B_p^\omega + \Lambda B_p c_p\omega^2[\cdot] + b_r$$

$$c \equiv -i\rho_0^2\omega^3 c_p B_p\,[\cdot]$$

$$[\cdot] \equiv \left[\, C_p^\omega B_p^\omega - \frac{\gamma-1}{\gamma}\left(B_\theta^\omega\right)^2\right]$$

The visco-reactive contributions are

$$a_r \equiv k_\theta\,\alpha_{13}$$

$$b_r \equiv i\rho_0\omega\alpha_{13} c_p C_p^\omega + i\omega\beta_\theta\,\alpha_{33}\, B_\theta^\omega$$

$$E_\rho \equiv M^{-1}_{\rho\sigma}\,\overset{+}{R}_\sigma\,(\overset{+e}{\xi}_\sigma)^{-1/2}$$

$$\alpha_{13} \equiv i\rho_0\omega\sum_{\rho=1}^{R}\frac{E_\rho V_\rho \tau_\rho}{1 + i\omega\tau_\rho} \qquad \alpha_{33} \equiv i\rho_0\omega\sum_{\rho=1}^{R}\frac{E_\rho H_\rho \tau_\rho}{1 + i\omega\tau_\rho}$$

The characteristic equation for χ may be written in terms of dimensionless groups of parameters including a frequency number X, thermoviscous number Y, a non-Newtonian viscosity number P, as well as additional parameters due to the presence of reactions as follows:

$$a'\left(\frac{\chi}{k_0}\right)^4 + b'\left(\frac{\chi}{k_0}\right)^2 + c' = 0$$

$$a' \equiv XY\left[i(1 + P\gamma X) - \gamma X B_p^{\omega} \right]$$

$$b' \equiv k_o^2 \left\{ C_p^{\omega} + i\gamma XY B_p^{\omega} + \gamma X [\bullet](i + P) \right\}$$

$$c' \equiv k_o^4 \gamma [\bullet]$$

$$[\bullet] \equiv \left[C_p^{\omega} B_p^{\omega} - \frac{\gamma - 1}{\gamma} (B_\theta^{\omega})^2 \right]$$

This characteristic equation represents a systematic generalization and unification of the acoustical classes of materials discussed in sections 3.0 and 5.0. Both chemical and viscoelastic relaxations are explicitly accounted for. This proposed constitutive theory, however, remains to be derived from formal continuum thermodynamics and verified by experiments.

7. Conclusions

Acoustic wave propagation is directly dependent upon the properties of the transmitting medium. In particular, measurements of attenuation and dispersion of ultrasonic waves in liquids and gases have provided useful estimates of material properties, assuming a set of constitutive relations for the characteristic responses of the fluid. We have systematically examined and solved the viscothermal forced plane wave acoustic problem for a variety of acoustical classes of materials, starting with the linearly viscous and Fourier heat-conducting fluid case. Theoretical extensions have been presented for acoustic wave propagation in: 1) a reacting fluid mixture in equilibrium (without diffusion); 2) a binary diffusive mixture (without reaction); 3) a general linearized compressible fluid with memory, as mathematically modelled by Coleman and Noll; and finally 4) the case of a viscoelastic fluid with simultaneous coupled chemical reactions. For all the cases examined, an exact general solution for $\mathcal{X}$ (the complex propagation constant) was found. High (or low) frequency expansions were developed to interpret the contributions of the various phenomena separately. Calculations of sound absorption and dispersion were made and comparisons to available experimental data were discussed whenever possible.

All in all, the linear acoustic wave problem provides interesting research results to measure several material properties complementing other experimental approaches, to test constitutive theories, and to provide a check on the consistency among quantitative estimates of physico-chemical parameters which may have been independently measured.

Moreover, the acoustic wave propagation problem provides a means to introduce and teach valuable information about balance laws for single and multicomponent mixtures of fluids and constitutive relations.

8. List of Symbols

A_j	Chemical affinity of jth reaction $[ML^2T^{-2}]$
A_{ij}	Coefficient of change of affinity of ith reaction with degree-of-advancement variable $[M\ mol^{-1}]$
a_α	Activity of αth - component [1]
b_j	Components of external body-force vector per unit mass $[L^2T^{-2}]$
B_p^ω	Frequency-dependent isothermal coefficient of expansion [1]
B_θ^ω	Frequency-dependent isobaric coefficient of expansion [1]
c_0, c	Reference and frequency-dependent speeds of sound respectively $[LT^{-1}]$
c_p, c_p^e	Instantaneous and equilibrium heat capacities at constant pressure $[L^2T^{-2}\Theta]$
c_v, c_v^e	Instantaneous and equilibrium heat capacities at constant volume $[L^2T^{-2}\Theta]$
c_α	Mass fraction of constituent α [1]
C_p^ω	Frequency-dependent heat capacity [1]
$C_{p\sigma}^\omega$	Frequency-dependent heat capacity for σth orthonormal reaction [1]
C_α	Chemical constituent [1]
$\frac{d(\ )}{dt}$	Total or material derivative $[T^{-1}]$
d_{ij}	Rate of deformation tensor $[T^{-1}]$
D_{ij}	Rate of diffusion tensor $[T^{-1}]$
D_{AB}	Binary diffusion coefficient
f	Frequency of sound wave $[T^{-1}]$
g	Specific Gibbs free energy $[L^2T^{-2}]$
G_{ij}	Matrix coefficients of kinetics equations $[T^{-1}]$
h_i	Heat-flux vector $[HL^{-2}T^{-1}]$
h_σ	Heat of σth reaction at constant temperature and pressure $[H\text{-}mol^{-1}]$
H_σ	Heat of σth orthonormal reaction $[H\ mol^{-1}]$
J_{ij}	Symmetrized matrix coefficients of kinetics equations $[T^{-1}]$

k	Number of constituents [1]
k	Wave number [L^{-1}]
K_D	Thermo-diffusion coefficient
K_σ	Equilibrium constant of σth reaction [1]
K_0	Bulk viscosity [$ML^{-1}\ T^{-1}$]
k_σ^F, k_σ^R	Forward and reverse reaction rate coefficients respectively of σth reaction [mol $M^{-1}\ T^{-1}$]
K_T	Thermal diffusion coefficient
k_θ	Thermal conductivity of mixture [$MLT^{-3}\ \theta^{-1}$]
m_α	Molar density of αth constituent [mol (of α) L^{-3}]
M_w	Mean molecular weight [mol M^{-1}]
M_a	Molecular weight of ath component [mol (of a) M^{-1}]
M_e	Molecular weight of equilibrium mixture [mol M^{-1}]
p, p_0, p^a	Thermodynamic, thermostatic and acoustic pressures respectively [$ML^{-1}\ T^{-2}$]
p_α	Partial pressure of α-component
$\mathcal{P}$	Diffusive pressure $\mathcal{P} = c_B p_A - c_A p_B$
r	Number of reactions [1]
R	Number of independent reactions [1]
R	Universal gas constant [$ML^2\ T^{-2}\ \theta^{-1}\ mol^{-1}$]
$\bar{R}_\sigma$	Visco-reactive coefficient for σth reaction
η	Specific entropy [H $M^{-1}\ \theta^{-1}$]
S_{ij}	Total stress tensor [$ML^{-1}\ T^{-2}$]
$S_{\sigma\alpha}^*$	Signed stoichiometric numbers [1] $S_{\sigma\alpha}^* = S_{\sigma\alpha}^- - S_{\sigma\alpha}^+$
S_σ	Sum of signed stoichiometric numbers of σth reaction [1]
T_{ij}	Extra-stress tensor $T_{ij} = S_{ij} + p\,\delta_{ij}$
t_{xx}^a	x,x-component of linearized extra-stress tensor
t	Time variable [T]
$\underline{u}$	Relative diffusion velocity $\underline{u} = \underline{u}_A - \underline{u}_B$
v, v_0, v^a	Total, static and acoustic velocities in x-direction [LT^{-1}]
υ	Specific volume of equilibrium mixture [$L^3\ M^{-1}$]
v_σ	Volumetric change of σth reaction at constant temperature and pressure [$L^3\ mol^{-1}$]

V_σ — Volumetric change of σth orthonormal mode at constant temperature and pressure [L^3 mol^{-1}]

$\underline{x}$, x_i — Position vector and components [L]

x_α — Mole-fraction of α-constituent

X — Frequency number [1]

Y — Thermoviscous number [1]

Z_σ — Extent-of-reaction of σth orthonormal reaction

α — Absorption coefficient [L^{-1}]

β_θ, β_θ^e — Instantaneous and equilibrium isobaric coefficients of thermal expansion [θ^{-1}]

β_p, β_T^e — Instantaneous and equilibrium isothermal coefficients of compressibility [T M L]

β_ω — Volumetric change due to concentration (component A) $\beta_\omega = \frac{1}{v}\frac{\partial \bar{v}}{\partial \omega_A}$

γ_α — Activity coefficient of th chemical components [1]

$\gamma = c_p/c_v$ — Ratio of instantaneous heat capacities

δ_{ij} — Kronecker delta symbol [1]

ε — Specific internal energy [$L^2 T^{-2}$]

$\underline{\xi}$, ξ_σ — Degree-of-advancement vector and σth component [mol M^{-1}]

$\dot{\xi}_\sigma^e$ — Reaction velocity at equilibrium [mol M^{-1} T^{-1}]

η — Specific entropy [$L^2 T^{-2} \theta^{-1}$]

η_0 — Shear viscosity [$ML^{-1} T^{-1}$]

θ, θ_0, θ^a — Absolute equilibrium and acoustic temperature respectively [θ]

λ — Wavelength of acoustic wave [L]

λ_0 — Volume coefficient of viscosity [$ML^{-1} T^{-1}$]

λ_σ — Eigenvalue of σth reaction [T^{-1}]

μ_α — Chemical potential of αth component [$L^2 T^{-2}$ mol^{-1}]

$\mu_{AA}^\dagger$ — Reduced chemical potential $\mu_{AA}^\dagger \equiv \frac{\partial \hat{\Psi}}{\partial \omega_A} = \frac{\partial \bar{g}}{\partial \omega_A} = M_A\left[\frac{\mu_A}{M_A} - \frac{\mu_B}{M_B}\right]$

ρ_0 — Density of equilibrium mixture [ML^{-3}]

τ_ρ, τ_ρ^* — Relaxation times of ρth reaction at constant (p,θ) and (η,p) respectively [T]

$\varkappa$ Complex propagation coefficient [L]

ψ Helmholtz free energy $[L^2 T^{-2}]$

ω Angular frequency $[T^{-1}]$

$\underline{\omega}$, ω_α Molar concentration vector and αth component $[\mathrm{mol}\ M^{-1}]$

ω Total molar concentration of mixture

$\overset{+}{\underline{\omega}}$, $\overset{+}{\omega}_\alpha$ Molar production vector and αth component $[\mathrm{mol}\ M^{-1} T^{-1}]$

Ω_σ Dimensionless chemical relaxation time of σth reaction [1]

Ω_{ij} Antisymmetric diffusive tensor $\Omega_{ij} \equiv \frac{1}{2}(u_{i,j} - u_{j,i})$

[ ] = 'dimensions of '; M = mass (g), L = length (cm),

T = time (sec), Θ = absolute temperature (°K),

H = heat (cal), mol = g mole

ACKNOWLEDGEMENT

We dedicate this paper to Stan Corrsin -- whose inspirational teaching, significant research accomplishments, and warm, caring personality will always be remembered.

References

1. B. D. Coleman and V. J. Mizel, J. Chem. Phys. 40, 1116-1125, 1964.
2. C. A. Truesdell, J. Rational Mech. and Analysis 2, 643-741, October, 1953.
3. E. Buckingham, Phys. Rev. 4, 345, 1914; Phil. Mag. (6) 42, 696, 1921.
4. P. Biquard, Ann. Phys. (11) 6, 195-304, 1936.
5. J. Thoen, E. Vangeel and W. Van Dael, Physica, 45, 339-356, 1969.
6. M. Eigen and L. deMaeyer, "Relaxation Methods," in Technique of Organic Chemistry, 2nd ed., edited by S. L. Friess, E. S. Lewis and A. Weissberger, (Interscience, New York, 1963), Vol. VIII/2, pp. 895-1054.
7. R. Haase, Thermodynamics of Irreversible Processes, Addison-Wesley Publishing Co., Reading, Mass., 1969; S. R. deGroot and P. Mazur, Nonequilibrium Thermodynamics, North-Holland, Amsterdam, 1969.
8. L. S. Garcia-Colin and S. M. T. De La Selva, Physica, Utrecht 75, 37, 1974.

9. T. S. Margulies and W. H. Schwarz, J. Chem. Phys. 77, No. 2, 1005, 15 July 1982.
10. W. Jost, Z. Naturforsch. Teil A2, 159, 1947; Z. Phys. Chem. 195, 317, 1950.
11. G. W. Castellan, Ber. Bunsenges Phys. Chem. 67, 898, 1963.
12. R. M. Mazo, J. Chem. Phys. 28, 1225, 1958.
13. T. S. Margulies and W. H. Schwarz, paper presented at the 107th Acoustical Society of America Meeting, Norfolk, VA 6-10 May 1984; "Acoustic Wave Propagation in Fluids with Coupled Chemical Reaction," U.S. Nuclear Regulatory Commission, NUREG-0935, August 1984.
14. O. J. Heilman, Mat.-Fys. Medd. K. Dans. Vidensk. Selsk. 38, 1, 1972.
15. M. Kohler, Z. Physik 127, 40-48, 1949.
16. J. Meixner, Acustica 2, 101-109, 1952.
17. E. Goldman, J. Acoust. Soc. Am. 41, 93-99 (1967).
18. G. S. Wang Chang and G. E. Uhlenbeck, in Studies in Statistical Mechanics, edited by J. de Boer and G. E. Uhlenbeck, North-Holland, Amsterdam, 1970. Vol. V, 1.
19. J. D. Foch and G. W. Ford, in Studies in Statistical Mechanics, edited by J. de Boer and G. E. Uhlenbeck, North-Holland, Amsterdam 1970, Vol. V, 103.
20. J. D. Foch, Jr., G. E. Uhlenbeck and M. F. Losa, Phys. Fluids 13, 1224-1232, 1972.
21. A. Fick, Ann. Physik, 94, 59, 1855.
22. Ming-Nan Huang, Ph.D. Thesis, The Johns Hopkins University, 1973.
23. E. Goldman and L. Sirovich, Phys. Fluids 10, 1928-1940.
24. G. J. Prangsma, R. M. Jonkman and J. J. M. Beenakker, Physica 48, 323-330, 1970.
25. E. Goldman and L. Sirovich, Phys. Fluids 12, 245-247, 1969.
26. M. Greenspan, J. Acoust. Soc. Amer. 28, 644-648, 1956.
27. M. Greenspan, in Physical Acoustics, edited by W. P. Mason, Academic Press, 1965, Vol. II.
28. E. Meyer and G. Sessler, Z. Physik. 149, 15-39, 1957.
29. B. D. Coleman and W. Noll, "Simple Fluids with Fading Memory," in Second-Order Effects in Elasticity, Plasticity, and Fluid Dynamics," edited by M. Reiner and D. Abir, International Symposium, Haifa, Israel, April 23-27, 1962.
30. B. D. Coleman and W. Noll, "Foundations of Linear Viscoelasticity," Rev. of Modern Physics 33, 239, 1961.

31. J. D. Ferry, Viscoelastic Properties of Polymers, John Wiley and Sons, Inc. (1980).
32. R. B. Bird, R. C. Armstrong and O. Hassager, Dynamics of Polymeric Liquids, Volume 2: Kinetic Theory, John Wiley and Sons (1977).
33. F. Bueche, J. Chem. Phys. 22, 603, 1954.
34. P. E. Rouse, Jr., J. Chem. Phys. 21, 1272, 1953.
35. B. H. Zimm, J. Chem. Phys. 24, 269, 1956.
36. A. J. Barlow, G. Harrison and J. Lamb, Proc. R. Soc. Lond. 282, 228, 1964.
37. J. L. Hunter and P. R. Derdul, J. Acoust. Soc. Am. 42, 1041, 1967.

Publications of Stanley Corrsin

1. Corrsin, S. 1943. Investigation of Flow in an Axially Symmetrical Heated Jet of Air, N.A.C.A. Adv. Confid. Rep. No. 3L23, (reissued as Wartime Rep. No. W-94).

2. Corrsin, S. 1944. Investigation of the Behavior of Parallel, Two-Dimensional Air Jets, N.A.C.A. Adv. Confid. Rep. No 4H24, (reissued as Wartime Rep. No. W-90).

3. Corrsin, S. 1947. Extended Applications of the Hot-Wire Anemometer, *Review Sci. Instr.*, 18,7: 469-71.

4. Corrsin, S. 1949. Extended Applications of the Hot-Wire Anemometer, N.A.C.A. Tech. Note 1864.

5. Corrsin, S. and Uberoi, M. S. 1950. Further Experiments on the Flow and Heat Transfer in a Heated Turbulent Jet, N.A.C.A. Report 998 (originally published as N.A.C.A. Tech. Note 1865, April 1949).

6. Corrsin, S. 1949. Diffusion of Submerged Jets, discussion in *Proc. Am. Soc. Civil Eng.*, pp. 912-14.

7. Corrsin, S. and Kovasznay, L. S. G. 1949. On the Hot-Wire Length Correction, *Phys. Rev.*, 75,12: 1954.

8. Corrsin, S., Uberoi, M. S. and Kovasznay, L. S. G. 1949. The Transformation Between One- and Three-Dimensional Power Spectra for an Isotropic Scalar Fluctuation Field, *Phys. Rev.*, 76,8: 1263-4.

9. Corrsin, S. 1949. An Experimental Verification of Local Isotropy, *J. Aero. Sci.*, 16,12: 757-8.

10. Corrsin, S. 1950. Hypothesis for the Skewness of the Probability Density of the Lateral Velocity Fluctuations in Turbulent Shear Flow, *J. Aero. Sci.*, 17,7: 396-8.

11. Corrsin, S. and Uberoi, M. S. 1951. Spectra and Diffusion in a Round Turbulent Jet, N.A.C.A. Rep. 1040, (originally published as N.A.C.A. Tech. Note 2124, Aug. 1950).

12. Corrsin, S. 1950. On the Derivation of Euler's Equation for the Motion of an Inviscid Fluid, *Am. J.Phys.*, 18,7: 467.

13. Corrsin, S. 1951. On the Spectrum of Isotropic Temperature Fluctuations in an Isotropic Turbulence, *J. Appl. Phys.*, 22,4: 469-73.

14. Corrsin, S. 1951. The Decay of Isotropic Temperature Fluctuations in an Isotropic Turbulence, *J. Aero. Sci.*, 18,6: 417-23.

15. Corrsin, S. 1951. A Simple Geometrical Proof of Buckingham's Π Theorem, _Am_. _J_. _Phys_., 19,3: 180-1.

16. Corrsin, S. 1951. An Integral Relation from the Turbulent Energy Equation, _J_. _Aero_. _Sci_., 18,11: 773-4.

17. Corrsin, S. and Kovasznay, L. S. G. 1951. The Energy Equation for Two Kinds of "Incompressible" Flow, _J_. _Aero_. _Sci_., 18,12: 843-844.

18. Corrsin, S. 1952. Heat Transfer in Isotropic Turbulence, _J_. _Appl_. _Phys_., 23,1: 113-8.

19. Corrsin, S. 1952. Patterns of Chaos, _The Johns Hopkins Magazine_, 3,4: 2-7.

20. Corrsin, S. 1952. Effect of Wind Tunnel Nozzle on Steady Flow Nonuniformities, _J_. _Aero_. _Sci_., 19,2: 135-136.

21. Corrsin, S. and Uberoi, M. S. 1953. _Diffusion of Heat From a Line Source in Isotropic Turbulence_, N.A.C.A. Rep. 1142 (originally published as N.A.C.A. Tech. Note 2710, June 1952).

22. Corrsin, S. 1952. Generalization of a Problem of Rayleigh, _Q_. _J_. _of_ _Appl_. _Math_., 10,2: 186-9.

23. Corrsin, S. 1953. Turbulent Channel Flow from a Variational Principle, _J_. _Aero_. _Sci_., 20,12: 853-4.

24. Corrsin, S. 1953. Intrepretation of Viscous Terms in the Turbulent Energy Equation, _J_. _Aero_. _Sci_., 20,12: 853-4.

25. Corrsin, S. and Kistler, A. L. 1955. _The Free Stream Boundaries of Turbulent Flows_, N.A.C.A. Rep. 1244 (originally published as N.A.C.A. Tech. Note 3133, Jan. 1954).

26. Corrsin, S. 1953. Remarks on Turbulent Heat Transfer, _Proc. First Iowa Sympos. on Thermodynamics_, pp. 5-30, State Univ. of Iowa, Iowa City.

27. Corrsin, S. 1955. Dimensional Analysis and Similarity, in _Fluid Models in Geopyhsics_, (Proc. Sympos. on Geophysical Models), ed. R. R. Long, U.S. Govt. Printing Office.

28. Corrsin. S. 1955. A Measure of the Area of a Homogeneous Random Surface in Space, _Q_. _J_. _of_ _Appl_. _Math_., 12,4: 404-8.

29. Corrsin, S., Kistler, A. L. and O'Brien, V. 1954. _Preliminary Measurements of Turbulence and Temperature Fluctuations Behind a Heated Grid_, N.A.C.A. Res. Memo. 54D19.

30. Corrsin, S. and Ruetenik, J. R. 1955. Equilibrium Turbulent Flow in a Slightly Divergent Channel, in "_50 Jahre Grenzschichtforschung_" (ed. H. Görtler and W. Tollmien), Braunschweig: F. Vieweg und Sohn.

31. Corrsin, S. 1955. On Turbulent Jet Mixing of Two Gases of Constant Temperature, discussion of paper by S. I. Pai, _J_. _Appl_. _Mech_., 603-604.

32. Corrsin, S., Kistler, A. L. and O'Brien, V. 1956. Double and Triple Correlations Behind a Heated Grid, _J_. _Aero_. _Sci_., 23,1: 96.

33. Corrsin, S. and Lumley, J. L. 1956. On the Equation of Motion for a Particle in Turbulent Fluid, Appl. Sci. Res., A,6: 114-16.

34. Corrsin, S. and Schwarz, W. H. 1957. Some Effects of Turbulence on a Pendulum at Moderate Reynolds Number, in Proc. Heat Trans. and Fluid Mechs. Inst., pp. 421-39,, Palo Alto, Stanford U. Press.

35. Corrsin, S. 1957. Simple Theory of an Idealized Turbulent Mixer, Am. Inst. Chem. Engr. J., 3,3: 329-30.

36. Corrsin, S, 1958. Statistical Behavior of a Reacting Mixture in Isotropic Turbulence, Phys. of Fluids, 1,1: 42-7.

37. Corrsin, S. 1958. Local Isotropy in Turbulent Shear Flow, N.A.C.A. Res. Memo RM 58B11.

38. Corrsin, S. 1957. Some Current Problems in Turbulent Shear Flow, in Symposium on Naval Hydrodynamics (ed. F. S. Sherman), pp. 373-400, Publ. 515, Wash., D.C.: Nat. Acad. Sci./Nat. Res. Council.

39. Corrsin, S., Mills, Jr., R. R., Kistler, A. L. and O'Brien, V. 1958. Turbulence and Temperature Fluctuations Behind a Heated Grid, N.A.C.A. Tech. Note 4288.

40. Corrsin, S. and Mills, Jr. R. R. 1959. Effect of Contraction on Turbulence and Temperature Fluctuations Generated by a Warm Grid, N.A.S.A., Memo. 5-5-59W (condensed version in Proc. 1958 Heat Transfer and fluid Mechanics Inst., pp.15-24, Stanford University Press, June 1958).

41. Corrsin, S. 1959. Progress Report on Some Turbulent Diffusion Research, in Atmos. Diffusion and Air Pollution (ed. F. N. Frenkiel and P. A. Sheppard), pp. 161-4, Advances in Geophysics, vol. 6, , New York & London: Acad. Press.

42. Corrsin, S. and Lumley, J. L. 1959. A Random Walk with Both Lagrangian and Eulerian Statistics, in Atmos. diffusion and Air Pollution (ed. F. N. Frenkiel and P. A. Sheppard), pp. 179-83, Advances in Geophysics, vol. 6, New York & London: Acad. Press.

43. Corrsin, S. 1959. Lagrangian Correlation and Some Difficulties in Turbulent Diffusion Experiments, Atmos. Diffusion and Air Pollution (ed. F. N. Frenkiel and P. A. Sheppard), pp. 441-6, Advances in Geophysics, vol. 6, , New York & London: Acad. Press.

44. Corrsin, S. 1959. Outline of Some Topics in Homogeneous Turbulent Flow, J. Geophys. Res. 64,12: 2134-50.

45. Corrsin, S. 1961. Comments on "Rarefaction Effects in Low-Speed Turbulence," J. Aero. Sci., 28,4: 333-4.

46. Corrsin, S. and Kennedy, D. A. 1961. Spectral Flatness Factor and "Intermittency" in Turbulence and in Non-Linear Noise, J. Fluid Mech., 10,3: 366-70.

47. Corrsin. S. 1961. Turbulent Flow, Am. Scientist 49,3: 300-25.

48. Corrsin. S. and Phillips, O. M. 1961. Contour Length and Surface Area of Multiple-Valued Random Variables, J. Soc. Indust. and Appl. Math., 9,3: 395-404.

49. Corrsin, S. 1962. Discussion of Paper by L. M. Brush, Jr. "Exploratory Study of Sediment Diffusion," J. Geophys. Res., 67,4: 1435.

50. Corrsin, S. 1961. Reactant Concentration Spectrum in Turbulent Mixing with a First-Order Reaction, J. Fluid Mech. 11,3: 407-16.

51. Corrsin, S. 1962. Turbulent Dissipation Fluctuations, Phys. of Fluids, 5,10: 1301-2.

52. Corrsin, S. 1962. Measurement of Turbulence, in Encyclopaedic Dictionary of Physics, (ed. J. Thewlis) C 59, pp. 1-4, Oxford: Pergamon Press.

53. Corrsin, S. 1962. Discussion of "Turbulent Diffusion and Anemometer Measurements" by L. V. Baldwin and W. R. Mickelsen, J. Eng'g Mech. Div., Amer. Soc. Civ. Eng., EM6: 151-3.

54. Corrsin, S. 1963. Turbulent Flow, Experimental Methods, in Handbuch der Physik, Vol. 8, part 2, pp. 524-90 (ed. S. Flügge and C. Truesdell), Berlin: Springer-Verlag.

55. Corrsin, S. 1962. Some Statistical Properties of the Product of a Turbulent First-Order Reaction, in Fluid Dynamics and Applied Mathematics (Proc. Univ. of Md. 1961 Sympos., ed. J. B. Diaz and S. I. Pai), pp. 105-24, New York: Gordon & Breach.

56. Corrsin, S. 1962. Theories of Turbulent Dispersion, in Mécanique de la Turbulence, Colloques Internat. du C.N.R.S., Paris, (distrib. in U. S. by Gordon & Breach).

57. Corrsin, S. 1963. Estimates of the Relations Between Eulerian and Lagrangian Scales in Large Reynolds Number Turbulence, J. Atmos. Sci., 20,2: 115-19.

58. Corrsin, S. 1964. (7) Evalulation of a Toy Ground Effect Machine, pp. 14-15; (22) Effect of Disturbances on the Transition from Laminar to Turbulent Flow, pp. 56-7; (35) Windmill Powered Propellor Driven Boat, pp. 97-98; (36) Toy Ornithopter, pp. 99-100, four experiments in Laboratory Experiments and Demonstrations in Fluid Mechanics and Heat Transfer (ed. F. Landis): New York University: Mechanical Eng'g Dept.

59. Corrsin, S. 1964. Further Generalization of Onsager's Cascade Model for Turbulent Spectra, Phys. of Fluids 7,8: 1156-9, (erratum 8,10, Oct, 1965).

60. Corrsin, S. 1964. The Isotropic Turbulent Mixer: Part II. Arbitrary Schmidt Number, A. I. Ch. E. J., 10,6: 870-7. 1964.

61. Corrsin, S. and Patterson, Jr., G. S. 1965. Computer Experiments on Random Walks with Both Eulerian and Lagrangian Statistics, in Dynamics of Fluids and Plasmas (Proc. of a Univ. of Md. Sympos. honoring J. M. Burgers, (ed. S. I. Pai et al.), pp.275-307, Academic Press, New York.

62. Corrsin, S. and Comte-Bellot, G. 1966. The Use of a Contraction to Improve the Isotropy of Grid-Generated Turbulence, J. Fluid Mech. 25,4: 657-82.

63. Corrsin, S. 1966. Two Comments on Interdisciplinary Research, in Annals of the New York Acad. Sci. (ed. S. Jakowska), 130,3: 967-8.

64. Corrsin, S. 1968. Effect of Passive Chemical Reaction on Turbulent Dispersion, Am. Inst. Aero. and Astro. J. 6,9: 1797-98.

65. Corrsin, S. 1968. Conjecture on a Connection between Lift and Particle Displacement, Am. Inst. Aero- and Astro. J. 6,9: 1811-12.

66. Corrsin, S. and Morton, J. B. 1969. Experimental Confirmation of the Applicability of the Fokker-Planck Equation to a Nonlinear Oscillator, J. Math. Phys., 10,2: 361-8.

67. Corrsin, S. and Karweit, M. J. Fluid Line Growth in Grid-generated isotropic turbulence. J. Fluid Mech. 39,1: 87-96.

68. Corrsin, S., Champagne, F. H. and Harris, V. G. 1970. Experiments on nearly homogeneous turbulent shear flow, J. Fluid Mech. 41,1: 81-139.

69. Corrsin, S. and Morton, J. B. 1970. Consolidated expansions for estimating the response of a randomly driven nonlinear oscillator, J. Statistical Phys. 1,2: 153-94.

70. Corrsin, S. and Comte-Bellot, G. 1971. Simple Eulerian time correlation of full- and narrow-band velocity signals in grid-generated "isotropic" turbulence, J. Fluid Mech. 48,2: 273-337.

71. Corrsin, S. and Riley, J. J. 1971. Simulation and computation of dispersion in turbulent shear flow, Proc. of Conf. on Air Pollution Meteorology, Amer. Meteorol. Soc.

72. Corrsin, S. and Kuo, A. Y.-S. 1971. Experiments on internal intermittency and fine-structure distribution functions in fully turbulent fluid, J. Fluid Mech., 50,2: 285-319.

73. Corrsin, S. 1972. Random geometric problems suggested by turbulence, in "Statistical Models and Turbulence," vol. 12 of Lecture Notes in Physics (ed. M. Rosenblatt & C. Van Atta), pp. 300-316, Berlin: Springer-Verlag.

74. Corrsin, S. and Karweit, M. J. 1972. Simple and compound line growth in random walks, in "Statistical Models and Turbulence" Notes in Physics (ed. M. Rosenblatt & C. Van Atta), pp. 317-326, Berlin: Springer-Verlag.

75. Corrsin, S. and Karweit, M. J. 1972. The mixing of scalar stripes by an isotropic ensemble of single velocity modes, in "Statistical Models and Turbulence" Notes in Physics (ed. M. Rosenblatt & C. Van Atta), pp. 327-332, Berlin: Springer-Verlag.

76. Corrsin, S. 1972. "Backward diffusion" in turbulent flow, Phys. of Fluids, 15,6: 986-7.

77. Corrsin, S. 1972. Simple proof of fluid line growth in stationary homogeneous turbulence, Phys. of Fluids, 15,8: 1370-2.

78. Corrsin, S. and Karweit, M. J. 1972. A note on the angular dispersion of a fluid line element in isotropic turbulence, J. Fluid Mech., 55,2: 289-300.

79. Corrsin, S. and Kuo, A. Y.-S. 1972. Experiment on the geometry of the fine-structures regions in fully turbulent fluid, J. Fluid Mech., 56,3: 447-80.

80. Corrsin, S., Berger, R. E. and Karweit, M. J. 1972. Motion of the pre-corneal tear film after a blink, Proc. 25th Ann. Conf. on Eng'g in Med. & Biol.

81. Corrsin, S. Comment on "Transport equations in turbulence," Phys. of Fluids, 16,1: 157-8.

82. Corrsin, S. and Gad-el-Hak, M. 1974. Measurements of the Nearly Isotropic Turbulence Behind a Uniform Jet Grid, J. Fluid Mech.,62,1: 115-43.

83. Corrsin, S. and Shlien, D. J. 1974. A Measurement of Lagrangian Velocity Auto-correlation in Approximately Isotropic Turbulence, J. Fluid Mech., 62,2: 255-71.

84. Corrsin, S. and Riley, J. J. 1974. The Relation of Turbulent Diffusivities to Lagrangian Velocity Statistics for the Simplest Shear Flow, J. Geophys. Res., 79,12: 1768-71.

85. Corrsin, S. and Ross, S. M. 1974. Results of a Wavy Wall Analytical Model of Muco-ciliary Pumping, J. Appl. Physiol.,37,3: 333-40.

86. Corrsin, S. and Berger, R. E. 1974. A Surface-tension Gradient Mechanism for Driving the Pre-corneal Tear Film after a Blink, J. Biomech., 7: 225-38.

87. Corrsin, S., Erian, F. F., Yoshinaga, M. A. and Davis, S. H. 1974. A Model of Maternal Blood Flow in the Placenta, Proc. 27th Ann. Conf. on Eng'g in Med. & Biol.

88. Corrsin, S. 1974. Limitations of Gradient Transport in Random Walks and in Turbulence, in Advances in Geophysics, 18A: 25-60, "Turbulent Diffusion in Environmental Pollution," ed. by F. N. Frenkiel and R. E. Munn, New York: Academic Press.

89. Corrsin, S. and Karweit, M. J. 1975. Observation of Cellular Patterns in a Partly Filled, Horizontal, Rotating Cylinder, Phys. of Fluids, 18,1: 111-12.

90. Corrsin, S., Erian, F. F. and Davis, S.H. 1975. Influence of Variable Permeability and Intertia on a Model of Maternal Blood in the Placenta, Proc. 28th Ann. Conf. on Eng'g in Med. & Biol.

91. Corrsin, S. and Shlien, D. J. 1976. Dispersion Measurements in a Turbulent Boundary Layer, Int. J. Heat & Mass Transfer, 19: 285-295.

92. Corrsin, S., Harris, V. G. and Graham, J. A. H. 1977. Further Measurements in Nearly Homogeneous Turbulent Shear Flow, J. Fluid Mech., 81: 657-687. Corrigendum, J. Fluid. Mech., 86: 793-796.

93. Corrsin, S. and Kollmann, W. 1977. Preliminary Report on Suddenly Sheared Cellular Motion as a Qualitative Model of Homogeneous Turbulent Shear Flow, in Turbulence in Internal Flows, pp. 11-33 (ed. S. N. B. Murthy) Washington/London: Hemisphere Publ. Co.

94. Corrsin, S., Erian, F. F. and Davis, S. H. 1977. Maternal, Placental Blood Flow: A Model with Velocity-dependent Permeability, J. Biomech., 10: 807-814.

95. Corrsin, S. and Higdon, J. J. L. 1978. Induced Drag of a Bird Flock, The American Naturalist, 112: 727-744.

96. Tavoularis, S., Corrsin, S. and Bennett, J. C. 1978. Velocity Derivative Skewness in Small Reynolds Number, Nearly Isotropic Turbulence, *J. Fluid Mech.*, 88: 63-69.

97. Corrsin, S. and Bennett, J. C. 1978. Decay of Nearly Isotropic, Grid-Generated Turbulence at Small Reynolds Numbers, in a Straight Duct and a Slight Contraction, *Physics of Fluids*, 21,12: 2129-2140.

98. Corrsin, S. and Kellogg, R. M. 1980. Evolution of a Spectrally Local Disturbance in Grid-Generated Nearly-Isotropic Turbulence, *J. Fluid Mech.*, 94: 641-669.

99. Corrsin, S., Sreenivasan, K. R., Tavoularis, S. and Henry, R. 1980. Temperature Fluctuations and Scales in Grid-Generated Turbulence, *J. Fluid Mech.*, 100: 597-621.

100. Corrsin, S. and Tavoularis, S. 1981. Experiments in Nearly Homogeneous Turbulent Shear Flow with a Uniform Mean Temperature Gradient, *J. Fluid Mech.*, 104: 311-347.

101. Corrsin, S. and Tavoularis, S. 1981. Experiments in Nearly Homogeneous Turbulent Shear Flow with a Uniform Mean Temperature Gradient. Part 2. The Fine Structure. *J. Fluid Mech.*, 104: 349-367.

102. Tavoularis, S. and S. Corrsin 1981. Theoretical and Experimental Determination of the Turbulent Diffusivity Tensor in Homogeneous Turbulent Shear Flow, in *Third Symposium on Turbulent Shear Flows*, eds. L. J. S. Bradbury et al., pp. 15.24 - 15.27, Davis, CA: The University of California.

103. Corrsin, S., Sreenivasan, K. R. and Tavoularis, S. 1981. A Test of Gradient Transport and Its Generalizations, *Turbulent Shear Flow 3*, (ed L. J. S. Bradbury, F. Durst, B. E. Launder, F. W. Schmidt, and J. H. Whitelaw), pp. 96-112, Berlin/Heidelberg: Springer-Verlag.

104. Corrsin, S. and Fournier, J.-L. 1982. Viscous Dissipation Rates of Velocity Component Kinetic Energies, *Phys. of Fluids*, 25,4: 583-585.

105. Corrsin, S. 1982. A More Explicit Estimate for the "Implications of Athlete's Bradycardia on Lifespan", *J. Theo. Biol.*, 96.

106. Corrsin, S. and Tavoularis, S. 1984. Effects of Shear on the Turbulent Diffusivity Tensor, *Internat. J. Heat & Mass Trans.* (in press).

107. Corrsin, S., Budwig, R. and Tavoularis, S. 1984. Temperature Fluctuations and Heat Flux in Grid-Generated Isotropic Turbulence with Streamwise and Transverse Mean Temperature Gradients, (submitted to *J. Fluid Mech.*).

Index of Contributors